Materials Horizons: From Nature to Nanomaterials

Series Editor

Vijay Kumar Thakur, School of Aerospace, Transport and Manufacturing, Cranfield University, Cranfield, UK

Materials are an indispensable part of human civilization since the inception of life on earth. With the passage of time, innumerable new materials have been explored as well as developed and the search for new innovative materials continues briskly. Keeping in mind the immense perspectives of various classes of materials, this series aims at providing a comprehensive collection of works across the breadth of materials research at cutting-edge interface of materials science with physics, chemistry, biology and engineering.

This series covers a galaxy of materials ranging from natural materials to nanomaterials. Some of the topics include but not limited to: biological materials, biomimetic materials, ceramics, composites, coatings, functional materials, glasses, inorganic materials, inorganic-organic hybrids, metals, membranes, magnetic materials, manufacturing of materials, nanomaterials, organic materials and pigments to name a few. The series provides most timely and comprehensive information on advanced synthesis, processing, characterization, manufacturing and applications in a broad range of interdisciplinary fields in science, engineering and technology.

This series accepts both authored and edited works, including textbooks, monographs, reference works, and professional books. The books in this series will provide a deep insight into the state-of-art of Materials Horizons and serve students, academic, government and industrial scientists involved in all aspects of materials research.

Review Process

The proposal for each volume is reviewed by the following:

1. Responsible (in-house) editor
2. One external subject expert
3. One of the editorial board members.

The chapters in each volume are individually reviewed single blind by expert reviewers and the volume editor.

Bibhu Prasad Swain
Editor

Advances in Mechanical Coating

Processing, Properties and Applications

Springer

Editor
Bibhu Prasad Swain
Department of Physics
National Institute of Technology Manipur
Imphal, Manipur, India

ISSN 2524-5384 ISSN 2524-5392 (electronic)
Materials Horizons: From Nature to Nanomaterials
ISBN 978-981-96-7483-1 ISBN 978-981-96-7484-8 (eBook)
https://doi.org/10.1007/978-981-96-7484-8

This Springer imprint is published by the registered company Springer Nature Singapore Pte Ltd.
The registered company address is: 152 Beach Road, #21-01/04 Gateway East, Singapore 189721, Singapore

Preface

In recent decades, significant advancements in mechanical coating technologies—encompassing processing methods, material properties, and diverse applications—have profoundly influenced the implementation of surface coatings across a broad spectrum of engineering disciplines. These technologies have evolved into essential tools for addressing complex challenges in mechanical engineering, materials science, tribology, corrosion science, biomedical engineering, aerospace engineering, and beyond.

In many engineering applications, critical structural and functional components are routinely exposed to hostile environments that can compromise their integrity and operational lifespan. For example, marine propellers and wind turbine bearings must endure high temperatures, mechanical friction, and corrosive media. Likewise, synthetic biodegradable materials used in aerospace applications are vulnerable to wear and abrasion in cryogenic or marine-launch environments. High-velocity motion through the atmosphere subjects component surfaces to severe aerodynamic heating and pressure. These diverse service conditions necessitate material surfaces with specialized attributes such as low friction, enhanced wear resistance, corrosion resistance, thermal stability, and superhydrophobicity.

However, achieving a robust correlation between the intrinsic structure of materials and their surface-dependent properties remains an ongoing scientific challenge. The direct interaction of material surfaces with the surrounding environment often leads to premature degradation unless mitigated through effective surface engineering strategies.

One of the most reliable and widely adopted solutions is the application of mechanical coatings, which function as protective interfaces between the base material and the external environment. Over the last several decades, a concerted effort by researchers and industry practitioners has led to the development of high-performance coating materials and advanced deposition techniques. These developments aim not only to enhance the functional properties of the coatings themselves but also to improve their adhesion, surface finish, and overall reliability in service.

Figure 1a, b derived from Scopus data as of October 21, 2024, illustrate the increasing scholarly attention devoted to mechanical coating research—highlighting

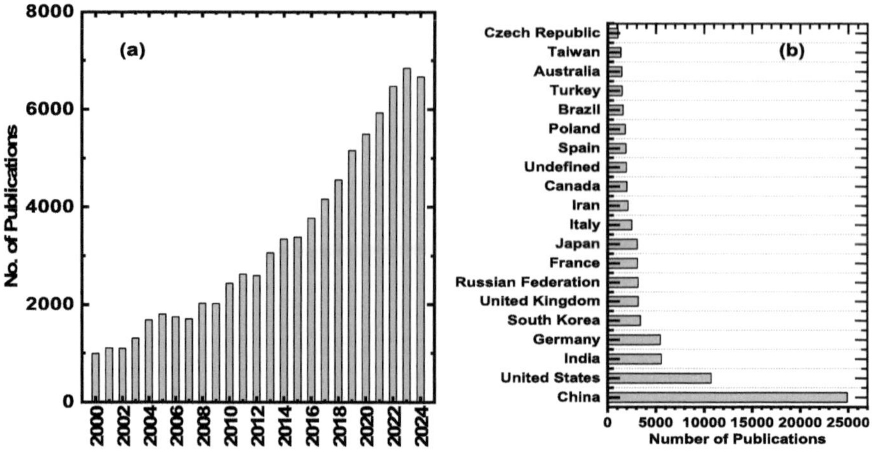

Fig. 1 **a** number of publications from the years 2000 to 2024 and **b** countries publishing articles on mechanical coating [scopus on October 21, 2024]

both the temporal growth in publications and geographical contributions from leading research-active nations.

The current volume, titled *Advances in Mechanical Coating: Processing, Properties, and Applications*, is conceived as a comprehensive reference for scientists, researchers, and engineers working in surface engineering and related fields. It is intended to serve both as a technical guide and as a resource for academic instruction and applied research. The selection of appropriate coating materials, tailored to specific operational and environmental constraints, remains a formidable task for design engineers. Given that surface phenomena predominantly dictate material-environment interactions, surface coatings are increasingly utilized to enhance durability, reduce maintenance costs, and enable novel functionalities.

Mechanical coatings, as a subset of surface modification techniques, encompass a wide range of metallic and non-metallic systems, each with unique performance characteristics. This book offers a detailed exposition of state-of-the-art coating techniques, coating materials, and industrial applications, aiming to bridge fundamental science and practical implementation. The inclusion of both theoretical insights and empirical findings enhances its value for a multidisciplinary audience, including graduate students, academic researchers, and industry professionals.

The book is structured into eleven authoritative chapters, each authored by subject matter experts from academia and industry:

- Chapter 1 provides an in-depth review of the historical development and scientific principles underlying modern coating technologies. It emphasizes their transformative role in industries such as healthcare, electronics, aerospace, and renewable energy.

- Chapter 2 addresses the mechanical characterization of coatings, with emphasis on parameters such as adhesion strength, hardness, and elastic modulus—essential for evaluating performance and durability in industrial applications.
- Chapter 3 explores high-performance coating systems, including diamond-like carbon (DLC), metal carbides, metal nitrides (e.g., TiN, TiC), molybdenum disulfide (MoS_2), aluminum oxide (Al_2O_3), and carbon nitrides (C_3N_4). The chapter discusses their compositional features, deposition methodologies, and functional attributes.
- Chapter 4 details surface preparation techniques, encompassing mechanical (e.g., abrasive blasting), chemical (e.g., acid etching), and physical methods (e.g., plasma and laser treatments). These processes are critical to achieving optimal coating adhesion and interfacial strength.
- Chapter 5 investigates surface wettability and its underlying mechanisms, highlighting the influence of surface energy, topography, chemical compatibility, and surface modifications on adhesion and coating behavior in diverse applications.
- Chapter 6 presents a comprehensive treatment of residual stresses in mechanical coatings. It covers their origins, measurement techniques, mechanical implications, and mitigation strategies to ensure structural integrity and long-term reliability.
- Chapter 7 explores analytical and computational modeling approaches, including finite element analysis, for simulating coating behavior under mechanical loading, film deposition dynamics, and ion beam interactions in MEMS and NEMS systems.
- Chapter 8 focuses on fracture mechanics of coatings, examining methods for evaluating fracture toughness via indentation and scratch testing, along with qualitative and quantitative fractographic analysis.
- Chapter 9 provides a rigorous overview of tribology, and the science of interacting surfaces in relative motion, and discusses the critical roles of friction, wear, and lubrication in coated systems.
- Chapter 10 outlines the multifunctional benefits of coatings, such as improved surface hardness, corrosion resistance, and tribological performance. The chapter also examines their role in reducing fuel consumption, environmental impact, and operational costs.
- Chapter 11 emphasizes surface passivation and its relevance in enhancing the durability and chemical stability of coatings. It covers both organic and inorganic passivation strategies tailored to specific industrial requirements.

Collectively, this volume provides a unified and detailed treatment of surface coating science and technology, combining foundational theory, experimental research, and industrial practice. This book will serve as a valuable resource for advancing innovation, education, and practical implementation in surface engineering and mechanical coatings.

Imphal, India Bibhu Prasad Swain

Contents

Chapter 1
Introduction to Coating Technology

Ningthoujam Bidyapani Chanu and Bibhu Prasad Swain

Abstract Coating technology enriches ordinary surfaces by depositing thin layers of material for both aesthetic and practical purposes. This advanced technique relies on various procedures to reinforce, improve, and augment existing objects, increasing their resistance and lifetime. Coatings have played a significant role throughout history, from ancient craftsmanship to present advances, and are vital today in various industries, from medicine to aerospace. The essential ideas, benefits, and applications of many coating methods, including electroplating, thermal spraying, chemical vapour deposition, and physical vapour deposition, are discussed here. The unique applications in the solar energy and electronics sectors, magnetron sputtering, and cathodic arc evaporation deposition methods are also presented here. This chapter covers the development of coating technologies, their practical applications, and the underlying scientific theories governing their functionality across various industrial sectors. Further growth in coating technology will enable new opportunities in large-scale industries such as healthcare, electronics, and energy sources.

Keywords Coating · Magnetron sputtering · HVOF · Electroplating · ALD

1 Introduction

Coating is the term used to describe a thin film of material applied to a substrate's surface. It can serve decorative and functional purposes in various forms: liquids, gases, or solids, like powder coatings. Coating technology encompasses methods of deposition of an exterior film layer on the surface of an object to protect, improve, or add new features and properties to the surface of the base object.

This technology significantly reduces or eliminates corrosion, fouling, biofouling, physiological and morphological deterioration, and surface damage caused by exposure to environmental elements such as humidity, climatic conditions, and UV rays [1–5]. It helps preserve and prolong the longevity of materials by enhancing

N. B. Chanu · B. P. Swain (✉)
Department of Physics, National Institute of Technology, Langol, Imphal, Manipur, India
e-mail: bibhuprasad.swain@gmail.com; bpswain@nitmanipur.ac.in

B. P. Swain (ed.), *Advances in Mechanical Coating*, Materials Horizons: From Nature to Nanomaterials, https://doi.org/10.1007/978-981-96-7484-8_1

the durability, steadiness, and lifespan of treated surfaces. People have used it for artistic expression and practical protection for ages. Evidence of decorative coatings spans approximately 30,000 years, while robust historical evidence indicates the use of protective coatings as early as 5,000 BC [6, 7]. Tar, beeswax, gelatin, asphalt, and clay were used to safeguard maritime vessels and human dwellings.

The efficiency of the mechanical systems is usually reduced by higher stress and damage in the more significant components or materials [8]. So, in worldwide production, applying coatings to products has become a standard practice to reduce production expenses and boost profits, which is essential for maintaining economic competitiveness in the industry [9]. Today, protective coatings are ubiquitous, serving diverse purposes, from ships and aircraft to buildings and cars. The evolution of these coatings reflects a journey marked by enduring challenges and innovations. The tribomechanical surface properties of materials across various microelectronics and transportation industries drive these durable protective coatings [10–15]. This chapter delves into the principles, working mechanisms, advantages, and applications of various coating methods.

2 Types of Coating Technology and Their Applications

Depending on the purpose of the coating, various coating processes are used, such as anodising, thermal spraying, cold spraying, laser processing, chemical vapour deposition (CVD), and physical vapour deposition (PVD) [16–20]. This chapter discusses a few of these coating processes.

2.1 Magnetron Sputtering

Magnetron sputtering is an advanced deposition method that uses a specially designed magnetic field around a diode-sputtering target. The cathode surface is enveloped by this magnetic field, generating electron traps that produce ($E \times B$) drift currents, forming closed loops [21]. This field confines primary and secondary electrons near the cathode in a continuous racetrack pattern. The resulting higher electron density boosts the ionisation of gas atoms, leading to enhanced ionisation efficiency and lower plasma impedance. As a result, magnetron sputtering operates at significantly lower (500–600) V compared to traditional diode systems, which require several kilovolts. This higher ionisation efficiency leads to more significant ion current density and faster target erosion.

Various magnetron designs exist, such as post magnetrons, planar magnetrons, and S-guns, all based on the same principle of controlled electron drift and trapping. The planar magnetron is popular due to its simplicity and scalability, with cathodes that can be extended to any required size [23]. Large cathodes are ideal for continuous processing and can be fabricated into various shapes as long as the racetrack exists.

Figure 1 shows a planar magnetron in more detail. Electromagnets or permanent magnets can also be used in magnetron design, but permanent magnets are preferred for simplicity. These magnets can be in various geometries, but must create a tunnel-shaped magnetic field before the target surface. The magnetic field strength must be around 20 mT to effectively trap electrons, though it is often much stronger. The most straightforward magnet system uses a single high-strength magnet with a ferromagnetic pole component to accurately shape the magnetic field. Higher field strengths improve electron trapping efficiency, allowing these systems to operate down to the 50 mPa range at lower pressures. The process resembles vacuum evaporation at these low pressures due to the long mean-free path length.

Over the years, magnetron sputtering technology has been continuously improved to address various challenges, including electrical irregularities like arcs, maximising metal ionisation, optimising target utilisation, preventing target toxicity in reactive sputtering, increasing deposition rates, and reducing operating costs [24, 25]. The growing demand for high-quality functional coatings and layered structures with diverse functionalities has driven this progress. Recent developments in sputter deposition techniques focus on enhancing the output of the sputter source (the cathode target), ensuring uniformity in the deposited layer, enabling directional metal deposition, and controlling the energy of ions impacting the substrate [26].

Magnetron sputtering has become crucial in various industrial applications, including metallisation in integrated circuits, coatings for anti-wear and protection against corrosion, optical protective coatings, wide area coatings on architectural

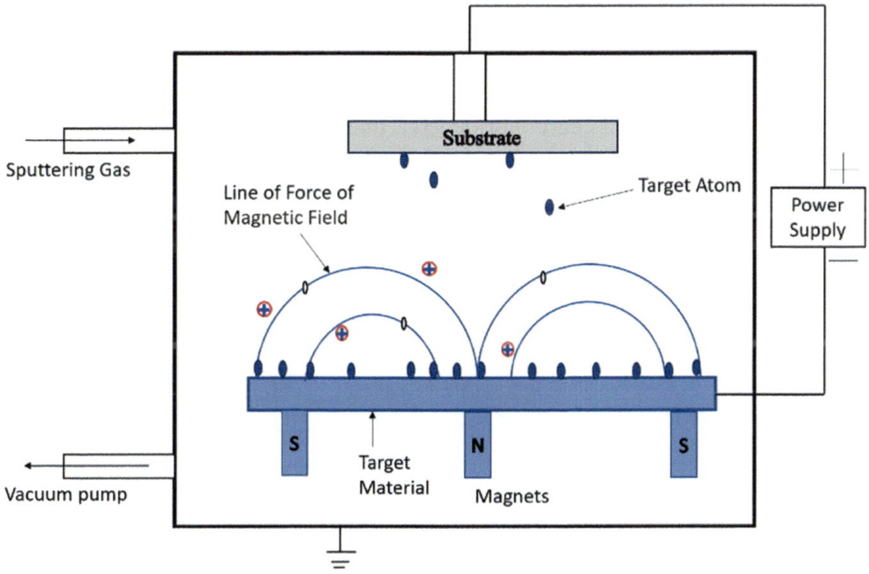

Fig. 1 Schematic diagram of magneton sputtering deposition process [22]

glass, photovoltaic solar cells, and displays [27]. Moreover, it is also used in roll-to-roll systems to deposit conductive layers in microelectronic devices for large sheets coating of architectural glass and polymeric web. The versatility of this technology is reflected in the wide range of cathode target sizes and configurations, as well as the different waveforms used for the cathode voltage or discharge current. In semiconductor applications, low-temperature magnetron sputtering deposits the components for transistors and positive-intrinsic-negative diodes. Balanced magnetrons are suitable for low-energy ion bombardment and highly uniform thin film deposition for optical coatings and semiconductors.

Window and Savvides [28] first used an unbalanced magnetron while studying seven planar magnetrons. They are grouped into three magnetic field configurations, as shown in Fig. 2. It was turned up only during the 80s and is now extensively helpful for the deposition of anti-corrosive, low friction, wear-resistant, and ornamental films. The unbalanced magnetron differs from the balanced magnetron in its applications, construction, and functionality. The outer ring of magnets in the unbalanced magnetron is stronger than the core pole, creating an asymmetry in magnetic field strength. Magnetic field lines will also be distributed toward the substrate, channelling some electrons. As a result, this design causes the plasma to flow out towards the substrate, facilitating high ion current without needing external biasing. An unbalanced magnetron is a highly efficient ion source that provides a high flux of coating atoms. The ion-to-atom arrival ratio remains constant even as the deposition rate rises since the ion current drawn at the substrate is precisely proportional to the target current. In another way, an unbalanced magnetron can be constructed with a stronger magnetic field in the central magnet than in the outer magnet. This configuration is sometimes considered an unbalanced magnetron assembly of Type 1.

Highly unbalanced magnetrons are used to deposit titanium nitride [29], zirconium nitride [30], titanium carbonitride [31], titanium aluminium nitride [32], and chromium nitride, which are used in machining and cutting tool applications. For instance, Madhuri et al. [33] used DC magnetron sputtering to synthesise AlN films

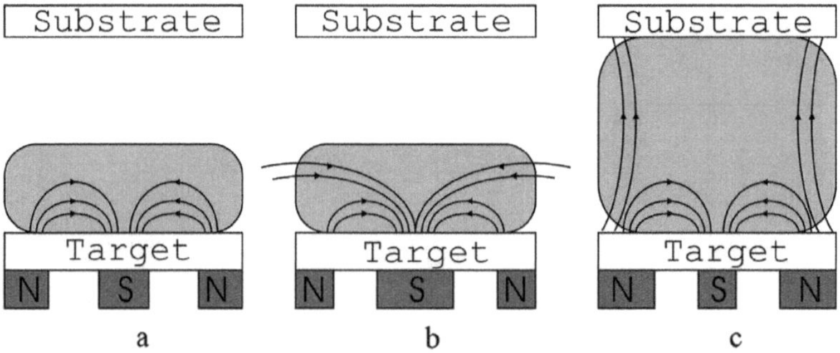

Fig. 2 Three configurations of magnetron: **a** conventional, **b** type 1 unbalanced magnetron, **c** type 2 unbalanced magnetron [28]

on p-Si (100), quartz, and glass substrates, investigating the role of nitrogen flow rate. Similarly, Jena et al. [34] utilised DC magnetron sputtering to produce TiN_{xnm}/Ti_{xnm} multilayers on p-type silicon wafers and amorphous quartz substrates. Txintxurreta et al. [35] deposited indium titanium oxide with a 550 × 125 mm^2 target size unbalanced magnetron.

Double magnetrons, often called dual magnetrons or dual cathodes, are employed for rapid reactive deposition of oxides such as silicon dioxide, titanium dioxide, and tantalum pentoxide for display or plastic web coatings. It also offers a versatile method for enhancing the surface of medical metal implants. This process enables the creation of thick, tightly bonded films with a well-monitored composition of elements. Hermans et al. [36] used a dual magnetron sputtering setup to deposit $ZnWO_4$ thin films and studied their structural and morphological properties. Titanium or titanium alloy implants are often used as long-term tooth or bone injury fixes [37].

2.2 Cathodic Arc Evaporation Deposition

Cathodic arc evaporation deposition is a PVD technique that utilises a high-energy electric arc to vapourise cathode material, producing an ionised plasma that condenses on the surface of a substrate to create a thin, adherent coating. The cathodic arc represents a plasma discharge with low voltage but high current. It occurs in a vacuum between the cathode target and the anode, which is usually part of the coating chamber, as shown in Fig. 3. Its purpose is to evaporate particles from the target and deposit them on the surface of a substrate. Since a tiny section of the cathode is vapourised quickly in arc vapourisation, this process consumes more energy than sputtering (Fig. 4).

Understanding the sequence of steps and the fundamental principles of the deposition process is crucial for enhancing the quality of coatings and their resulting performance in terms of friction and wear. The cathodic arc evaporation deposition begins by placing properly prepared substrates into the chamber. Typically, the target material and the substrate are maintained at a gap of (18–25) cm to minimise radiant heating effects during evaporation [39]. Subsequently, the chamber is evacuated to create a vacuum, ranging from 10^{-3} to 10^{-4} Pa, to prevent gas contamination during coating. Subjecting it to a temperature of 150–500 °C is done before etching. During etching and coating, the substrate rotates at speeds of (1.5–2) rpm or higher to ensure uniformity. Etching aims to remove residual contaminants from decontamination, often achieved using inert gas or metal ion plasma. High-speed plasma containing inert gas or metal ions is directed towards a negatively charged substrate, causing the sputtering of surface contaminants and activating a superficial layer on the substrate's surface. Metal ion etching is quicker and provides better coating adhesion than inert gas etching. However, a combination of both methods may be employed. The etching process further follows the evacuation of gaseous contaminants and the re-establishing of vacuum conditions.

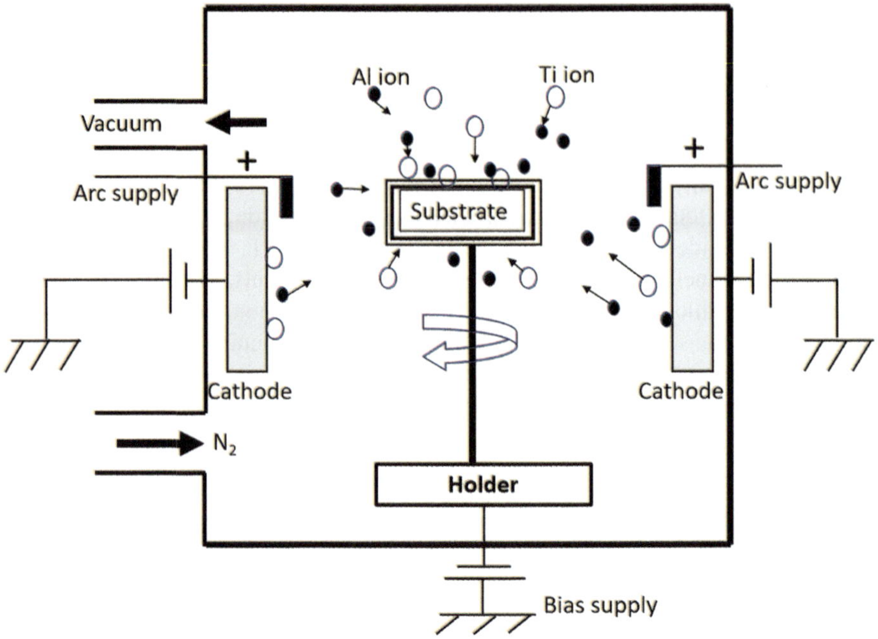

Fig. 3 Schematic diagram of the principle of the cathodic arc deposition technique of (Ti, Al)N coating [38]

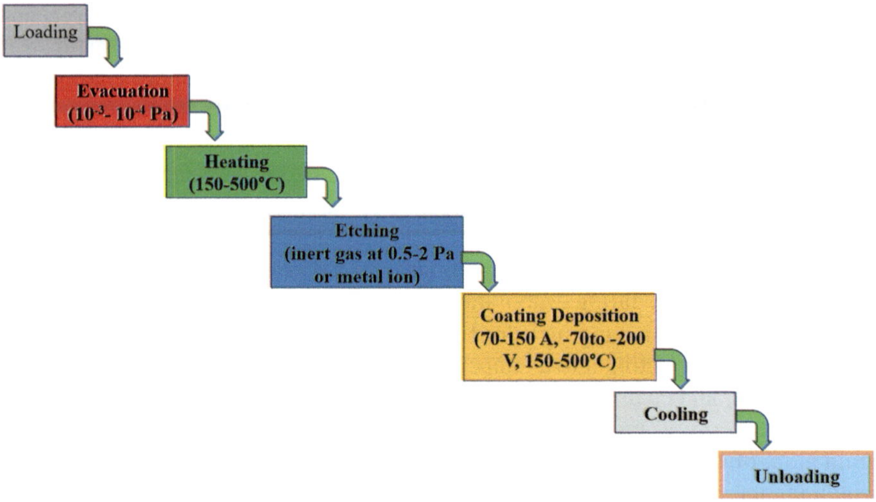

Fig. 4 Flowchart of Cathodic arc evaporation process

The coating deposition begins with the evaporation of coating material from a target, usually by creating an arc on the cathode. Various methods can initiate this arc, such as high voltage on an auxiliary electrode or low-voltage electric current between electrodes in a vacuum. A magnetic field may be used to regulate it by a direct current power supply. The characteristics of the cathode, like its geometry and material, also affect the behaviour of the arc. The concentrated arc on the cathode surface creates micro spots known as cathode spots, characterised by high current density, short lifespan, and a characteristic small size of 1 μm [40]. This spot erodes the cathode, causing localised melting and evaporation of material as well as the formation of metal plasma. The number and properties of these spots depend on the arc current. Steinhorst et al. [41] investigated how deposition temperature affects the properties of carbon-based coatings created through cathodic arc evaporation. The total thickness of the films they produced was approximately 60–66 nm. The film deposited at 200 °C exhibited only minor peaks. In contrast, the films deposited at 300 and 400 °C displayed similar spectra with distinct D and G peaks, as shown in Fig. 5. Consequently, the properties and performance of the deposited coatings are mainly influenced by factors such as arc current, substrate bias, and the current temperature.

The microparticulate formed from the low-melting materials is the primary drawback during cathodic evaporation. Several researchers have been working to mitigate this disadvantage by dispersing arc discharge over a cathode designed with thermal insulation subjected to the plasma stream to an electric field to reduce macro particle formation. Another researcher used a straight filter for particles to lower particulate matter formation and enhance plasma intensity. In some systems, an additional anode is added to improve ionisation efficiency. The resulting gaseous plasma, consisting of

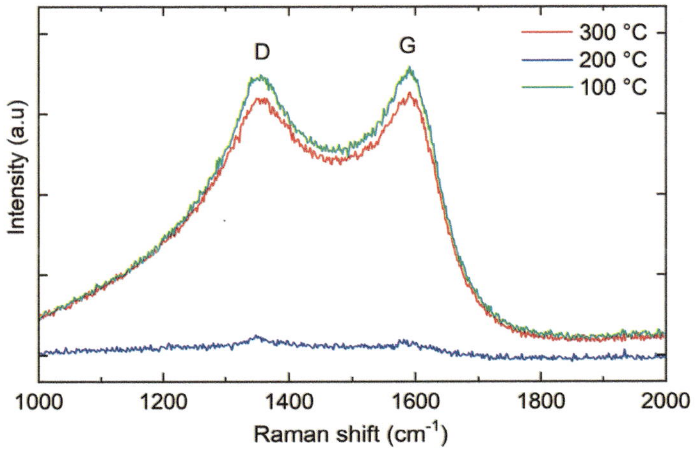

Fig. 5 Variation of Raman spectra of the coated sample with a carbon top layer deposited at 300, 200, and 100 °C [41]

electrons, metal plasma, and ions, deposits coatings. Reactive arc evaporation deposition involves reactive gases like nitrogen to enhance dissociation and produce more ions. The plasma loses some species when directed toward the substrate, creating a potential difference. So, applying a negative bias to the substrate accelerates positive ions towards it, facilitating faster deposition. Consequently, the plasma species condenses on the surface of the substrate. The coated substrate is cooled, typically with water cooling, and finally unloaded from the deposition chamber.

Let us consider an example of the application of CAE in TiAlN coating! TiAlN coatings can be generated through various methods, including ion beam sputter deposition, magnetron sputtering, arc ion plating, and cathodic arc evaporation (CAE). CAE stands out for its ability to produce dense, defect-free coatings by removing droplet particles, producing high-quality coatings with extremely hard properties. For example, Koller et al. [42] studied CAE-produced TaAlN, TiAlN, TiAlTaN, and multilayered TiAlN/TaAlN. All their coatings except the quaternary $Ti_{0.45}Al_{0.46}Ta_{0.19}N$ coating and the TiAlN/TaAlN multilayers have a hardness of nearly 32 ± 2 GPa. Alhafian et al. [43] achieved a 31.4 ± 0.7 GPa hardness on carbide substrates using CAE, enhancing wear resistance. Sveen et al. [44] studied the scratch adhesion of TiAlN coatings on different substrates, such as high-speed steel and cemented carbide. They determined that hardness is vital in yield strength, wear resistance, and overall mechanical performance. Long et al. [45] conducted a comparison between $Ti_{0.5}Al_{0.5}N$ and $Cr_{0.3}Al_{0.7}N$ coatings on silicon, revealing that $Ti_{0.5}Al_{0.5}N$ displayed superior wear resistance due to its greater hardness and finer grain size. These findings highlight the critical role of high hardness in enhancing resistance to mechanical degradation. However, TiAlN coatings produced via cathodic arc evaporation (CAE) typically exhibit moderate hardness, often less than 40 GPa. The ratio of aluminium content significantly influences hardness, with the highest value of 31.4 GPa (3200 HV) observed at an Al/(Ti + Al) ratio of 60%.

Consequently, optimising the aluminium content is essential for achieving high hardness in these coatings. Large particles can compromise the coating's strength, uniformity, and smoothness during the CAE process. In contrast, filtered cathodic vacuum arc (FCVA) deposition can eliminate these particles, resulting in a smooth, hard, and dense structure even at low temperatures. Adjusting parameters such as cathode current, bias voltage, nitrogen gas pressure, magnetic filter angle, and magnetic field intensity can precisely control the composition, hardness, wear resistance, and TiAlN coatings morphology. Building on this idea, Zhang et al. [46] applied a $Ti_{0.33}Al_{0.67}$ coating to silicon wafer substrates and 304L stainless steel, adjusting the nitrogen flow rate in the filtered cathodic arc deposition system, as illustrated in Fig. 6. In their research, all TiAlN coatings demonstrated hardness values above 40 GPa, categorising them as ultrahard as the nitrogen flow rate increased from 40 to 60 sccm, and hardness improved from 43.15 to 46.39 GPa. The maximum hardness of 46.39 GPa and Young's modulus of 505.4 GPa, observed at 60 sccm, were credited to TiAlN's preferred (111) orientation and intensified structure.

Cathodic arc evaporation deposition serves various applications beyond enhancing cutting tool durability in machining industries [47]. For instance, it can produce hard and wear-resistant coatings, making it useful for tooling, cutting, and machining

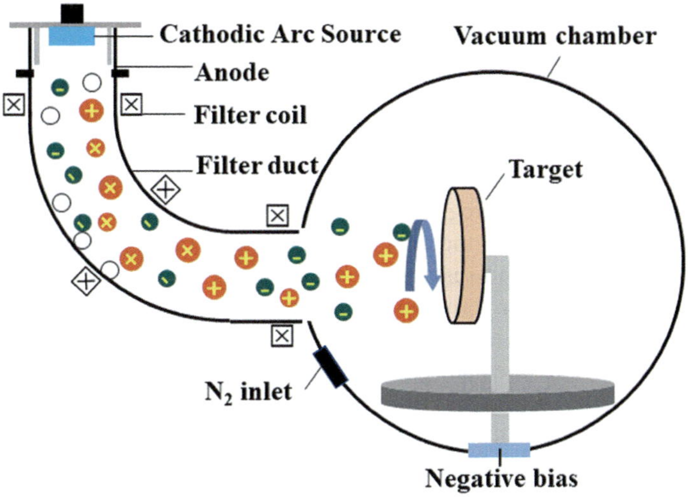

Fig. 6 Diagrammatic representation of the Filtered cathodic vacuum arc deposition (FCVA) setup [46]

applications. These coatings can extend the lifespan of tools and improve their performance. It also finds utility in producing high-performance coatings for aerospace components, where coatings must withstand extreme temperatures, wear, and corrosion. CAE deposition is also instrumental in depositing thin films for diffusion barriers, interconnects, and passivation layers in microelectronics and integrated circuits in the semiconductor industry.

Moreover, it is pivotal in producing decorative coatings for metal, glass, and plastic surfaces, offering aesthetic appeal and protection against corrosion and abrasion. Furthermore, cathodic arc deposition is employed in fabricating thin-film solar cells, enabling the deposition of layers with specific optical and electrical properties to enhance solar energy conversion efficiency. These diverse applications underscore the versatility and importance of cathodic arc evaporation deposition across various industries and research domains.

2.3 Radiofrequency and Pulsed DC Sputtering

Sputtering is a widely used PVD method for thin-film coatings, essential in various industrial and scientific applications. Based on the power supply, it is classified as DC (direct current) and AC (alternating current) sputtering. DC sputtering is a cost-effective deposition process for thin film coatings on electrically conductive metallic targets. However, when non-conductive dielectric materials are bombarded with positive ions, their surfaces become charged, which repels additional positive ions and halts the sputtering process, often causing arcing in the plasma. Radiofrequency (RF) sputtering is preferred for depositing films from electrically non-conductive target materials to address these limitations [48, 49]. So, among the sputtering methods, RF and pulsed DC sputtering have gained prominence due to their ability to deposit high-quality films with precise control over composition and properties.

The RF sputtering process begins with selecting a suitable solid target based on the desired properties of the coating. The target and the substrate are placed within the vacuum chamber to ensure precise deposition, as shown in Fig. 7. Here, the DC power source is replaced by the alternating current at radio frequencies (typically 13.56 MHz) to generate and sustain the plasma inside the vacuum chamber, which is maintained at (0.5–10) mTorr. The radio frequency source is capacitively coupled with the target, creating a time-varying electric field that ionises the gas, typically argon, to form plasma. The inert gas is a medium for transferring momentum from the ionised gas particles (plasma) to the target. The alternating nature of the RF field prevents charge accumulation on the target and substrate, making RF sputtering suitable for non-conductive dielectric materials. This power supply leads to neutralise the charge accumulation on the non-conductive plate. Since the target is non-conducting, it is bombarded by positive ions when a negative potential is connected. The positive charge accumulates on it until it deflects any additional positive bombardment. Subsequently, during the other half of the cycle, electrons bombard it, thereby neutralising the positive ions collected on its surface. This collision process effectively ejects atoms from the target material. These ejected atoms travel through the vacuum chamber, navigating the controlled environment and settling onto the strategically placed substrate. As a result, a coating forms on the substrate, transferring the desired properties of the initially chosen target material to the substrate.

RF sputtering offers precise control over the deposition process, producing extremely uniform and consistently reproducible thin films. It can handle a broad spectrum of target materials, including metals, alloys, and compounds, enabling

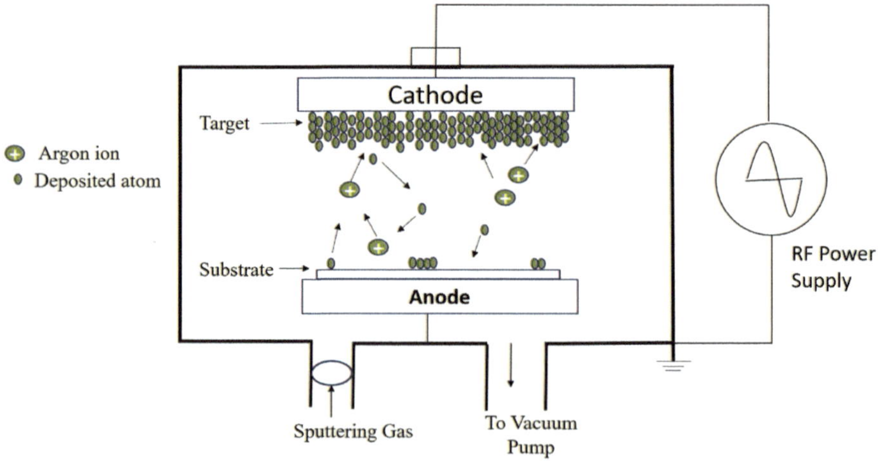

Fig. 7 Radio frequency sputtering system [50]

the deposition of thin films tailored to precise specifications. As an illustration, Nur-E-Alam et al. [51] deposited the 100 nm thick RuO_2 layer on polyimide and alumina substrates by RF magnetron sputtering at a 4 mTorr pressure of the chamber. Their study revealed an unusual magnetic hysteresis pattern with a middle saturation phase in garnet trilayers and extremely sensitive magnetic switching in garnet-oxide composites (GOC). Additionally, it highlighted the exceptional thermal stability found in certain nanoengineered metal-dielectric multilayers. Singh et al. [52] observed a remarkable increase in photocatalytic efficiency of the nanostructured 40nm thick film TiO_2 deposited on the silica glass substrates by RF magnetron sputtering. They attributed it to its reduced band gap and more excellent light absorption in the UV-visible spectrum. Garcia-Valenzuela et al. [53] discovered that optimal surface passivation is achieved using low RF power and high argon pressure when depositing Al_2O_3 thin films. This improvement is linked to decreased surface damage caused by the sputtering process. Furthermore, the ability of RF magnetron sputtering to operate at low voltage and pressure levels further distinguishes it as superior to alternative sputtering methods such as DC sputtering.

Pulsed DC sputtering involves applying a periodic pulse of voltage to the sputtering target material [55]. The pulsing helps resolve arcing and target poisoning, particularly when sputtering reactive materials. The pulsed DC power supply switches between a negative voltage (sputtering phase) and a positive or zero voltage (deposition phase), enhancing the stability and quality of the sputtered films. During the negative voltage phase, ions bombard the target, causing sputtering, and the positive or zero voltage phase helps neutralise charge accumulation, reducing arcing and enhancing film quality. Sputtered atoms are deposited on the substrate, forming a uniform thin film. The pulsed direct current sputtering process is typically carried out with 50 to 90 % duty cycles and 10–350 kHz pulse frequencies. Figure 8 shows a diagram of a typical voltage sequence for single magnetron pulsed DC-reactive sputtering. A negative voltage pulse of a few hundred volts is applied to the target for some time, τ_{on}.

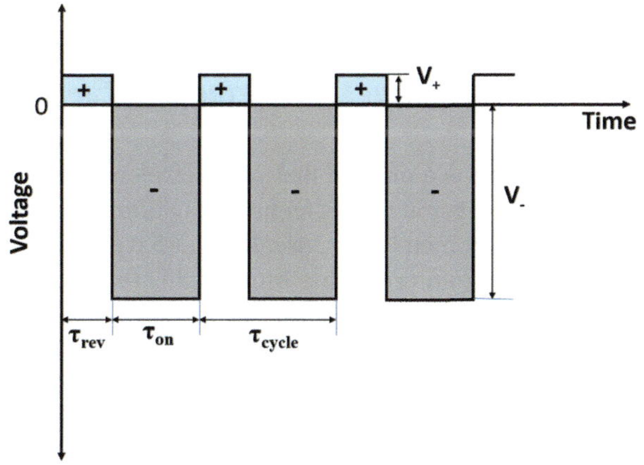

Fig. 8 Ideal voltage sequence applied to pulsed DC sputtering of dielectrics [54]

The power is either regulated to a low positive voltage (nearly 20 V) or turned off for τ_{off} after τ_{on}. Subsequently, the charged dielectric surfaces must undergo complete discharge during the reverse time, τ_{rev}. To ensure arc-free operation, τ_{on} must be short enough to prevent charge accumulation that may lead to breakdown and arcing, while τ_{rev} must be sufficiently long to fully discharge the surfaces, thus preventing charge build-up in successive cycles. Generally, τ_{on} is ten times of τ_{rev} [56]. τ_{on} and τ_{rev} determine the minimum pulsed frequency ($1/\tau_{cycle}$), the critical frequency, and the maximum duty cycle to achieve arc-free pulsed DC reactive sputtering. This critical frequency increases with current and decreases with reverse time and pressure.

Consequently, a corresponding increase in the reverse time is necessary to maintain an arc-free continuous discharge with a longer on-time. The advantages of pulsed DC sputtering include arc suppression, critical when sputtering reactive or compound materials, and improved density. It has been widely employed for depositing thin films of dielectric materials like alumina, Al_2O_3, and titania, TiO_2. The smooth structure of these films is attributed to the absence of particles generated by micro-arcs. In 2019, Erfurt et al. [57] developed hydrogen-doped indium oxide using pulsed DC magnetron sputtering, improving its electrical properties. They discovered that this deposition process can achieve electron mobilities greater than 100 cm^2/Vs.

While studying the sputtering process, comparing these two methods is beneficial. Therefore, let's consider the work of Lelis et al. They analysed the structural and morphological difference between the pulsed DC and RF magnetron sputtering of TiO_2 film and found that at peak nominal power without sample heating or calcination, pulsed-DC magnetron sputtering typically produces smooth, amorphous TiO_2 films. Conversely, due to increased energy during physical vapour condensation, RF magnetron sputtering produces rough films with highly oriented crystalline rutile and anatase TiO_2 phases, specifically in (110) and (101) orientations. Implementing a pulsed-DC and RF magnetron co-sputtering technique made it feasible to control the nucleation and growth of TiO_2 crystalline domains. This method formed purely anatase phase TiO_2 films with a relatively high proportion of the most photocatalytically active {001} facets [58].

2.4 *Electroplating, Anodizing and Electroless Plating*

Electroplating is a method that involves electrochemically depositing a metal layer onto a solid substrate. This technique is often used in nano and microfabrication to achieve thicker metal films, usually over 5 μm [59]. Other methods, like evaporation or sputtering, find producing films of this thickness challenging due to the internal thermal stress developed during deposition. It can be carried out in a room or at a lower temperature, resulting in less thermal stress than other methods. As a result, this finishing process is extensively applied in aerospace, electronics and automotive applications to ward off corrosion and enhance aesthetics. Electroplating works on the principle of electrolysis, where both an anode (metal to be deposited) and a cathode (the base metal) are immersed in an electrolytic solution containing the metal ions intended for deposition. This setup is arranged within a non-conducting vessel, typically made of glass, to prevent unwanted electrical interactions with the surroundings. Upon applying the external power source, the voltage across the electrodes will cause them to become polarised. The presence of metal ions at the cathode surface diminishes, forming a diffusion layer on its surface. The thickness of this diffusion layer is influenced by the temperature, the movement of the solution, and the rate at which metal ions are deposited.

The electrochemical reactions involved in this process are as follows:

At the anode: $M(s) \rightarrow M^{n+}(aq) + ne^-$

At the cathode: $M^{n+}(aq) + ne^- \rightarrow M(s)$

M^{n+} represents the metal ion in these reactions, and n is the number of electrons involved. During electrolysis, a metallic coating is formed. Generally, electrolyte solutions are aqueous, though molten salt or solid electrolytes may occasionally be used.

Electroplating and electroless plating primarily differ because the former involves an external current source for electrolysis. Figure 9 shows the basic difference between the electrolytic and electroless plating processes. Redox reactions occur without an external current supply, with the metallic surface catalysing the reduction of the coating metal from its salts during electroless plating. This process is generally referred to as chemical plating or autocatalytic plating. It was first coined by Abner Brenner and Grace E. Riddell in 1946, even though it was serendipitously discovered by Charles Wurtz in 1846 while performing a nickel phosphorous bath [61]. When Wurtz left the nickel phosphorous bath on the benchtop, it broke down spontaneously and a black powder-coated surface.

Here, the metal ions are preferentially reduced at the surface of the catalytic substrate placed in the electroless plating solution [62]. This substrate continues

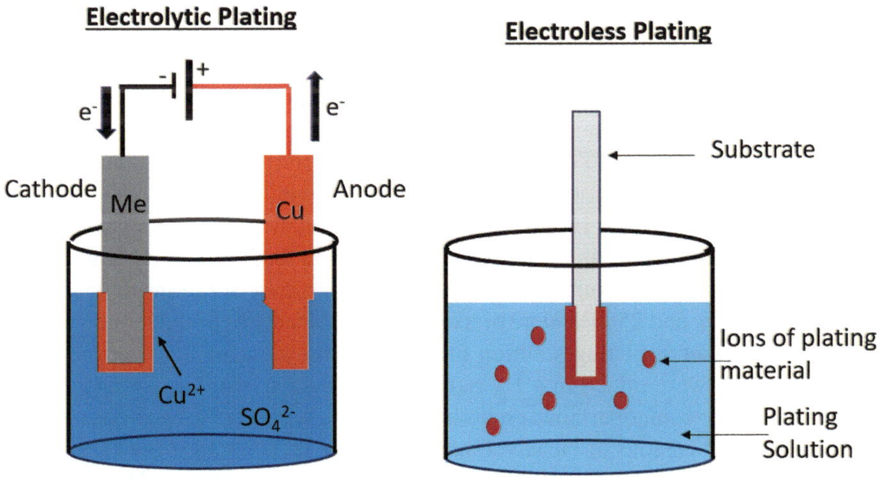

Fig. 9 Comparison of electrolytic and electroless plating processes [59, 60]

to catalyse the deposition of metal ions, and the deposit facilitates further reduction reactions. Above all, the substrate metal atoms retain electrons, ensuring the substrate remains intact and is not depleted during the electroless plating process. It produces coatings with a thickness range of 1 to 120 μm. It is used only when its unique advantages, such as uniform thickness distribution over complex shapes (hard-to-reach areas like corners and cavities), are necessary, since it requires more precise control and is costlier than electroplating. The plating solution contains metal salts for plating, metal complex builders, buffering agents, reducing agents, stabilisers, and brighteners to achieve a glossy finish if desired. The rate at which deposition occurs is influenced by factors like the concentration of metal and reducing agents, pH levels, and temperature.

Additionally, the solution's presence and type of complexing agents can impact plating speed. While typical plating rates range from (3–5) μm per hour, electroless nickel plating can achieve speeds as high as 20 μm per hour. Electroless nickel plating is standard for this type of deposition. Various composition percentages of phosphorus in nickel-phosphorus alloy are coated on metals and plastics without applying an electric current. Electroless plating has three categories: Electroless metallic coating, Electroless composite coating, and Electroless alloy coatings. Nickel plating can be performed both in electrolytic plating and electroless plating, but the obtained products show slightly different properties, including hardness, surface roughness, microstructure, and reduced modulus of elasticity. Electrolytic plating is performed in a Brenner-type electrolytic bath at 50 °C, while a commercial bath is used for electroless nickel-phosphorous plating at 90 °C. Therefore, according to the application, we must consider which plating is more suitable.

Bhujel et al. [63] prepared SiNW arrays using Ag-assisted chemical etching on p-Si (100) wafers. The samples underwent RCA cleaning and drying to eliminate organic contaminants and the surface oxide layer. The cleaned wafers were electroplated by immersing them in a solution of 10% HF and 0.35 M $AgNO_3$ for three different durations at room temperature. The Ag nanoparticle-coated samples were placed in an etching solution of 10% HF and 0.882 M H_2O_2, with etching times of 15, 30, 45, and 60 minutes. After the etching process, the Ag nanoparticles were dissolved by immersing the etched Si substrate in nitric acid for 30 min. Lastly, the wafers were washed and dried. In one paper, they [64] further produced AgNPs through electroless deposition. The silicon wafers were initially treated with acetone and ultrasonic agitation for 10 min, then rinsed multiple times with deionised water. They were subsequently immersed in a piranha solution for 30 min and washed again with deionised water to ensure cleanliness. The silver nitrate solutions of concentrations 5, 10, 15, 20, and 25 mM were prepared. The cleaned silicon substrate was then dipped into an electroplating solution of 5% hydrofluoric acid. The various silver nitrate solutions were mixed in a 1:1 ratio for 10 seconds, resulting in a significant deposition of silver nanoparticles on both sides of the wafer. It was found from their study that the dimensions of the silver nanoparticles (AgNPs) range between 55 and 100 nm, with an average size fluctuating between 57.32 and 80 nm. The crystallite dimensions of the AgNPs ranged from 16.27 to 7.63 nm, accompanied by a corresponding reduction in microstrain within the AgNPs film, which ranged from 0.01 to 0.012, as the concentration of $AgNO_3$ varied from 5 to 25 mM.

In contrast to electroplating, anodising involves an exclusive electrochemical surface treatment to thicken the natural oxide layer on metal parts, typically aluminium, enhancing corrosion resistance and surface hardness and allowing for decorative finishes [65]. Here, the metal is first cleaned, submerged in an electrolyte solution (usually sulfuric acid), and subjected to an electrical current, which forms a thick oxide layer. This layer can be sealed and coloured for additional benefits. There are three primary categories of anodising techniques: chromic anodising, where the solution used is chromic acid; sulfuric anodising, which employs sulfuric acid as the solution; and hard anodising, which involves sulfuric acid either on its own or with additional substances. Less common methods involve sulfuric acid with oxalic acid, phosphoric acid, sulfosalicylic, sulfophthalic acid, and tartaric acid. Most anodic coatings typically have thicknesses ranging from 5 to 18 μm [66]. Anodising is primarily used for aluminium, creating a stable oxide layer, whereas electroplating applies a thin metallic coating to various metals. A notable example is the anodising of aluminium components in consumer electronics, such as Apple's MacBook and iPhone casings, which result in sleek, durable, and colourful finishes that protect against scratches and corrosion. The characteristics of the anodised coating layer depend on the type of aluminium alloy and the anodising conditions.

2.5 Combustion Spray Process

Thermal spray coating is a functional surface coating method where molten or heated materials are released onto a surface to preserve or enhance the efficiency of a substrate [67]. Based on the energy source used for heating or melting the coating materials, the thermal spray process is broadly categorised into three groups: combustion, electric wire, and plasma. The combustion spray process is a thermal spray coating technique that involves applying materials onto surfaces. It requires a combustion flame to melt coating materials and spray them onto a substrate to form a coating. This method enhances surface properties like wear resistance, corrosion resistance, and thermal insulation. The flame spray powder, flame spray wire, detonation, and High Velocity Oxygen Fuel (HVOF) are combustion spray methods.

2.5.1 High-Velocity Oxygen Fuel Spray (HVOF)

The introduction of high-velocity oxygen fuel (HVOF) spraying in the 1980s revolutionised thermal spraying technology and its industrial applications. HVOF coating significantly enhances sprayed coatings' binding strength, hardness, and density thermally. It also lowers or eliminates the oxide mass fraction in the coatings. HVOF

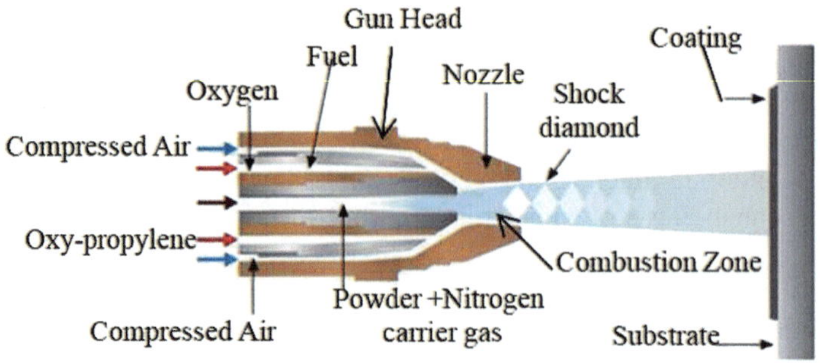

Fig. 10 Schematic diagram of the HVOF spray process [69]

coating features an ultra-high flame velocity of 2000 m/s with a comparatively low-temperature range of (1300–3000) °C, resulting in an optimal thermal and kinetic energy blend [68] (Fig. 10).

In this process, oxygen and propane, kerosene, or acetylene as fuel are delivered into the combustion chamber, mixed, and ignited [69]. The combustion generates high pressure, propelling the gas through an aperture to create a supersonic stream. Typically, carrier argon or nitrogen gas introduces particles into this supersonic stream radially or axially. When particles are heated and accelerated, they undergo collision with the substrate in a solid, semi-molten, or molten state, leading to plastic deformation and forming a dense and firmly adhered coating on the substrate. The three generations of this gun system are designed to boost gas pressure and particle velocity while lowering the flame and particle temperature. The first generation consists of a large combustion chamber with a straight nozzle or a convergent barrel. The second stage of the HVOF spray gun improves upon this by incorporating a Laval nozzle.

Both first and second generations use equal-section nozzles to produce a combustion pressure of 3–5 bars, with spray particles introduced axially into the combustion chamber. The third generation of this system operates at a higher combustion chamber pressure of (6–10) bars, allowing spray particles to be injected either radially or axially. A Laval nozzle is attached here to accelerate the flame flow by (30–50) %. It facilitates better control over particle heating, thereby enhancing deposition efficiency. Typically, the HVOF method utilises low-melting ceramics, metals, and cermets (a composite material consisting of ceramic (cer) and metallic (met) materials).

HVOF Spray Process in Tungsten Carbide Coatings

Tungsten carbide (WC) particles enhance coating hardness, while binders like Ni, Co, or Co-Cr provide toughness. Co improves WC wetting during sintering and maintains strength, existing in stable α (FCC) and ε (HCP) forms depending on temperature. Ni is used instead of Co due to concerns about Co toxicity, cost, and corrosion resistance.

Conventional thermal spraying can convert WC to W_2C, compromising resistance to wear, oxidation, cavitation, and corrosion. The attributes of the feedstock powder, including particle size, shape, and production method, are crucial in determining the quality of the coatings. Carbide particles measuring between 1 and 5 μm are typically used in agglomerated feedstock for applying WC-Co coatings. Achieving particle sizes smaller than 1 μm is challenging due to carbide dissolution and carbon loss, though smaller sizes are preferred. An innovative processing method addresses these issues, with nano-sized particles offering better hardness and toughness. Using nano-powders with HVOF is challenging due to rapid decomposition and increased carbon loss. Despite this, finer powders improve coating quality and thickness. Nano-carbide particles achieve better coating thickness than medium-sized particles. The liquid HVOF method also reduces porosity by about 40% compared to the gas HVOF method.

Song et al. [70] examined the wear behaviour of WC-based coatings and found no phase changes in either type of coating, likely due to the low flame temperature in the liquid HVOF process. The high velocities in the HVOF process provide sufficient deformation energy during splat impacts, minimising porosity and producing high-quality coatings. Process parameters are crucial for enhancing coating properties. Murugan et al. [71] optimised these parameters to reduce porosity and increase hardness, which is essential to wear and corrosion resistance. The oxygen flow rate significantly impacts hardness and porosity, followed by the liquefied petroleum gas flow rate, powder feed rate, and spray distance. Higher cobalt content lowers particle surface temperature without affecting velocity, while finer carbide particles result in lower particle velocity than coarser ones. Hardness improves with a lower stand-off distance and fuel flow rate in the HVOF process. Particle temperature and velocity decrease as particle size increases, with a 30% reduction observed as size increases from 10 to 60 μm. Increased methane flow rate raises particle velocity but lowers temperature, while oxygen flow rate does not significantly impact velocity or temperature. Higher chamber pressure enhances particle velocity due to increased gas viscosity. Ethanol as a fuel increases particle velocity and temperature with higher flow rates. Higher particle feed rates reduce velocity and temperature and increase stresses.

Adding elements like carbon nanotubes (CNTs) and TiC can improve performance. CNTs enhance microhardness and reduce fracture toughness, while TiC forms clusters that do not significantly alter microstructure but grow more considerably with increased TiC content. High cooling rates in the HVOF process led to amorphous and metastable phases, but adding chromium to the binder can enhance coating performance, particularly for wear and erosion resistance.

HVOF Spray Process in Chromium Carbide Coatings

Coatings based on Cr_3C_2 are highly resistant to corrosion but less durable against wear and erosion. Although they do not match the hardness of WC coatings, they are often used in environments where corrosion is a significant problem, especially

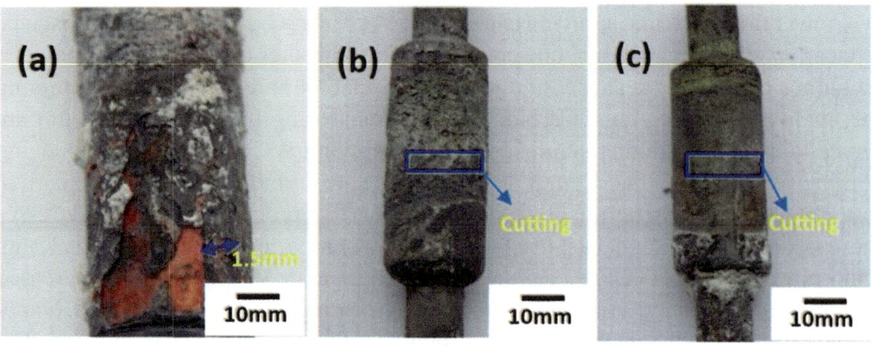

Fig. 11 Physical appearance of **a** P91 steel, **b** Cr_3C_2-NiCr coating, and **c** Cr_3C_2-NiCrMoNbAl coating after 10 hot corrosion cycles testing up to 650 °C [73]

when combined with high temperatures and erosion. When depositing larger carbide particles, using a low-velocity process minimises bouncing. The carbide particles' size and molten speed are critical factors for their retention in the coating. HVOF deposition can cause interdiffusion, leading to increased chromium levels, but also results in oxidation and decarburisation, which may create surface cracks. The hardness of the coating is primarily influenced by the bond strength within and between layers.

Studies have shown that spraying conditions significantly impact the microstructure and performance of carbide coatings. Fine-tuning process parameters is essential, as their interplay affects the quality of the coating. For example, Varis et al. [72] improved the fatigue resistance of Cr_3C_2-NiCr coatings by adjusting process settings, noting that the high-impact energy from HVOF creates compressive stresses that help prevent crack formation and growth. In the study of Zhou et al. [73], the microstructures of HVOF produced-Cr_3C_2-NiCr and Cr_3C_2-NiCrMoNbAl coatings, after undergoing cyclic hot corrosion tests up to 650 °C, revealed that both coatings effectively protected the substrate materials without any corrosive phenomena at the interface, as shown in Fig. 11. The Cr_3C_2-NiCr coating showed a maximum thickness loss of approximately 200 μm due to corrosion.

In comparison, the Cr_3C_2-NiCrMoNbAl coating exhibited a loss of about 70 μm, indicating that the corrosion loss for Cr_3C_2-NiCr was nearly three times higher. The Cr_3C_2-NiCrMoNbAl coating displayed excellent corrosion resistance with no spallation, surpassing both the uncoated P91 steel and the Cr_3C_2-NiCr coating in performance. Xie et al. [74] observed that the feed rate of powder significantly affects the toughness, adhesion, and porosity of Cr_3C_2–25% NiCr coatings. A lower feed rate enhances the melting of particles, leading to increased hardness and reduced porosity. Achieving optimal coating properties relies on carefully adjusting thermal spray parameters to produce coatings with high carbide content and minimal defects, resulting in superior wear and corrosion resistance (Fig. 12).

The main benefits of the HVOF process include very low porosity (less than 1%) due to higher particle impact speeds, strong bonding (over 80 MPa), smoother spray

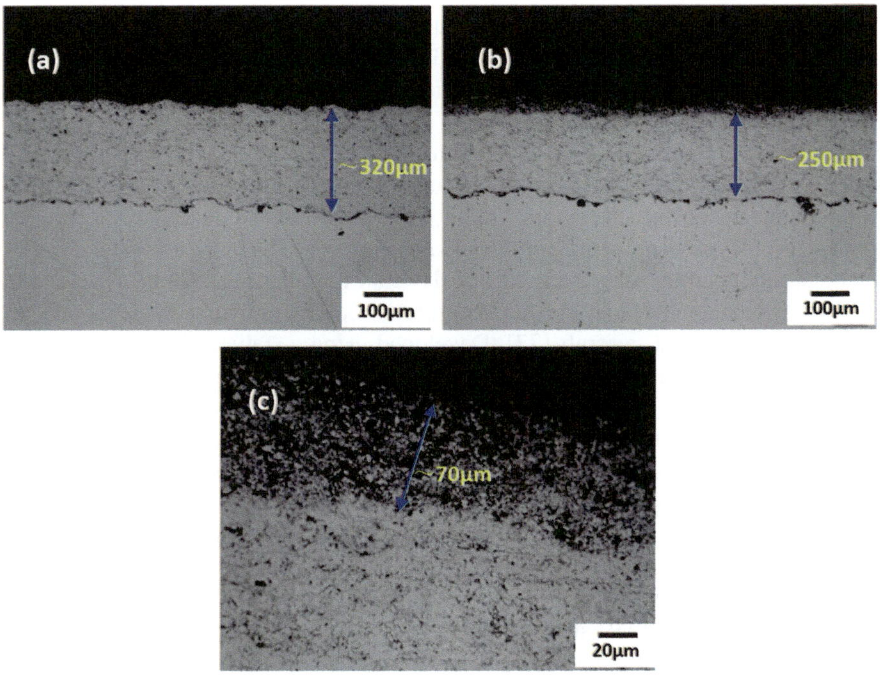

Fig. 12 **a** Optical micrograph of the polished cross-section of the sprayed Cr_3C_2-NiCrMoNbAl coating, **b** and **c** optical micrographs of the polished cross-section of Cr_3C_2-NiCrMoNbAl coating after hot corrosion [73]

surfaces from higher impact speeds and finer powder sizes, excellent wear resistance with more rigid, harder coatings up to 1400 Hv, and dense, thick coatings with minimal residual stresses. Compared to electroplating, HVOF offers less environmental impact, downtime, and enhanced density with fewer cracks and pores. One of its drawbacks is that it is complicated since different processing factors affect its characteristics and microstructure. Narrow size distributions are required due to the restriction of powder sizes to a range of about 5–60 μm. Additionally, it is expensive and demands skilled personnel to provide continuous coating quality and safe functioning. HVOF spraying needs a clear line of vision to the surface and a spraying range of 150–300 mm, which makes it challenging or impossible to deposit coatings onto the inner surfaces of tiny cylindrical components or other regions with limited access. Meticulous adjustment of spray parameters and selection of powder feedstock is necessary to ensure adequate particle velocity for effective coating while avoiding excessive particle heating that could lead to decarburisation. Pulsford et al. focused on demonstrating that WC-Co-Cr HVOF thermal-sprayed coatings with effective tribological properties can be applied to internal surfaces with diameters as narrow as 70 mm. The 70 mm internal diameter coating outperformed the 110 mm during wear tests at 96 N. However, increasing the internal diameter to 110 mm led to lower fracture toughness and higher wear rates under all test loads. The coating applied at

90 mm showed decreased specific wear rates at both test loads because of its micro-hardness and superior fracture toughness. In contrast, the coating applied at 110 mm exhibited the highest specific wear rates under both load conditions due to its inferior fracture toughness.

Despite its drawbacks, the HVOF coating technique is extensively used in marine, paper, petroleum products, aircraft, automotive, power generation, mineral extraction, chemical, and other sectors [75, 76]. STOVL aircraft face particle erosion damage in dusty environments, compromising engine performance and safety. HVOF nano-carbide coatings are proposed by Ma et al. to enhance erosion resistance on significant components like airfoils, as current coatings for helicopter engines are unsuitable for STOVL aircraft [77]. Developed using a statistical design of experiments, these coatings show 3 to 10 times greater erosion resistance than titanium alloys, depending on impact angles. Stainless steel and D2 tool steel substrates of varying depths have been effectively repaired using this process, achieving a total accumulated thickness of nearly 5.5 mm [78]. The fabrication of superior-quality wear-resistant coatings for carbides and satellites also widely employs the HVOF process. Corrosion in boiler steel tubes, caused by corrosive agents, poses a significant danger. To mitigate this, Singh et al. [79] applied Cr_3C_2-25(Ni-20Cr) and Ni-20Cr coatings to T91 boiler tube steel using the HVOF process. Hot corrosion studies on uncoated and coated T91 steel specimens were conducted in a Na_2SO_4–60% V_2O_5 environment at 900 °C under cyclic conditions. Each cycle included 1 h of heating and 20 min of air cooling. They found that the Cr_3C_2-25(Ni-20Cr) coating provided superior protection compared to the Ni-20Cr coating.

2.6 *Vacuum Deposition Process*

A vacuum, defined as space free of matter, decreases the number of atoms in the environment by increasing the average distance an atom can travel before colliding with another. Creating a vacuum can eliminate unwanted gas atoms from the deposition environment, helping to achieve the desired chemical composition of the layers. Vacuum deposition applies thin material films onto a substrate within a low-pressure (vacuum) environment [80]. Sputtering and sputter deposition became known in oil-sealed piston pumps in the middle of the 1800s. However, vacuum deposition was not practical until Sprengel invented Mercury-column vacuum pumps. Edison evacuated the first carbon-filament incandescent bulbs in 1879 using this pump, and the first vacuum evaporation was carried out in 1887 by Nahrwold. The widespread vacuum deposition of metallic thin films did not become prevalent until the 1920s.

Here, the purity of the film primarily depends on the vacuum pressure deposition rate and the materials' reactivity. Vacuum pressure higher than 10^{-4} Torr allows atoms to travel longer distances on average before colliding in the vacuum during deposition. The substrate is contaminated even at 10^{-4} Torr due to significant amounts of unwanted trace gases. So, either an ultrahigh vacuum ($<10^{-9}$ Torr) or a high vacuum (10^{-7} Torr) environment is necessary to improve the purity of the film for vacuum deposition [81].

The equilibrium vapour pressure describes the vapour phase of a substance in equilibrium with its solid or liquid phase at a specific temperature. It is accurately measured under controlled circumstances using a Knudsen cell. This effusion cell is comprised of a sealed chamber with a small opening of accurately defined conductance. The vapour pressures of various substances can vary significantly at a particular temperature. Achieving a feasible deposition rate in vacuum deposition requires a sufficiently high vaporisation rate. The optimal vapor pressure for practical deposition is 10^{-2} Torr. High-vapour-pressure surface impurities evaporate first when the materials are heated. The material evaporates freely from its surface when it vaporises. This rate of evaporation, which depends directly on the vapour pressure, is governed by the Hertz-Knudsen equation.

When alloys are vaporised, the selective vaporisation creates a gradient in the composition, which can be beneficial or problematic. For instance, when a copper-gold alloy film is evaporated to coat the surface of polymers, gold, which has a lower vapour pressure than copper, evaporates afterwards. This causes the accumulation of copper at the interface, which enhances the bond between the polymer and the deposited layer. Feeding sources, such as a wire or pellet feeder, are sometimes used to restore the depleted component of such deposition. Combining an element with another material can also modify its vaporisation properties. The material's surface is first required to achieve the vapour pressure via resistive heating, inductive (RF) heating, and high- and low-energy electron beams for thermal vaporisation.

There are two ways to evaporate the material to be deposited: chemical and physical methods. In chemical vapour deposition, solid materials are agglomerated from a gas phase, and physical vapour deposition is where solid precursors are vapourised from a target and deposited onto the substrate. An electron beam, resistive heating, and an electric arc are often used in physical vapour deposition to heat the target material in a vacuum until it evaporates and sputters off the target. The vapourised material is then deposited onto the surface of the substrate. Ultimately, the metal vapour condenses and forms a film on the surface of the substrate. Thermal evaporation deposition is one of the oldest physical vapour deposition methods. Its phenomenon is schematically shown in Fig. 13. This method can deposit films at a rate of 10–100 nm/min, depending on the sublimation of the target material in the vacuum chamber. However, due to the higher rate of deposition, superior step coverage, a more comprehensive selection of materials, and dense and well-adherent film, the sputtering process is frequently chosen over thermal evaporation in the commercial fields. Moreover, the thickness of PVD deposition depends on the deposition time and the vapour density.

The process of CVD engages the chemical reaction of precursors to coat the desired material onto the surface of the base material. CVD processes are usually performed at low pressure to prevent contamination in the gas phase reactions. Vacuum deposition is widely used in various industries, including electronics, optics, and materials science, to create coatings with precise thickness and composition for applications such as ceramic metallisation, semiconductor metallisation, capacitor

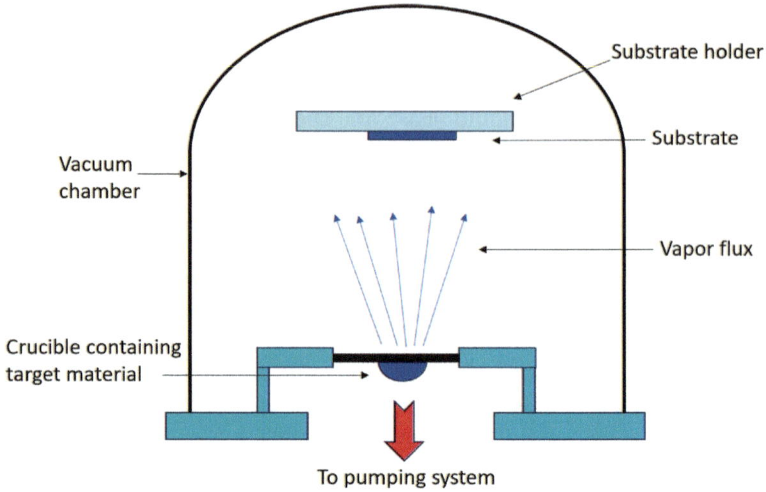

Fig. 13 Schematic diagram of the thermal evaporation process [82]

foils metallisation, optical coatings, decorative aluminium, and gold coatings on plastics. The vacuum environment ensures high purity and control over the deposition process, producing high-quality films with uniform properties. Each technique offers distinct advantages and limitations. The desirable quality of the end product is generally the deciding factor when choosing between these two deposition methods. PVD is ideal for hard coatings and optical applications and for creating durable and precise coatings. However, it may be limited by its lower deposition rates and requirement for line-of-sight application.

In contrast, chemical vapour deposition (CVD) can create films with excellent conformality and uniformity over complex shapes and offers higher deposition rates. Nevertheless, CVD might involve hazardous chemicals and higher temperatures, which can limit its use with specific substrates. Thus, selecting between PVD and CVD depends on specific coating requirements, including thickness, composition, uniformity, and the nature of the substrate.

Unlike solution-based coating methods, vacuum deposition has established itself as the premier technology in the thin film semiconductor industry due to its ability to enable large-area coverage, high throughput, contamination-free and solvent-free deposition, continuous production compatibility, and outstanding robustness. Feng et al. developed a scalable vacuum deposition technique for fabricating large-area perovskite films using low-temperature annealing [83]. The films deposited on glass and flexible substrates showed excellent uniformity and quality. When annealed at 60 °C under vacuum, the films exhibited large grain sizes, long carrier lifetimes, and minimal trap densities.

2.7 *Chemical Vapor Deposition Process*

Chemical deposition processes are methods used to deposit a material onto a substrate through chemical reactions. These processes have become popular in various fields, such as electronics, optics, and materials science, for creating high-performance, high-quality thin films and coatings. J.M. Blocher proposed at the 1960 Electrochemical Society (ECS) conference that chemical reactions based on thin film deposition methods be distinguished from those based on physical processes, leading to the adoption of the terms CVD and PVD [84]. Many individuals consider him to be the founder of contemporary chemical vapour deposition. The earliest CVD processes, patented in the late 19th century, produced carbon powder for pigments and carbon fibers for lamp filaments. CVD of metals also began in this period, with one of the first examples being Wohler's 1855 deposition of tungsten from WCl_6 in a hydrogen atmosphere.

Gaseous precursors in CVD react or decompose on the base surface to coat a solid film. The procedure for this deposition involves heating the substrate in a reaction chamber, introducing gaseous precursors into the chamber, and initiating chemical reactions on the surface of the base material to deposit a thin coating. Energy is delivered to the substrate inside the chamber to assist the reaction with the carrier gas. The fundamental processes include creating a reactive gas mixture, transporting the reactant gas to the substrate, adsorbing the reactants on the substrate surface, and reacting with the adsorbed materials to form the coating. The substrate must be mechanically or chemically cleaned before deposition, and the chamber must be free of leaks, dust, and moisture.

Common materials used in CVD coatings include nickel, tungsten, chromium, and titanium carbide, which enhance corrosion and wear resistance and provide specific properties that are challenging to achieve using other methods. Chemical Vapor Deposition (CVD) is one of the earliest methods to produce industrial pigments like Al_2O_3, TiO_2, Si_3N_4, SiO_2, and carbon black in nano or micron sizes. It can also deposit pure metals on substrates. The quality of layers produced by CVD is influenced by substrate temperature, reactant concentration, gas pressure, and gas flow rate, similar to the Physical Vapor Deposition (PVD) process.

2.7.1 Types of CVD Process

This includes techniques like Atomic Layer Deposition, Metal-Organic Chemical Vapor Deposition, Plasma-Enhanced Chemical Vapor Deposition, Vapor Phase Epitaxy, and Laser Chemical Vapor Deposition. Chemical Vapor Deposition processes, particularly ALD, are renowned for producing highly uniform and conformal coatings with precise atomic-scale thickness control.

Atomic Layer Deposition

The development of ALD, patented by Suntola in 1977, is considered the latest breakthrough in the field [85]. ALD was first commercially applied in thin film electroluminescent displays, producing superior luminescence and protective layers compared to the best thin film synthesis techniques of the 1970s. However, research on this remained scarce until the late 1990s, when it emerged as a potential solution for high-permittivity thin films in memory devices and transistors. Intel recently achieved a notable milestone in the miniaturisation of integrated circuit chips by developing large-permittivity metal gate transistors using ALD (atomic layer deposition). Its ability to reduce the size of microelectronic devices has been a key factor in the increasing interest in this field. ALD is a film deposition technique characterised by sequential self-terminating gas-solid reactions.

When applying thin films of metals, oxides, polymers, or other materials to different substrates, a gas-phase chemical process naturally regulates itself to deposit one atomic layer at a time. It is known as atomic layer deposition technology. Before initiating this process, the surface is prepared to a known and controlled state, often through methods such as heat treatment. The growth of this process typically involves a repetitive sequence of the following four characteristic steps:

- The self-terminating reaction of the first reactant.
- Evacuation removes reactants that have not undergone reaction and gaseous by-products from the chamber.
- Self-terminating reaction of the second reactant or another surface activation of the treatment for the first reactant.
- Evacuation to prepare the chamber for the next cycle of reactions [86].

Each reaction cycle deposits a specific amount of material onto the surface, called the growth per cycle. Material layers are grown by repeating these cycles until the desired thickness is achieved. Due to meticulous surface control, ALD-grown films exhibit exceptional conformity and uniformity in thickness. These processes typically occur above 100 °C, except Al_2O_3 (200 °C). A particularly successful ALD process spanning a broad temperature range is the deposition of Al_2O_3. Al_2O_3 ALD growth can be achieved at temperatures as low as 33 °C. Despite slower reaction rates and more extended purging periods with H_2O reactants, it exhibits a high dielectric constant, low leakage currents, and low surface roughness, similar to films grown at higher temperatures. Figure 14 shows the schematic diagram for the Al_2O_3 atomic layer deposition cycle [87]. In this, trimethylaluminium, $Al_2(CH_3)_6$, and water introduce hydroxyl groups to the metal surface. As a result of partial breakdown, $Al_2(CH_3)_6$ combines with the hydroxyl groups to generate a surface terminated with $–CH_3$, creating one monolayer. Excess precursors are purged using inert gas. The $–CH_3$-ended surface then interacts with water, similarly resulting in the creation of one monolayer of Al_2O_3. This ALD cycle is repeated to reach the desired film thickness, leveraging the self-limiting features of the reactions. As Al_2O_3 material is highly impermeable to oxygen gas and water vapour, it is ideal for effectively blocking

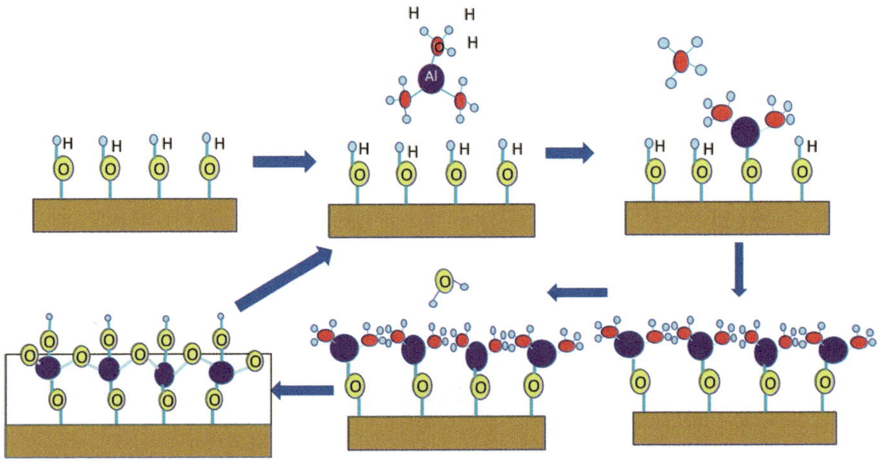

Fig. 14 The schematic diagram for the Al_2O_3 atomic layer deposition cycle [87]

these gases. This coating protects the mechanical and optical properties of the coated object.

Thermal Chemical Vapour Deposition

The deposition of thin films onto a substrate by chemically reacting gaseous precursors at high temperatures (500–1000 °C) is termed Thermal Chemical Vapor Deposition (CVD). Here, the precursor is introduced into a heated chamber and subjected to reduction or thermal decomposition. The reaction causes material to deposit onto the substrate, creating a thin film or coating. It also produces additional compounds through interactions with other gases present.

Kumar et al. [88] utilised a CVD process to deposit SiCN thin films on a Si (100) substrate. SiC and Si_3N_4 nanopowders were used as sources, while a mixture of N_2 and H_2 served as precursors. The N_2 flow rate was adjusted between 40 cm to 100 sccm, with a constant H_2 flow rate of 80 sccm. The powders and the cleaned Si (100) substrate were placed in a ceramic boat at the center of a furnace. The deposition occurred at a base pressure of 10^{-6} Torr and atmospheric pressure, with a processing temperature of 1000 °C. The film was deposited over 3 hours with a heating rate of 5 °C/min, and the furnace was then cooled gradually at a rate of 3 °C/min. Figure 15 shows that as the N_2 flow rate increases, the peaks for the (200) and (102) planes become narrower, indicating improved intermediate-range order. Additionally, the Si (400) plane shifts from 72.94° to 71.92° with an increasing N_2 flow rate from 40 to 120 sccm. This shift towards the standard diffraction angle of 69.13° suggests that higher N_2 flow rates reduce strain in the Si crystal within the SiCN thin film. In the CVD chamber, titanium aluminium nitride (TiAlN) thin films were deposited on p-type Si (100) substrates with varying N_2 flow rates by Das et al. [89]. FESEM

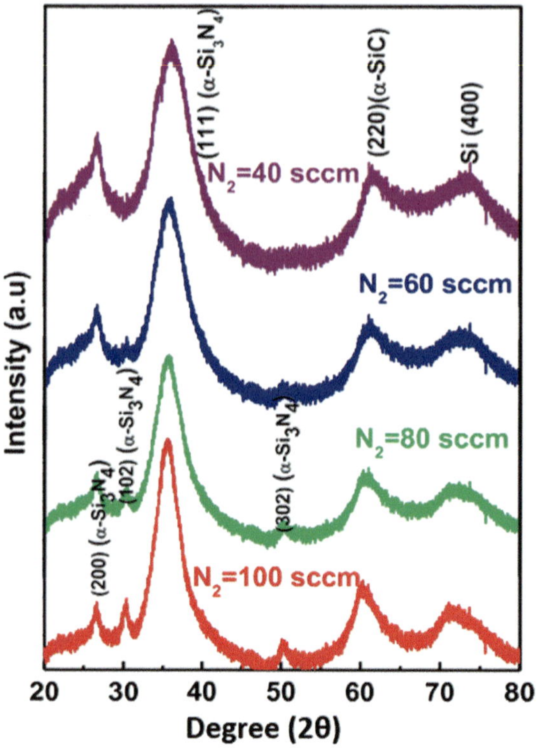

Fig. 15 XRD plot of SiCN thin films [88]

images showed increased porosity and uneven grain distribution with higher N_2 flow rates. XRD results identified TiAlN, AlN, and Al_2O_3 phases at specific diffraction angles. AFM images revealed an average surface roughness of 36.9 nm at a 30 sccm N_2 flow rate. Mechanical tests showed a hardness of 21 GPa, Young's modulus of 400 GPa, a plasticity index of 0.094, and a deformation-to-plastic resistance ratio of 0.318 GPa, demonstrating enhanced crack resistance with increased N_2 flow rates.

Plasma-Enhanced Chemical Deposition

Plasma is an energy source for plasma-enhanced chemical vapour deposition (PECVD). As shown in Fig. 16, it is a complex deposition process involving various species and reactions. The density of carbon-related species influences the morphology of as-grown graphene. Species like H_2 and Ar act as etchants for amorphous carbon, enhancing graphene quality. Electrical properties can be tuned by doping graphene with heteroatoms using NH_3 and N_2. Adhikari et al. [91] found that hydrogen improves graphene quality by shaping the grains and removing weak carbon-carbon bonds, including those in amorphous and oxidised carbon.

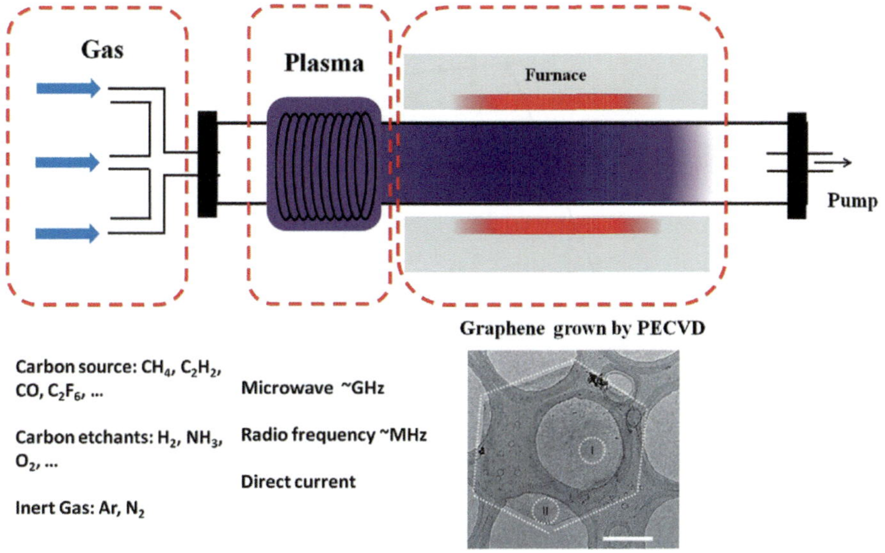

Fig. 16 Schematic of the experimental setup of plasma-enhanced chemical vapor deposition [90]

Vishwakarma et al. [92] investigated how temperature affects the CVD process for growing monolayer and few-layer graphene on thin silicon oxide/Si substrates using ERC-CVD. At 550 °C, graphene shows large grain sizes and high crystallinity but suffers from poor stitching due to random grain orientation. Increasing the temperature to 650 °C results in smaller grains with improved conductivity and enhanced material quality, while annealing treatments help remove hydrogen, making the low-temperature method scalable and suitable for various applications. Using the plasma-enhanced chemical vapour deposition method, Kulczyk-Malecka et al. [93] synthesised vertically aligned graphene on silicon substrates at approximately 300 °C. They observed that the I_D/I_G ratio increases with a deposition time of up to 25 min, as shown in Fig. 17, indicating an increase in defects as sp^2 carbon converts to sp^3 amorphous carbon. The 15-min sample has the highest defect density. Beyond 25 min, the ratio decreases, suggesting fewer defects and stabilised sp^2 carbon. The broadening of Raman spectral peaks, as indicated by the full width at half maximum (FWHM), is attributed to the random orientations of graphene layers. Deposition times in the range of 15–25 min were found to be optimal, with samples deposited for 15 min exhibiting the highest defect density (FWHM).

The plasma-assisted chemical vapour technique creates uniform reactions to form reactive radicals and ions, forming layers on a substrate [94]. A major advantage of this technique over thermal chemical vapour deposition involves its ability to occur at near-ambient temperatures, making it suitable for temperature-sensitive substrates. It is especially advantageous for carbon-based material deposition and applications requiring a low substrate temperature and the production of high-density films.

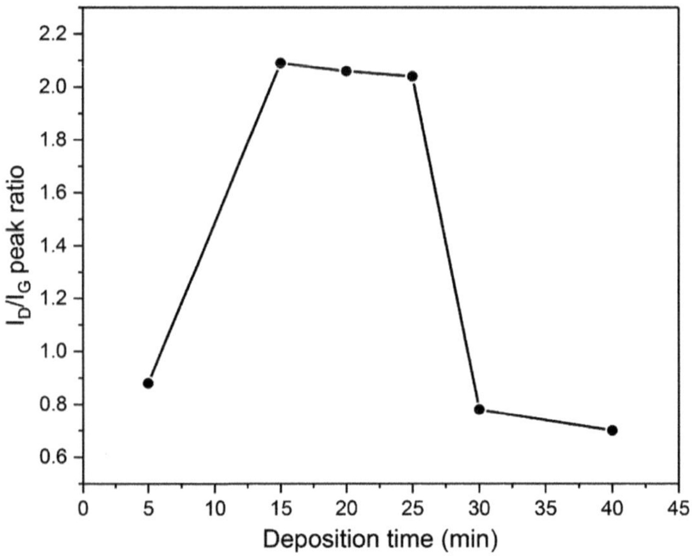

Fig. 17 The change in I_D/I_G Raman intensity ratio over deposition time [93]

Low-Pressure Chemical Vapor Deposition

Low-pressure chemical vapour deposition (LPCVD) involves low pressure (0.1–10 Torr) [95]. Since the late 1900s, LPCVD has been a successful approach for fabricating superior carbon nanotubes (CNTs). This technique has also shown great potential for synthesising graphene on transition metal bases. Li et al. [96] effectively utilised LPCVD to create a single layer of graphene on copper foil with a thickness of 25 μm. Yamauchi et al.'s work [97] on LPCVD of TiO_2 in an H_2-ambient shows a lower deposition rate and activation energy than an O_2-ambient, with improved electrical resistivity but inferior crystallographic quality. Optical absorption increases around 2.2 eV due to Ti^{3+} and Ti^{3+}-related complexes.

2.7.2 Comparative Study of Different CVD Techniques

The decomposition process of CVD can be enhanced or accelerated through heat, plasma, or other techniques. Ion plating, low-pressure CVD, plasma-enhanced CVD, laser-enhanced CVD, active reactive evaporation, ion beam, and laser evaporation techniques are included. The main distinction among these techniques is how chemical reactions are started. They are generally categorised, according to the working pressure, as atmospheric pressure, low pressure, and ultrahigh vacuum (below 10^{-8} Torr). The comparative study of some of the CVD techniques is presented in Fig. 18.

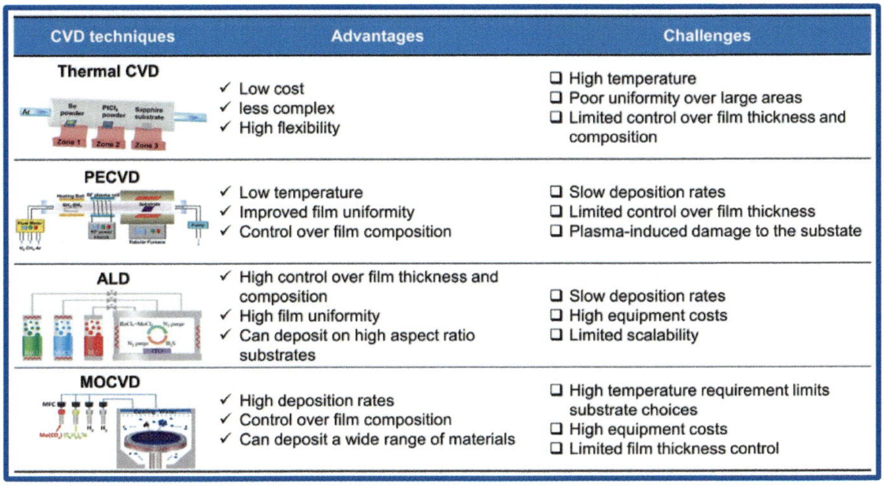

CVD techniques	Advantages	Challenges
Thermal CVD	✓ Low cost ✓ less complex ✓ High flexibility	❑ High temperature ❑ Poor uniformity over large areas ❑ Limited control over film thickness and composition
PECVD	✓ Low temperature ✓ Improved film uniformity ✓ Control over film composition	❑ Slow deposition rates ❑ Limited control over film thickness ❑ Plasma-induced damage to the substate
ALD	✓ High control over film thickness and composition ✓ High film uniformity ✓ Can deposit on high aspect ratio substrates	❑ Slow deposition rates ❑ High equipment costs ❑ Limited scalability
MOCVD	✓ High deposition rates ✓ Control over film composition ✓ Can deposit a wide range of materials	❑ High temperature requirement limits substrate choices ❑ High equipment costs ❑ Limited film thickness control

Fig. 18 Overview of different CVD techniques [98]

2.7.3 Advantages and Disadvantages of Chemical Vapour Deposition

CVD is an accurate and scalable method for depositing thin films, offering excellent control over thickness, composition, and properties by optimising reaction conditions. It can accommodate multiple substrates simultaneously, making it ideal for commercial applications, and it can deposit a wide range of materials, including metals, semiconductors, and ceramics [99]. CVD's adaptability to new materials, such as 2D and high-purity polymers, has broadened its applications. It produces high-quality films with superior uniformity, purity, and adhesion, and can conformally coat complex geometries. The substrate-independent nature of CVD allows bioactive surface chemistries to be transferred to various substrates, enhancing tissue engineering applications. CVD facilitates fundamental research into the effects of surface chemistry and topography on cell differentiation and is increasingly used to modulate cellular behaviour in tissue engineering [100]. With rapid deposition methods and advancements in real-time monitoring, CVD continues to expand its applications across consumer products, energy, sustainability, and healthcare. Its ability to produce mechanically and chemically stable films under diverse conditions further underscores its versatility and reliability.

CVD technology, much like a double-edged sword, offers numerous benefits but also faces several limitations, including the need for high temperatures and high vacuum conditions, particularly for the synthesis of graphene and h-BN [101]. Quartz tubes, commonly used for their thermal and mechanical strength, are scalable only up to 12 in. in diameter. In contrast, stainless steel chambers, used in industrial-scale reactors, require advanced chiller systems to manage heat. Contamination is a significant issue, as quartz reactors need periodic replacement when contaminated, and stainless-steel reactors require regular cleaning to maintain synthesis quality. Residual precursors can cause severe contamination in reaction chambers, necessitating replaceable quartz tubes and gaseous precursors for 2D material synthesis.

Additionally, plasma modules combined with thermal CVD systems can introduce defects, particularly in graphene. Roll-to-roll synthesis of materials like copper foils poses mechanical challenges at high temperatures, requiring vertical tension control to minimise deformation and promote high-quality synthesis. Quality control is also challenging, as current fast analysis methods are valid only for post-CVD graphene on insulating substrates, making it challenging to evaluate as-grown graphene on copper foil. Finally, achieving uniform deposition over large areas is still limited, and there is a tendency for the process to eject microsized particles, impacting the coating quality.

3 Conclusion

In this chapter, we have explored coating technology broadly, highlighting a significant range of methods, advancements, and applications across various sectors. It has evolved from ancient basic protective layers to robust, sophisticated systems that enhance materials and products' durability, functionality, and visual appeal. These wide varieties of coating methods discussed, including magnetron sputtering, cathodic arc evaporation deposition, radio frequency and pulsed DC sputtering, electroplating, anodising, electroless plating, combustion spray techniques like HVOF, and vacuum deposition, feature their flexibility and accuracy in addressing various industrial needs. Each process provides unique benefits to managing materials' deposits, improving material characteristics and ensuring they meet specific application requirements.

Further advancements in coating technologies are propelled by a continuous drive to enhance key performance indicators of wear protection, anticorrosion development, thermal regulation, and optical characteristics. These innovations serve in aerospace, automotive, and electronics and open new routes in renewable energy and healthcare. As we tackle challenges in environmental sustainability and technological integration, coating technology will remain invaluable to shaping the future of material design and manufacturing.

References

1. Huang JH, Tsai ZE, Yu GP (2008) Mechanical properties and corrosion resistance of nanocrystalline ZrN_xO_y coatings on AISI 304 stainless steel by ion plating. Surf Coat Tech 202:4992–5000
2. Kuprin AS, Ovcharenko VD, Gilewicz A, Tolmachova GN, Kolodiy IV, Vasilenko RL, Kuznetsova T, Lapitskaya V, Warcholinski B (2022) Structural, mechanical and tribological properties of Cr-V-N coatings deposited by cathodic arc evaporation. J Tribo Int 165:107246

3. Raman Singh RK, Banerjee PC, Lobo DE, Gullapalli H, Sumandasa M, Ashwin K, Lokesh C, Rachel T, Pulickel MA, Mainak M (2012) Protecting copper from electrochemical degradation by graphene coating. J Carbon 50:4040–4045
4. Liu C, Ma C, Xie Q, Zhang G (2017) Self-repairing silicone coating for marine anti-biofouling. J Matr Chem A 5:15855–15861
5. Newby BZ, Cutright T, Barrios CA, Xu Q (2006) Zosteric acid-an effective antifoulant for reducing fresh water bacterial attachment in coatings. J Coat Technol Res 3:69–76
6. Craughwell TJ, Craughwell TJ (2012) 30,000 years of inventions. Tess Press, New York
7. Edwards KN, Mislang HB (2000) History of coatings. In: Craver CD, Carraher CE (eds) Applied polymer science: 21st Century. Pergamon, Oxford, pp 439–447
8. Allcock BW, Lavin PA (2003) Novel composite coating technology in primary and conversion industry applications. Surf Coat Technol 163–164:62–66
9. Cherrington R, Marshall J, Alexander AT, Goodship V (2022) Exploring the circular economy through coatings in transport. J Sus Prod Consump 32:136–146
10. Jin Q, Tian G, Li J, Zhao Yan, Yan H (2019) The study on corrosion resistance of superhydrophobic magnesium hydroxide coating on AZ31B magnesium alloy. J Col Surf A 577:8–16
11. Khorsand S, Raeissi K, Ashrafizadeh F, Arenas MA (2015) Super-hydrophobic nickel-cobalt alloy coating with micro-nano flower-like structure. Chem Engg J 273:638–646
12. Fauchais P, Vardelle A (2012) Thermal sprayed coatings used against corrosion and corrosive wear. In: Jazi HS (ed) Advanced plasma spray applications. Intech, Rijeka
13. Chu PK (2007) Enhancement of surface properties of biomaterials using plasma-based technologies. Surf Coat Technol 201:8076–8082
14. Pikul JH, Zhang HG, Cho J, Braun PV, King WP (2013) High-power lithium ion microbatteries from interdigitated three-dimensional bicontinuous nanoporous electrodes. Nat Commun 4:1732
15. Kenneth K, Allan M (2009) Coatings tribology: properties, mechanisms, techniques, and applications in surface engineering. Elsevier, Amsterdam
16. Madhuri KV (2020) Thermal protection coatings of metal oxide powders. In: Al-Douri Y (ed) Metal oxide powder technologies. Elsevier, Amsterdam, pp 209–231
17. Singh TRR, Mcmillan H, Mooney K, Alkilani AZ, Donnelly RF (2013) Microneedles for drug delivery and monitoring. In: Zhou Y (ed) Li XJ. Microfluidic devices for biomedical applications, Woodhead Publishing, pp 185–230
18. Czettl C, Thurner J, Schleinkofer U (2018) Knowledge-based coating design of CVD TiN-TiBN-TiB_2 architecture. Int J Refract Met Hard Mat 71:330–334
19. Barber ZH (2006) The control of thin film deposition and recent developments in oxide film growth. J Mater Chem 16:334–344
20. Gracio JJ, Fan QH, Madaleno JC (2010) Diamond growth by chemical vapor deposition. J Phys D: Appl Phys 43:374107
21. Thornton JA (1978) Magneton sputtering: basic physics and application to cylindrical magnetrons. J Vac Sci Technol 15:171–177
22. Jain NK, Sawant MS, Nikam SH, Jhavar S (2016) Metal deposition: plasma-based processes. In: Shohet JL (ed) Encyclopedia of plasma technology, 1st edn. Taylor and Francis, New York, p p19
23. Waits RK (1978) Planar magnetron sputtering. J Vac Sci Technol 15:179–187
24. Gudmundsson JT (2020) Physics and technology of magnetron sputtering discharges. Plasma Sourc Sci Technol 29:113001
25. Baptista A, Silva FJG, Porteiro J, Miguez JL, Pinto G, Fernandes LL (2018) On the physical vapor deposition (PVD): evolution of magnetron sputtering processes for industrial applications. Procedia Manuf 17:746–757
26. Swann S (1988) Magneton sputtering. Phys Technol 19:67
27. Kay E (1963) Magnetic field effects on an abnormal truncated glow discharge and their relation to sputtered thin-film growth. J Appl Phys 34:760–768

28. Golosov DA (2017) Balanced magnetic field in magnetron sputtering systems. Vacuum 139:109–116
29. Mayrhofer PH, Kune F, Musil J, Mitterer J (2002) A comparative study on reactive and non-reactive unbalanced magnetron sputter deposition of TiN coatings. Thin Solid Films 415:151–159
30. Cubillos GI, Bethencourt M, Olaya JJ (2015) Corrosion resistance of Zirconium oxynitride coatings deposited via DC unbalanced magnetron sputtering and spray pyrolysis-nitriding. Appl Surf Sc 327:288–295
31. Hsieh JH, Wu W, Li C, Yu CH, Tan BH (2003) Deposition and characterisation of T(C, N, O) coatings by
32. Wuhrer R, Yeung WY, Phillips MR, McCredie G (1996) Study on d.c. magnetron sputter deposition of titanium aluminum nitride thin films: effect of aluminum content on coating. Thin solid films 290–291:339–342
33. Madhuri A, Jena S, Mukul G, Swain BP (2023) Nitrogen flow rate dependent atomic coordination, phonon vibration and surface analysis of DC magnetron sputtered nitrogen rich-AlN thin films. Physica B 666:415141
34. Jena S, Gupta M, Swain BP (2024) Effect of interlayer thickness on electrochemical properties and corrosion resistance of sputtered TiNxnm/Tixnm multilayer thin films. Mater Sci Semi Process 182:108729
35. Txintxurreta J, G-Berasategui E, Ortiz R, Hernandez O, Mendizabal L, Barriga J (2021) Indium tin oxide thin film deposition by magnetron sputtering at room temperature for the manufacturing of efficient transparent heaters. Thin Solid Films 11:92
36. Hermans Y, Mehmood F, Lakus-Wollny K, Hofmann JP, Mayer T, Jaegermann W (2021) Reactive dual magnetron sputtering: a fast method for preparing stoichiometric microcrystalline $ZnWO_4$ thin films. Surfaces 4:106–114
37. Constantin DG, Apreutesei M, Arvinte E, Marin A, Andrei OC, Munteanu D (2011) Magnetron sputtering technique used for coatings deposition; technologies and applications. Technol Appl Recent 12
38. PalDey S, Deevi SC, Alford TL (2004) Cathodic arc deposited thin film coatings based on TiAl intermetallics. Intermetallics 12:985–991
39. Warcholinski B, Gilewicz A, Myslinski P, Dobruchowska E, Murzynski D (2020) Structure and properties of AlCrN coatings deposited using cathodic arc evaporation. Coatings 10:793
40. Warcholinski B, Gilewicz A (2011) Multilayer coatings on tools for woodworking. Wear 271:2812–2820
41. Steinhorst M, Giorgio M, Teja R, Leyens C (2022) Influence of deposition temperature on the electrical and electrochemical properties of carbon-based coatings for metallic bipolar plates, prepared by cathodic arc evaporation. Adv Mater Phys Chem 12:45–57
42. Koller CM, Hollerweger R, Sabitzer C, Rachbauer R, Kolozsvari S, Paulitsch J, Mayrhofer PH (2014) Thermal stability and oxidation resistance of arc evaporated TiAlN, TaAlN, TiAlTaN, and TiAlN/TaAlN coatings. Surf Coat Technol 259:599–607
43. Alhafian MR, Chemin JB, Fleming Y, Bourgeois L, Penoy M, Useldinger R, Soldera F, Mucklich F, Choquet P (2021) Comparison on the structural, mechanical and tribological properties of TiAlN coatings deposited by HiPIMS and cathodic arc evaporation. Surf Coat Technol 423:127529
44. Sveen S, Andersson JM, Saoubi RM, Olsson M (2013) Scratch adhesion characteristics of PVD TiAlN deposited on high speed steel, cemented carbide and PCBN substrates. Wear 308:133–141
45. Long Y, Zeng J, Yu D, Wu S (2014) Microstructure of TiAlN and CrAlN coatings and cutting performance of coated silicon nitride inserts in cast iron turning. Ceramics Inter 40:9889–9894
46. Zhang Z, Zhang L, Yuan H, Qiu M, Zhang X, Liao B, Zhang F, Ouyang X (2022) Tribological behaviors of super-hard TiAlN coatings deposited by filtered cathode vacuum arc deposition. Mater 15:2236
47. Sanchette F, Ducros C, Schmitt T, Steyer P, Billard A (2011) Nanostructured hard coatings deposited by cathodic arc deposition: from concepts to applications. Surf Coat Technol 205:5444–5453

48. Mattox DM (2010) Physical sputtering and sputter deposition (sputtering). In: Mattox DM (ed) Handbook of physical vapour deposition (PVD) processing, 2nd edn. William Andrew Publishing, Boston, pp 237–286
49. Bishop CA (2011) Magnetron sputtering source design and operation. In: Bishop CA (ed) Vacuum deposition onto webs, films and Foils, 2nd edn. William Andrew Publishing, Oxford, pp 337–362
50. Srinivasan K (2004) L1(0) iron-platinum on nanocrystalline HITPERM soft magnetic underlayers for perpendicular recording media. Dissertation, Carnegie Mellon University
51. NurEAlam M, Lonsdale W, Vasiliev M, Alameh K (2019) Application-specific oxide-based and metal-dielectric thin-film materials prepared by radio frequency magnetron sputtering. Materials 12:3448
52. Singh J, Khan SA, Shah S, Kotnala RK, Mohapatra S (2017) Nanostructured TiO_2 thin films prepared by RF magnetron sputtering for photocatalytic applications. Appl Surf Sci 422:953–961
53. Garcia-Valenzuela JA, Rivera R, Morales-Vilches AB, Gerling LG, caballero A, Asensi JM, Voz C, Bertomeu J, Andreu J (2016) main properties of Al_2O_3 thin films deposited by magnetron sputtering of an Al_2O_3 ceramic target at different radio-frequency power and argon pressure and their passivation effect on p-type c-Si wafers. Thin Solid Films 619:288–296
54. Belkind A, Freilich A, Scholl R (1998) Pulse duration effects in pulse-power reactive sputtering of Al_2O_3. Surf Coat Technol 108–109:558–563
55. Vlcek J, Pajdarova AD, Musil J (2004) Pulsed DC magnetron discharges and their utilisation in plasma surface engineering. Contrib Plasma Phys 44:426–436
56. Belkind A, Freilich A, Lopez J, Zhao Z, Zhu W, Becker K (2005) Characterisation of pulsed dc magnetron sputtering plasmas. New J Phys 7:90
57. Erfurt D, Heinemann MD, Korner S, Szyszka B, Klenk R, Schlatmann R (2018) Improved electrical properties of pulsed DC magnetron sputtered hydrogen doped indium oxide after annealing in air. Mater Sci Semi Process 89:170–175
58. Lelis M, Tuckute S, Varnagiris S, Urbonavicius M, Laukaitis G, Bockute K (2019) Tailoring of TiO_2 film microstructure by pulsed-DC and RF magnetron co-sputtering. Surf Coat Technol 377:124906
59. Zhang JXJ, Hoshino K (2019) Fundamentals of nano/microfabrication and scale effect. In: Zhang JXJ, Hoshino K (eds) Molecular sensors and nanodevices, 2nd edn. Academic Press, pp 43–111
60. Yli-Pentti A (2014) Electroplating and electroless plating. In: Hashmi S, Batalha GF, Tyne CJ, Yilbas B (eds) Comprehensive materials processing. Elsevier, Oxford, p 277-306
61. Ratan M, Mainak G, Avijit P (2024) Electroless nickel coating on magnesium substrates: fabrication method and applications. In: Hashmi S (ed) Comprehensive materials processing, 2nd edn. Elsevier, Oxford, pp 547–564
62. Loto CA (2016) Electroless Nickel plating-a review. Silicon 8:177–186
63. Bhujel R, Rizal U, Agarwal A, Swain BS, Swain BP (2018) Synthesis and characterisation of silicon nanowires by electroless etching. J of Materi Eng Perform 27:2655–2660
64. Bhujel R, Swain BP (2019) Investigation of electronic environments and electrical characterisation of functionalised AgNPs synthesised by electroless process. Appl Phys A: Mater Sci Process 125
65. Abdel-Gawad SA, Osman WM, Fekry AM (2018) Characterisation and corrosion behavior of anodised aluminium alloys for military industries applications in artificial seawater. Surf Interf 14:314–323
66. Stevenson MF (1994) Anodising. In: Sprague JA, Smidt FA (eds) Cotell CM. Surface engineering, ASM International, pp 482–493
67. Dorfman MR (2018) Thermal spray coatings. In: Kutz M (ed) Handbook of environmental degradation of Materials, 3rd edn. William Andrew Publishing, pp 469–488
68. Yu J, Liu X, Yu Y, Li H, Liu P, Huang K, Sun R (2022) Research and application of high-velocity oxygen fuel coatings. Coatings 12:828

69. Wu M, Pan L, Duan H, Wan C, Yang T, Gao M, Yu S (2021) Study on wear resistance and corrosion resistance of HVOF surface coating refabricate for hydraulic support column. Coatings 11:1457–1468.
70. Song B, Murray JW, Wellman RG, Pala Z, Hussain T (2019) Dry sliding wear behaviour of HVOF thermal sprayed WC-Co-Cr and WC-Cr_xC_y-Ni coatings. Wear 442–443:203114
71. Murugan K, Ragupathy A, Balasubramanian V, Sridhar K (2014) Optimising HVOF spray process parameters to attain minimum porosity and maximum hardness in WC–10Co–4Cr coatings. Surf Coat Technol 247:90–102
72. Varis T, Suhonen T, Calonius O, Cuban J, Pietola M (2016) Optimisation of HVOF Cr_3C_2-NiCr coating for increased fatigue performance. Surf Coat Technol 305:123–131
73. Zhou W, Zhou K, Deng C, Zeng K, Li Y (2017) Hot corrosion behaviour of HVOF-sprayed Cr3C2-NiCrMoNbAl coating. Surf Coat Technol 309:849–859
74. Xie M, Zhang S, Li M (2013) Comparative investigation on HVOF sprayed carbide-based coatings. Appl Surf Sci 273:799–805
75. Pulsford J, Venturi F, Pala Z, Kamnis S, Hussain T (2019) Application of HVOF WC-Co-Cr coatings on the internal surface of small cylinders: effect of internal diameter on the wear resistance. Wear 432–433:202965
76. Gerard B (2006) Application of thermal spraying in the automobile industry. Surf Coat Technol 201:2028–2031
77. Ma X, Ruggiero P, Bhattacharya R, Rai AK (2023) Optimisation of HVOF-applied erosion-resistant coatings for large compressor and fan airfoils in STOVL aircrafts. J Therm Spray Tech 33:367–380
78. Bakan E, Mack DE, Mauer G, Vaben R, Lamon J, Padture NP (2020) High-temperature materials for power generation in gas turbines. In: Guillon Olivier (ed) Advanced ceramics for energy conversion and storage. Elsevier, pp 3–62
79. Tan JC, Looney L, Hashmi MSJ (1999) Component repair using HVOF thermal spraying. J Mater Process Technol 92–93:203–208
80. Quintino L (2014) Overview of coating technologies. In: Miranda R (ed) Surface modification by solid state processing. Woodhead Publishing, pp p1-24
81. Mattox DM (2010) Vacuum evaporation and vacuum deposition. In: Mattox DM (ed) Handbook of physical vapor deposition (PVD) processing, 2nd edn. William Andrew Publishing, Boston, pp 195–235
82. Bashir A, Awan TI, Tehseen A, Tahir MB, Ijaz M (2020) Interfaces and surfaces. In: Bashir A, Tehseen A (eds) Awan TI. Chemistry of nanomaterials, Elsevier, pp 51–87
83. Feng J, Jiao Y, Wang H, Zhu X, Sun Y, Du M, Cao Y, Yang D, Liu SF (2021) High-throughput large-area vacuum deposition for high-performance formamidine-based perovskite solar cells. Energy Environ Sci 14:3035–3043
84. Pedersen H, Elliot SD (2014) Studying chemical vapor deposition processes with theoretical chemistry. Theor Chem Acc 133:1476
85. Johnson RW, Hultqvist A, Bent SF (2014) A brief review of atomic layer deposition: from fundamentals to applications. Mater Today 17:236–246
86. Jussila T, Philip A, Tripathi T, Nielsch K, Karppinen M (2023) Atomic layer deposition of magnetic thin films: Basic processes, engineering efforts, and road forward. Appl Phys Rev 10:041313
87. Marquardt AE, Breitung EM, Drayman-Weisser T, Gates G, Phaneuf RJ (2015) Protecting silver cultural heritage objects with atomic layer deposited corrosion barriers. Heritage Sci 3:37
88. Kumar D, Ghadai RK, Das S, Sharma A, Swain BP (2019) Effect of nitrogen flow rate on the mechanical properties of CVD-deposited SiCN thin films. Bull Mater Sci 42:251
89. Das S, Ghadai R, Guha S, Sharma A, Swain BP (2020) Correlation of microstructural and mechanical properties of CVD deposited TiAlN coatings. Arab J Sci Eng 45:967–975
90. Li M, Liu D, Wei D, Song X, Wei D, Wee ATS (2016) Controllable synthesis of graphene by plasma-enhanced chemical vapor deposition and its related applications. Adv Sci 3:1600003

91. Adhikari S, Aryal HR, Uchida H, Umeno M (2016) Catalyst-free growth of graphene by microwave surface wave plasma chemical vapor deposition at low temperature. J Mater Sci Chem Eng 4:10–14
92. Munoz R, Martinez L, Lopez E, Munuera C, Huttel Y, Garcia-Hernandez M (2018) Direct synthesis of graphene on silicon oxide by low temperature plasma enhanced chemical vapor deposition. Nanoscale 10:12779–12787
93. Kulczyk-Malecka J, Santos IVJ, Betbeder M, Rowley-Neale S, Gao Z, Kelly PJ (2021) Low-temperature synthesis of vertically aligned graphene through microwave-assisted chemical vapour deposition. Thin Solid Films 733:138801
94. Souza FM, Gupta RK, Yasin G, Nguyen TA (2022) Plasma technology for nanostructures. In: Nguyen TA, Amrane A, Assadi AA, Yasin G (eds) Song H. Plasma at the nanoscale, Elsevier, pp 265–281
95. Eessaa AE, El-Shamy AM (2023) Review on fabrication, characterisation, and applications of porous anodic aluminum oxide films with tunable pore sizes for emerging technologies. Microelectron Eng 279:112061
96. Li X, Cai W, An J, Kim S, Nah J, Yang D, Piner R, Velamakanni A, Jung I, Tutuc E, Banerjee SK, Colombo L, Ruoff RS (2009) Large-area synthesis of high-quality and uniform graphene films on copper foils. Science 324:1312–1314
97. Yamauchi S, Ishibashi K, Hatakeyama S (2014) Low pressure chemical vapor deposition of TiO_2 layer in hydrogen-ambient. J Crystallization Process Technol 4:185–192
98. Zhang X, Lai J, Gray T (2023) Recent progress in low-temperature CVD growth of 2D materials. Oxford Open Mater Sci 3
99. Koudoumas E, Le KT, Vernardou D (2023) Recent advances of chemical vapor deposited thermochromic vanadium dioxide materials. Energy Nexus 11:100237
100. Khlyustova A, Cheng Y, Yang R (2020) Vapor-deposited functional polymer thin films in biological applications. J Mater Chem B 8:6588–6609
101. Sun L, Yuan G, Gao L, Yang J, Chhowalla M, Gharahcheshmeh MH, Gleason KK, Choi YS, Hong BH, Liu Z (2021) Chemical vapour deposition. Nat Rev Methods Primers 1:5

Chapter 2
Fundamentals of Mechanical Property Evaluation in Coating Materials

Niranjan Patra, Prathipati Ramesh, and Omar Cedillos-Barraza

Abstract This chapter presents a comprehensive study on the mechanical property evaluation of coating materials, focusing on critical aspects such as adhesion, hardness, and elastic modulus. The significance of coatings in enhancing the performance and durability of materials in various industrial applications is emphasised. A detailed analysis of the different testing methodologies, including scratch, indentation, and nanoindentation tests, is provided to assess the mechanical properties of coatings. The chapter also explores the influence of factors such as coating thickness, substrate material, and deposition techniques on the mechanical behaviour of coatings. Experimental results are discussed to illustrate the effectiveness of these evaluation techniques, offering valuable insights for developing and optimising advanced coatings with improved mechanical properties.

1 Introduction

The evaluation of mechanical properties in coating materials is a critical aspect of materials science and engineering, particularly given the widespread use of coatings in various industrial applications [1–3]. Coatings serve multiple functions, including protection against corrosion, wear resistance, aesthetic enhancement, and provision of specific functionalities such as electrical conductivity or thermal insulation [4–6]. Given their importance, understanding the mechanical properties of these coatings is essential to ensure their performance and longevity in real-world applications [7–9].

Coating materials are applied as thin layers on surfaces to enhance the substrate's properties or protect it from environmental and mechanical damage. The effectiveness of a coating is primarily determined by its mechanical properties, which include

N. Patra (✉) · P. Ramesh
Department of Chemistry, Koneru Lakshmaiah Education Foundation, Greenfield, Vaddeswaram, Andhra Pradesh, India
e-mail: patraji@gmail.com

O. Cedillos-Barraza
Department of Aerospace and Mechanical Engineering, University of Texas El Paso, at El Paso, TX 79968, USA

B. P. Swain (ed.), *Advances in Mechanical Coating*, Materials Horizons: From Nature to Nanomaterials, https://doi.org/10.1007/978-981-96-7484-8_2

hardness, elasticity, plasticity, toughness, fracture toughness, adhesion, cohesion, wear resistance, and fatigue resistance [10]. These properties govern how a coating reacts to mechanical stresses, resists deformation, and maintains its integrity over time. For example, hardness affects a coating's ability to resist scratching and indentation, while toughness and fracture toughness determine its capacity to absorb energy and resist crack propagation [11–13].

Evaluating coatings' mechanical properties involves theoretical understanding and practical testing methods. Standardised tests such as nanoindentation, tensile, compressive, impact, and wear tests provide critical data on how coatings perform under various conditions. For instance, nanoindentation is widely used to measure hardness and elastic modulus at the microscale, providing insights into the coating's resistance to deformation and indentation [14, 15]. Tensile and compressive tests, on the other hand, help determine the coating's ability to withstand stretching and compressive forces without failing [4]. These tests are essential for applications where coatings are subjected to mechanical loads, such as in the automotive and aerospace industries [16].

Several factors influence the mechanical properties of coating materials. The composition of the coating, including the base material and any additives or fillers, plays a significant role in determining its mechanical behaviour [17–19]. For instance, adding ceramic particles to a polymer matrix can significantly enhance its hardness and wear resistance [5, 6]. The microstructure and morphology of the coating, such as grain size and porosity, also impact its mechanical properties [20]. Finer grain sizes typically lead to higher hardness and strength due to the grain boundary strengthening mechanism [21].

Processing and deposition techniques are another critical factor affecting the mechanical properties of coatings. Methods such as chemical vapour deposition (CVD), physical vapour deposition (PVD), electroplating, and thermal spraying can produce coatings with different microstructures and bonding characteristics, which in turn affect their mechanical performance [5, 22, 23]. Additionally, the characteristics of the substrate, including its material composition and surface roughness, can influence the adhesion and overall performance of the coating [24–28]. Proper surface preparation and treatment are essential to enhance the bonding between the coating and the substrate [18, 29, 30].

Environmental factors such as temperature, humidity, and chemical exposure also play a significant role in the performance of coatings [31, 32]. High temperatures can lead to thermal expansion and stress, potentially causing cracks or delamination [31, 33, 34]. Humidity and moisture can accelerate corrosion processes, especially in metallic coatings, while chemical exposure can lead to degradation of the coating material [35]. Therefore, understanding the environmental conditions to which a coating will be exposed is crucial for selecting and evaluating the appropriate coating material.

In recent years, advances in testing methodologies and the development of new coating materials have significantly improved our ability to evaluate and enhance the mechanical properties of coatings. Innovations such as in-situ testing techniques, advanced microscopy, and spectroscopy provide more detailed and accurate data on coating performance [36]. Moreover, the advent of nanostructured coatings and smart coatings that can respond to environmental changes has opened new possibilities for improving the mechanical properties and functionality of coatings [37, 38].

This chapter aims to provide a comprehensive overview of the fundamental principles in evaluating coating materials' mechanical properties. We will explore the key mechanical properties and examine the factors that influence these properties. By understanding these fundamentals, researchers can better assess and optimise the mechanical performance of coatings for various applications.

2 Theoretical Background and Mechanical Properties Evaluation of Coating

The basic concepts in mechanical properties are essential for evaluating coating materials. These properties provide insights into how coatings behave under various mechanical stresses and environmental conditions. The key mechanical properties typically assessed include hardness, elasticity, plasticity, toughness, fracture toughness, adhesion, cohesion, wear resistance, and fatigue resistance. Each property provides valuable information about the material's performance and suitability for specific applications.

2.1 *Stress and Strain*

Stress and strain are fundamental concepts in mechanics that describe how materials respond to external forces. Stress (σ) is defined as the force (F) applied per unit area (A) of the material:

$$\sigma = \frac{F}{A}$$

Strain (ϵ) is the measure of deformation representing the displacement between particles in the material body relative to a reference length (Lo):

$$\epsilon = \frac{\Delta L}{Lo}$$

where (ΔL) is the change in length. The relationship between stress and strain is depicted in the stress–strain curve, which provides valuable information about the material's mechanical properties [39].

2.2 *Elastic Modulus (Young's Modulus)*

Elastic modulus (E), often referred to as **Young's modulus**, is a fundamental mechanical property that quantifies the stiffness of a material. It defines the relationship between stress and strain in the elastic region of a material's deformation curve, where the material returns to its original shape after removing an applied force. In the context of coating materials, the elastic modulus is critical in determining how coatings will react under mechanical loads, including how much they will stretch or compress when subjected to stress.

Young's modulus (E) is defined as the ratio of stress (σ) to strain (ε) in the linear elastic region of the stress–strain curve: $E = \frac{\sigma}{\epsilon}$

This parameter provides insight into the rigidity of a coating material. A higher Young's modulus indicates a stiffer material that deforms less under a given load, while a lower modulus signifies a more compliant material that deforms more readily. The elastic modulus is particularly important for coatings that must maintain their integrity and functionality under mechanical stress, such as those used in structural applications or on components subjected to bending, compression, or stretching.

2.2.1 Measurement of Elastic Modulus and Hardness by Nanoindentation

Nanoindentation is a widely used technique for measuring the elastic modulus and hardness of thin films and coatings. In this method, a sharp indenter is pressed into the coating surface, and the load–displacement response is used to calculate the elastic modulus. Nanoindentation is particularly useful for characterising the mechanical properties of thin coatings, as it allows for localised measurement with high precision [14].

Naseer et al. studied the hardness and elastic modulus of α-C:H:Si coatings by nanoindentation, which indicates significant improvements in the mechanical properties of silicon and oxygen-doped hydrogenated amorphous carbon (α-C:H:Si) coatings compared to those deposited by plasma-enhanced chemical vapour deposition (PECVD) [40]. The doped coatings showed enhanced hardness and reduced elastic modulus, with values of 24.4 ± 3.1 GPa and 212.9 ± 6.1 GPa, respectively, at a 5 mN load (Fig. 1). These values are notably higher than the typical ranges reported for PECVD-deposited coatings. The improvements are due to increased sp^3 bonding from silicon doping, which replaces less stable carbon–carbon bonds and reduces structural disorder. However, oxygen doping slightly increased sp^3 bonding while decreasing coating density and hydrogen doping further increased sp^3 fraction but reduced hardness; the magnetron sputtering method proved effective in producing coatings with superior mechanical properties compared to traditional PECVD techniques.

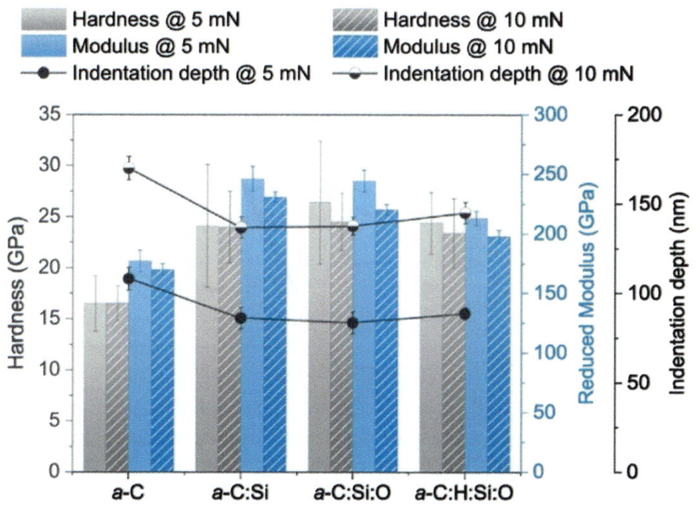

Fig. 1 Hardness and reduced elastic modulus and indentation depth were measured at 5 and 10 mN loads using Nanoindentor [40] (Adapted with permission from Elsevier)

2.2.2 Ultrasonic Testing

Ultrasonic techniques involve sending high-frequency sound waves through the coating and measuring the wave velocity. The elastic modulus can be determined from the wave velocity and material density, making this a non-destructive method ideal for evaluating coatings on large or complex surfaces [41].

Chai et al. studied the elastic modulus of laser cladding coatings by the ultrasonic method [42]. This method not only measures the elastic modulus but also evaluates the uniformity of the layer. Laser ultrasonic testing was then employed to collect surface wave signals, allowing for the determination of wave amplitude and delay time across the coating. The laser ultrasonic C-scan method effectively mapped the modulus distribution by correlating the surface wave velocity with the elastic modulus.

2.2.3 Three-Point Bending Test

The three-point bending test is often employed for thicker coatings or bulk materials. In this method, a beam coated with the material of interest is subjected to bending, and the modulus is derived from the stress–strain response (Fig. 2). This method is useful for coatings that are bonded to substrates, as it can account for the combined response of the coating-substrate system [43].

Luka et al. investigated the effectiveness of three-point bending tests in evaluating the adhesion of protective coatings applied to bronze substrates [44]. The findings reveal that this method is particularly sensitive to detecting the onset of failure at the

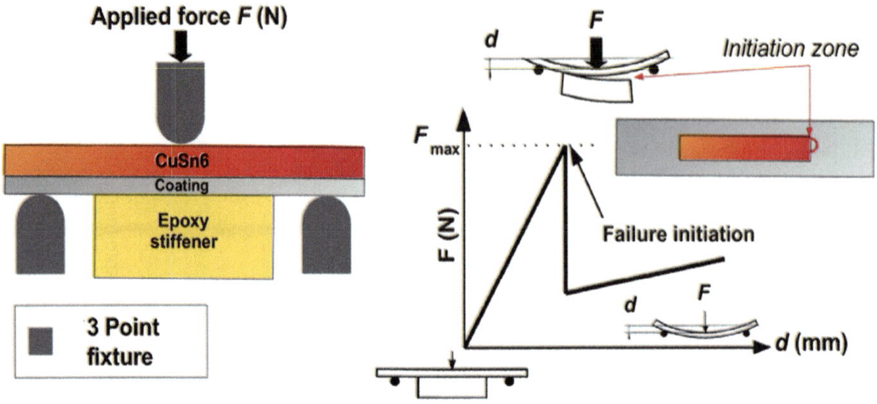

Fig. 2 Three-point bending test for assessing practical adhesion: (left) setup showing the sample's placement during testing and (right) representative load-deformation curve. The distinct change in slope indicates the onset of failure in the coated surface at the stiffener block's edge [44]. (Adapted with permission from Springer)

interface between the coating and the substrate. The tests showed that the initiation of failure typically occurs at the edges of the stiffener block, as indicated by a marked change in the slope of the load-deformation curve. This slope change corresponds to the coating's adhesion strength and is critical in identifying weak points in the coating system. The study demonstrates that three-point bending is a reliable technique for assessing practical adhesion, providing valuable insights into the performance and durability of protective coatings under mechanical stress.

Yilbas et al. investigated the mechanical behaviour of laser-treated and untreated Inconel 625 coatings on mild steel by three-point bending [45]. The results show that while the laser-treated and untreated coatings exhibit similar elastic behaviour initially, the laser-treated coatings have a reduced elastic limit compared to the untreated ones. This reduction is attributed to the laser treatment, which decreases the coating thickness and creates a more compact structure. However, despite this compactness, the laser treatment also introduces brittleness, particularly near the surface, due to splats with high oxygen content. The study further indicates that, despite the early termination of the elastic limit in laser-treated coatings, the overall load-bearing capacity of the coated samples is significantly higher than that of uncoated samples for a given displacement. This suggests that even a small thickness of the Inconel 625 coating substantially enhances the toughness of the underlying material despite the brittleness introduced by the laser treatment.

2.2.4 Resonance Frequency Testing

This technique measures a coated component's natural frequency of vibration. The elastic modulus is inferred from the resonance frequency, the mass of the component, and its geometric properties. This method benefits coatings on small or delicate components where traditional mechanical testing may be challenging [46].

Tarancon et al. studied the elastic modulus of coatings by recording the natural frequencies of coated specimens using a microphone, and the modulus can be calculated through the frequency equation specific to the coating's geometry[47]. For various W-Cu compositions, the RFA results align well with theoretical values derived from the rule of mixtures, except for pure copper. The variation in measured elastic modulus across different techniques underscores the influence of material texture and porosity. Specifically, the resonant frequency values are consistent with theoretical expectations, highlighting its effectiveness in capturing the elastic properties and uniformity of coatings despite the differences in measurements observed due to the scale and nature of each testing method.

The elastic modulus of a coating material has significant implications for its performance in various applications. In protective coatings, a high elastic modulus is often desirable to resist deformation and maintain a protective barrier under mechanical stress. For example, hard ceramic coatings used in cutting tools and wear-resistant surfaces must have high moduli to withstand the stresses encountered during use without failing. In biomedical applications, coatings must balance stiffness and compliance to match the mechanical properties of biological tissues. For instance, coatings used on medical implants need moduli that are not too high to avoid stress shielding but high enough to ensure mechanical stability.

Coatings are subjected to cyclic mechanical loads and thermal stresses in the aerospace and automotive industries. A well-defined elastic modulus ensures that these coatings can endure the mechanical demands without cracking or delaminating, thereby extending the service life of components. For coatings used in electronics and optics, the elastic modulus can affect the mechanical integrity and performance of devices. Stiff coatings are needed to protect delicate components while maintaining the required mechanical alignment and optical properties.

2.3 *Yield Strength and Plastic Deformation*

Yield strength (σ_Y) is the stress at which a material begins to deform plastically. Before reaching this point, the material deforms elastically, meaning it can return to its original shape once the stress is removed. However, once the yield strength is exceeded, the material undergoes plastic deformation, resulting in a permanent change in shape. Yield strength is a critical parameter in determining coatings' mechanical performance and durability, especially in load-bearing applications. Yield strength is typically determined through mechanical testing, such as tensile testing. A coated specimen is subjected to a uniaxial load in tensile testing until it deforms plastically. The yield strength is identified as the stress at which the material transitions from elastic to plastic deformation, often determined by the 0.2% offset method, where a line parallel to the elastic region is drawn from 0.2% strain, and the intersection with the stress–strain curve indicates the yield strength.

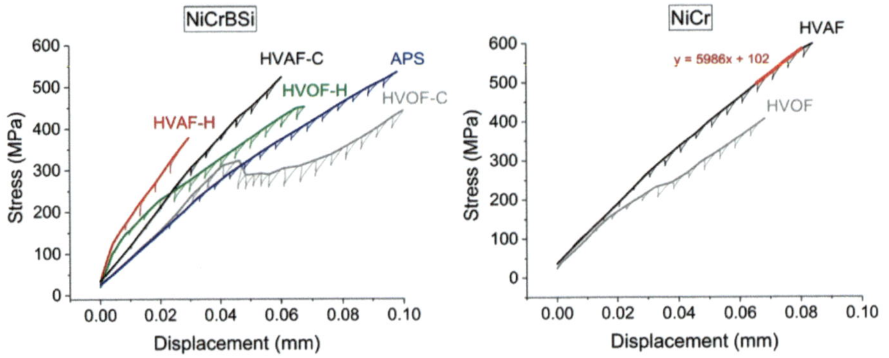

Fig. 3 Stress-displacement curves for the coatings, with the slope of $\Delta\sigma/\Delta\varepsilon$ fitted to the NiCr HVAF curve [48]

Varis et al. Studied the mechanical response of APS, HVOF, and HVAF-sprayed NiCr and NiCrBSi coatings by tensile testing [48]. The study reveals the non-linear stress-displacement behaviour of these coatings under tensile testing (Fig. 3). The stress-displacement curves, particularly for HVAF coatings, show less non-linear and inelastic deformation compared to HVOF and APS coatings, suggesting that HVAF coatings possess greater structural integrity and stiffness. The APS coating, however, was too fragile for tensile testing, indicating lower mechanical resilience. Notably, the NiCrBSi HVOF coating exhibited a crack at around 300 MPa, but it withstood additional stress before the final fracture, underscoring its durability despite initial failure. The study also emphasises that reducing the slope of the stress-displacement curve ($\Delta\sigma/\Delta\varepsilon$) near fracture is a critical parameter for evaluating the structural integrity of thermally sprayed coatings. This reduction is influenced by the extent of non-linear elastic and inelastic deformations, which are more pronounced in APS and HVOF coatings than in HVAF coatings. This finding suggests that HVAF coatings are more robust and less prone to deformation under stress, making them a superior choice for applications requiring high mechanical integrity.

Plastic deformation refers to the permanent change in shape that occurs in a material when it is subjected to stress beyond its yield strength. Unlike elastic deformation, which is reversible, plastic deformation results in a permanent alteration of the coating's microstructure and dimensions. In coatings, plastic deformation can significantly affect the material's performance, leading to changes in thickness, cracking, or even delamination from the substrate. Several testing methods are used to assess plastic deformation in coatings.

2.3.1 Indentation Testing

Indentation tests, such as Vickers or Rockwell hardness testing, involve pressing a hard indenter into the coating surface and measuring the size or depth of the resulting

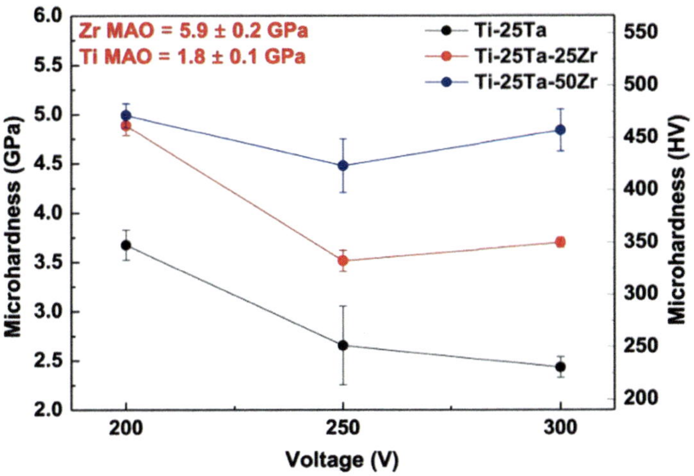

Fig. 4 Presents the microhardness values of the MAO coatings for Ti–25Ta–xZr alloys measured in GPa (left) and HV (right) across various applied voltages [49]. (Adapted with permission from Elsevier)

indentation. The degree of plastic deformation under the indenter provides insights into the coating's hardness and plasticity.

Kuroda et al. studied the Vickers microhardness values of oxide coatings using a Shimadzu HMV G20ST under a load of 0.5 kgf (4.90 N) for 15 s, adhering to the ASTM-E384 standard [49]. The results showed that increasing the zirconium (Zr) content in the $Ti_{25}T_{a-x}Zr$ alloy substrate led to higher hardness values in the oxidised layers. This increase is attributed to the higher hardness of ZrO_2 (5.9 GPa) compared to TiO_2 (1.8 GPa) (Fig. 4). Additionally, the study observed that higher voltage during the oxidation process reduced the hardness of the oxide layers, which is likely due to the formation of $CaCO_3$, a compound with lower hardness (0.75–5 GPa) compared to other detected oxides such as anatase, rutile, Ta_2O_5, and ZrO_2.

2.3.2 Scratch Testing

Scratch testing involves dragging a sharp stylus across the coating surface under a controlled load. The resulting scratch can reveal the coating's resistance to plastic deformation and its adhesion to the substrate [50].

2.3.3 Creep Testing

Creep testing measures the time-dependent plastic deformation of a coating under a constant load and temperature. This test is particularly relevant for coatings in high-temperature applications, where creep can lead to long-term deformation and failure (Fig. 5).

Jackson et al. conducted small punch creep (SPC) tests using a custom-built rig mounted on a Tinius Olsen H5KS single-column materials testing machine to evaluate the creep behavior of as-sprayed and heat-treated HVOF Stellite 6 coatings at 700 °C [51]. The tests applied a load through a 2.5 kN load cell, with punch head displacement monitored by two LVDTs, and utilised a three-tier, 3 kW furnace for controlled heating, regulated by three K-type thermocouples connected to a temperature controller. The study found that both the as-sprayed and heat-treated coatings exhibited typical creep behaviour, with time-displacement curves showing primary, secondary, and tertiary regions (Fig. 6). The heat-treated coating demonstrated greater resistance to creep, requiring higher loads to achieve similar times to failure as the as-sprayed coating. The results highlighted heat treatment's influence on enhancing the coatings' creep resistance under high-temperature conditions.

Yield strength and plastic deformation are fundamental mechanical properties that determine how coatings behave under mechanical loading conditions. Yield strength defines the stress level at which coatings begin to deform plastically, while plastic deformation represents the irreversible changes in shape that occur beyond this point. Understanding these properties is essential for designing coatings that can withstand the mechanical demands of their intended applications, whether in load-bearing structures, protective surfaces, or high-temperature environments.

2.4 *Toughness and Fracture Mechanics*

Toughness measures a material's ability to absorb energy and deform plastically before fracturing. In the context of coatings, toughness reflects the coating's ability

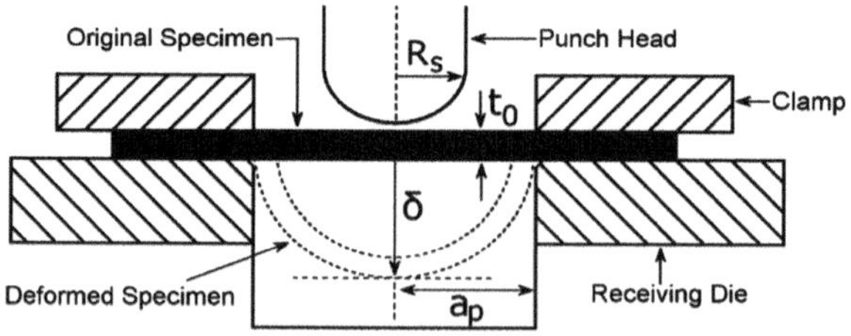

Fig. 5 The schematic cross-section of the small punch creep apparatus illustrates the load application via a hemispherical punch head and the resulting specimen displacement (δ). The diagram specifies key dimensions, including the receiving hole radius (ap = 2 mm), the punch head radius (Rs = 1 mm), and the original specimen thickness (t0 = 0.5 mm) [51]

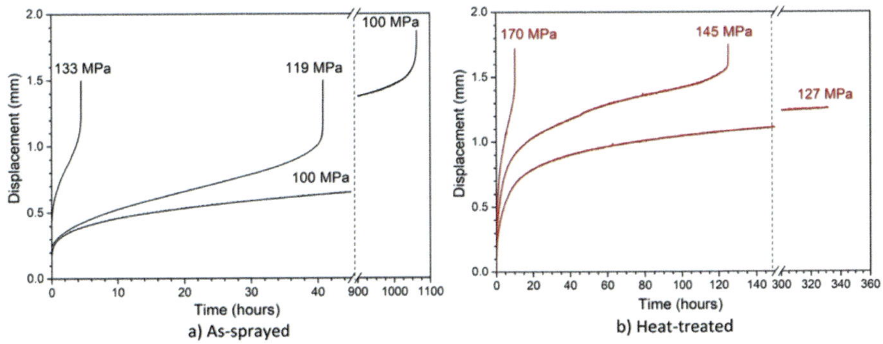

Fig. 6 Example SPC displacement–time curves for as-sprayed and heat-treated Stellite 6 coatings tested at 700 °C. The curves illustrate the typical creep behaviour of both coatings at this temperature [51]. (Adapted with permission from Elsevier)

to withstand impacts or other forms of mechanical stress without cracking or failing. A tough coating can absorb significant amounts of energy and distribute stress over a larger area, reducing the likelihood of localised failure. Testing the toughness of coatings typically involves measuring the amount of energy a coating can absorb before fracturing. Common testing methods include:

2.4.1 Impact Testing

Impact testing evaluates the coating's ability to resist crack initiation and propagation under sudden, high-energy loading conditions. The test involves striking a coated specimen with a pendulum or hammer, and the energy absorbed by the specimen before failure is recorded as a measure of toughness. Using an impact resistance testing machine, the impact resistance of polymer coatings was evaluated according to the PN-EN ISO 6272-1:2011 standard (falling weight impact test). Under standard conditions, this test determines the minimum drop height of a 20 mm diameter mass required to damage the coating under analysis.

Duda et al. [52] performed a comparative analysis of the impact resistance of multilayer polymer coatings. The evaluation involved determining the minimum height at which a 20 mm diameter mass needed to fall to cause damage to the coatings. In their tests, a 2 kg weight was dropped from a height of 0.9 m to induce material damage. This sample was then used as a reference for subsequent evaluations of other coatings. The impact testing revealed that the EP-GF-HF (epoxy resin with glass fabric hemp fiber) coating exhibited the highest delamination, while the EP (epoxy resin) coated sample showed the least delamination.

2.5 *Fracture Mechanics*

Fracture mechanics is the field of study that focuses on the behaviour of materials containing cracks and the conditions under which these cracks propagate and cause failure. In coatings, understanding fracture mechanics is crucial for predicting the onset of cracking and for designing coatings that can resist crack growth under various loading conditions. Fracture mechanics involves the analysis of several key parameters:

2.5.1 Stress Intensity Factor (K)

This parameter quantifies the stress concentration at the tip of a crack. It is influenced by the applied load, the crack's size and shape, and the coated component's geometry. When the stress intensity factor exceeds a critical value (K_{IC}), known as the fracture toughness, the crack will propagate, leading to failure.

2.5.2 Crack Growth Rate

In some cases, cracks may grow gradually under cyclic loading, a phenomenon known as fatigue. The crack growth rate measures how quickly a crack propagates per loading cycle. Understanding this rate is essential for predicting the service life of coatings under cyclic loading conditions.

2.5.3 Energy Release Rate (G)

This parameter represents the energy released as a crack propagates through the coating material. The energy release rate must be below a certain threshold for the crack to remain stable and not grow uncontrollably. This concept is closely related to toughness, as tougher materials can absorb more energy before a crack grows.

2.5.4 Fracture Toughness (K_{IC})

As mentioned earlier, fracture toughness is a critical property determining a material's ability to resist crack propagation. It is a material-specific value that reflects the combined effects of material composition, microstructure, and environmental conditions. The application of fracture mechanics to coatings involves understanding how cracks initiate, how they propagate, and what factors influence these processes. Factors such as the coating's microstructure, thickness, and adhesion to the substrate all determine how a coating will respond to cracks.

2.6 Adhesion and Cohesion

Adhesion refers to the ability of a coating to bond effectively to the substrate material. This property is crucial because it determines the coating's ability to remain attached to the surface under various conditions, including mechanical stress, thermal cycling, and exposure to environmental factors like moisture, chemicals, and UV radiation. Poor adhesion can lead to delamination, blistering, and peeling, which compromise the protective function of the coating and may result in substrate exposure and eventual failure. Several mechanisms contribute to adhesion between a coating and its substrate, including:

- **Mechanical Interlocking** occurs when the coating material physically locks into the surface roughness or porosity of the substrate. Surface preparation methods like sandblasting or etching are often employed to increase surface roughness, enhance mechanical interlocking, and improve adhesion.
- **Chemical Bonding**: Chemical bonds, such as covalent or ionic bonds, can form between the coating and the substrate at the interface. These bonds are often very strong and contribute significantly to the adhesion strength. Functional groups in the coating material that can react with the substrate surface are essential for forming chemical bonds.
- **Van der Waals Forces**: These are weak intermolecular forces that can contribute to adhesion, particularly in cases where chemical bonding is minimal. While van der Waals forces alone may not provide strong adhesion, they can play a supplementary role when combined with other adhesion mechanisms.
- **Electrostatic Forces**: In some cases, electrostatic attraction between the coating and substrate can enhance adhesion. This is particularly relevant in coatings applied through electrostatic spraying techniques, where the coating particles are electrically charged to attract to the substrate surface.

Testing adhesion is essential for predicting the long-term performance of coatings in their intended applications. Common methods for evaluating adhesion include:

2.6.1 Pull-Off Test

Pull-off test involves attaching a loading fixture to the coated surface and applying a tensile force perpendicular to the surface until the coating detaches. The force required to pull the coating off provides a measure of its adhesion strength. The pull-off adhesion strength is evaluated according to the ASTM D4541-09 (2009) standard. In this method, an adhesive with a known bond strength is applied to the surface of a dolly, which is then pressed onto the coated flat surface. Any excess adhesive squeezed out should be wiped away, and the sample with the attached dolly is allowed to dry for a specified period to ensure the adhesive has fully cured. To expedite the curing process, oven drying can be an alternative. After curing, the dolly is secured in a loading fixture designed to apply tension perpendicular to the test surface. The dolly is then subjected to a gradually increasing tensile load until it detaches from the coating. This load is applied using a hydraulic system, and the maximum perpendicular force the surface can withstand before detachment is recorded. Dollies come in various diameters, with 20 mm being the most commonly used. This method is distinct in that it applies tensile stress, unlike other methods that apply shear stress. Therefore, only results obtained using the same test method should be considered for comparative studies. To ensure that the perpendicular load is applied exclusively to the coated surface that matches the contact area of the dolly, the area around the dolly is cut using a specialised cutter. Various commercial systems are available for applying the normal load, either manually or with the assistance of a hydraulic pump system.

Dam et al. studied the adhesion properties of sol–gel films applied to differently pre-treated AA2024-T3 and AA7075-T6 aluminium alloys by pull-off adhesion testing, performed with an Elcometer® 106 tester following ASTM D4541–17[53]. Substrates were pre-treated and then coated with a sol–gel film, after which dollies were attached using a commercial two-part epoxy adhesive, Araldite® 2015–1, with a controlled bond line thickness of 300 μm. The pull-off testing revealed significant increases in adhesion strength after the application of sol–gel coatings, with the AA2024-T3 alloy showing a 28% increase and the AA7075-T6 alloy a 67% increase compared to untreated substrates (Fig. 7). Surface pre-treatments, such as HNO_3, KOH, and DI boiling water, further enhanced the adhesion, with the DI-treated surfaces showing the most significant improvement. The findings emphasise the importance of surface pre-treatment in optimising the adhesion of sol–gel coatings to aluminium alloy substrates.

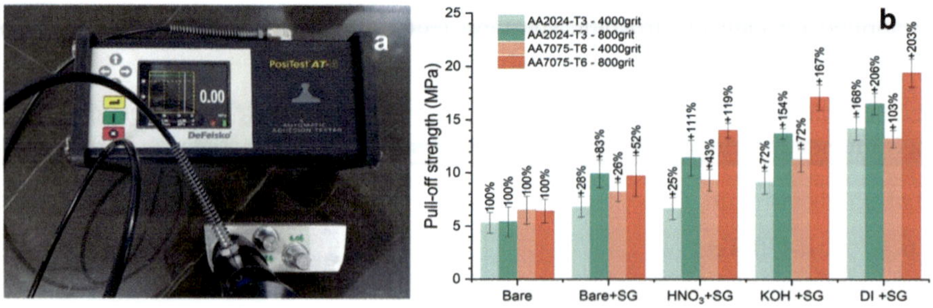

Fig. 7 **a**Commercial Ecometer for pull-off testing and **b** Comparison of average adhesion strength (pull-off strength) for AA2024-T3 and 7075-T6 aluminum alloys prepared with different surface finishes. The substrates were ground to 4000 or 800 grit, and Araldite epoxy-based adhesive was applied to all surfaces. "SG" represents sol–gel coatings, with the percentages indicating the relative improvement in adhesion strength compared to bare substrates with the same grit finish [53]

2.6.2 Cross-Hatch and Scotch Tape Test

A grid pattern is cut into the coating using a sharp blade, and adhesive tape is applied over the grid. The tape is then pulled off, and the amount of coating removed by the tape indicates the level of adhesion. The ASTM D3359-02 standard is used for evaluating adhesion through a tape test. This method uses a specialised cutter with sharp blades spaced 1 mm apart to create deep grooves in the coating. The cutter is pressed with enough force to penetrate the coating and reach the substrate, forming a grid pattern by making a second cut perpendicular to the first. A piece of sticky tape is then firmly applied to the grid and quickly removed at a right angle with a sudden motion. The surface is subsequently examined under a microscope using white light to determine the amount of material removed from the grid. Adhesion is assessed and ranked on a scale from 5 to 0B, where 5B indicates excellent adhesion with no material removed, and 0B indicates inferior adhesion with more than 65% of the material removed. The intermediate rankings reflect varying degrees of material removal: 4B for less than 5%, 3B for 5–15%, 2B for 15–35%, and 1B for 35–65% [54].

Verma et al. studied the adhesion of QGC/EP composite coatings on carbon steel, which was evaluated using the ASTM D3359-17 cross-cut tape test [55]. Coatings with different concentrations (0.01, 0.05, 0.1, and 0.5 wt.%) were tested by making a grid pattern on the coated steel and applying a 3 M® tape, which was then removed at a 180° angle. SEM analysis revealed that the 0.1 wt.% QGC/EP coating exhibited the highest adhesion, with less than 5% detachment (4B rating). In contrast, higher concentrations showed greater material removal, indicating reduced adhesion (Fig. 8). The findings suggest that lower concentrations of QGC in the epoxy matrix enhance adhesion to carbon steel.

Haritha et al. studied the adhesion of SiO_2–TiO_2/OTES-modified SiO_2 coatings on glass substrates, which was assessed using the Cross Hatch Cut method in accordance with ASTM D3359-17 standards [56]. A grid pattern was created on the coating, followed by applying and removing adhesive tape to evaluate coating removal. The test revealed no flaking or peeling, indicating excellent adhesion. Additionally, the durability of the double-layer anti-reflective (AR) coating was confirmed by stable transmittance over two years, demonstrating its long-lasting performance and suitability for various applications.

2.6.3 Scratch Test

A progressively increasing load is applied to a stylus that scratches the coating surface. The critical load at which the coating delaminates or shows signs of adhesion failure is used to assess adhesion strength [50]. The scratch test is a widely recognised method for assessing adhesion, largely due to its simplicity and broad applicability across various systems. It measures the adhesion strength of coatings on different substrates, including plastics, glass, and metals [50, 57]. This test is essential for evaluating coated surfaces' structural integrity and functionality. Organic

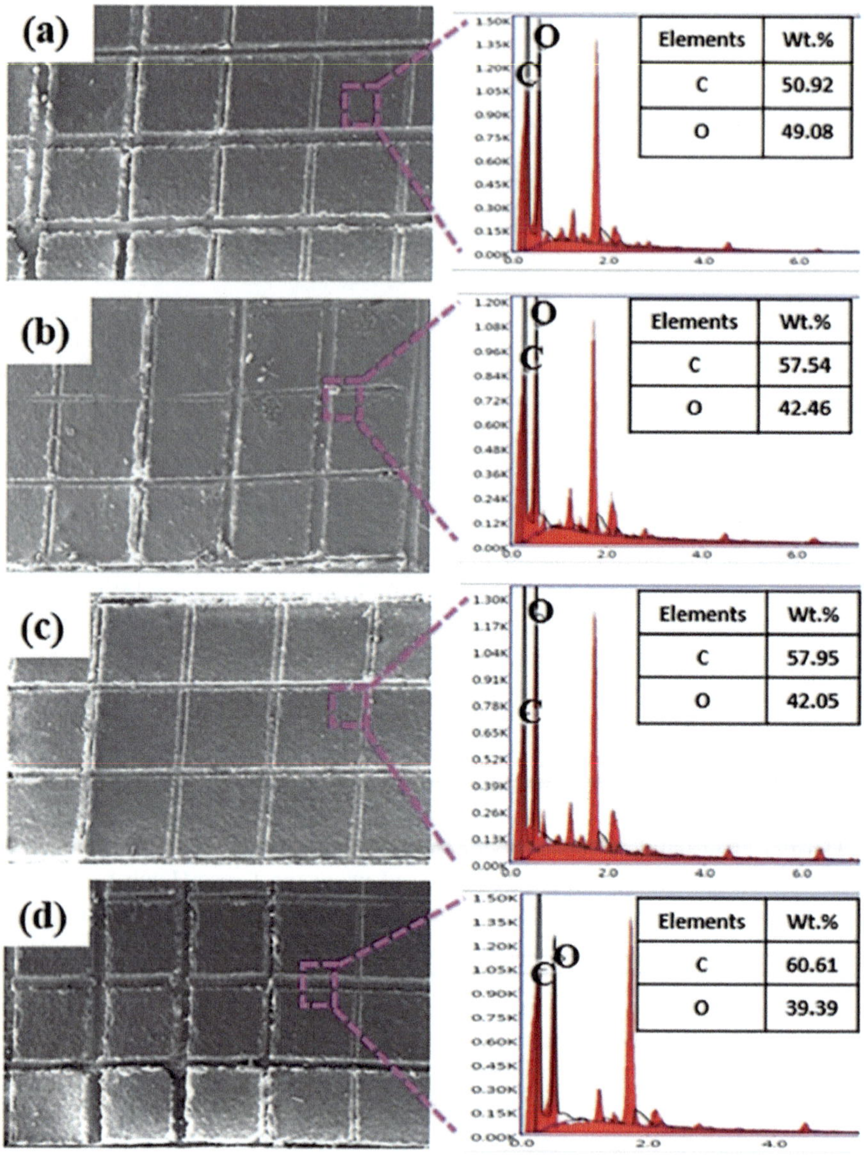

Fig. 8 SEM images after the cross-cut tape adhesion test, showing QGC/EP composite coatings on carbon steel at varying concentrations: **a** 0.01 wt.%, **b** 0.05 wt.%, **c** 0.1 wt.%, and **d** 0.5 wt.%. EDX data is also included for each sample [55]

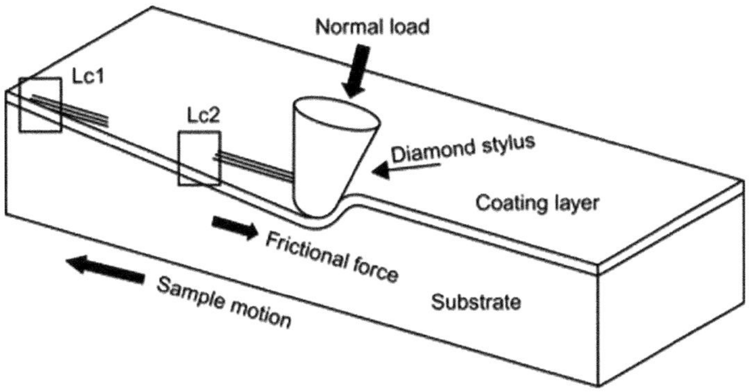

Fig. 9 Schematic showing an indenter scratching the workpiece. Lc1: Critical load for the first crack and Lc2: critical load for severe cracking [58]

coatings typically exhibit damage mechanisms such as elastic and plastic deformation, cracking, chipping, and delamination. Scratch resistance is typically evaluated using the knife test (ASTM D6677) and the pencil hardness test (ASTM D3363). During a typical scratch test, the workpiece remains stationary while an indenter is used to scratch the surface over a specified length with a predetermined normal load (as shown in Fig. 9). The load can either remain constant or increase progressively, depending on the specific testing requirements. The results are recorded in terms of acoustic emission, penetration depth, and tangential frictional force. Transducers are utilised to measure the frictional force exerted at the indenter tip. The point at which the coating begins to crack, commonly known as the critical load, corresponds to an increase in the coefficient of friction and acoustic emission, providing insight into the adhesion between the coating and substrate. The load at which the first crack forms is termed the "lower critical load," indicating the coating's cohesive strength. In contrast, the load at which the coating completely peels off from the substrate is referred to as the "upper critical load," serving as a measure of the adhesion strength between the coating and substrate [11].

Cohesion refers to the internal strength of the coating material itself, specifically the ability of the coating to resist internal forces that may cause it to fracture, crack, or delaminate within its own structure. Cohesion is essential for maintaining the integrity of the coating, particularly under conditions of mechanical stress, thermal cycling, or impact.

Cohesive strength is determined by the nature of the bonds within the coating material, including atomic, molecular, and crystalline structures. The type of material (e.g., polymer, ceramic, metallic) and its microstructure (e.g., grain size, phase distribution) play crucial roles in determining cohesion. For example, coatings with fine-grained microstructures often exhibit higher cohesion due to the uniform distribution of stresses across the material, reducing the likelihood of crack initiation and propagation.

2.7 Wear Resistance

Wear resistance is a critical property of coating materials, particularly when surfaces are subjected to repeated mechanical contact, friction, or abrasive conditions. It refers to the ability of a material to resist degradation, material loss, or damage caused by mechanical actions such as sliding, rolling, or impact. The wear resistance of a coating is essential for prolonging the life of components and maintaining their functional integrity in demanding environments. Wear can occur through various mechanisms, including:

- **Abrasive Wear** occurs when hard particles or asperities on a counter surface plow into the coating, leading to material removal. Abrasive wear is common in environments where particulate matter is present, such as mining, construction, and machining [11].
- **Adhesive Wear**: Also known as sliding wear, adhesive wear occurs when two contacting surfaces adhere to each other at a microscopic level, causing material transfer between surfaces. This type of wear is prevalent in applications involving metal-to-metal contact, such as in bearings, gears, and sliding components [30].
- **Erosive Wear**: Erosive wear is caused by the impact of high-velocity particles or fluid jets against the coating surface. It is typical in applications exposed to fluid flows containing suspended particles, such as in pumps, turbines, and pipeline systems [59].
- **Fatigue Wear**: Fatigue wear arises from repeated cyclic loading, forming and propagating cracks within the coating. This type of wear is common in applications where components are subjected to fluctuating stresses, such as in rolling-contact bearings and gear teeth [52].
- **Corrosive Wear**: This involves the combined action of mechanical wear and chemical or electrochemical reactions, often leading to accelerated material loss. It is prevalent in harsh chemical environments, such as chemical processing equipment and marine structures [12].

2.8 Measurement and Evaluation of Wear Resistance

Evaluating the wear resistance of coatings involves various testing methods, each tailored to simulate specific wear conditions:

2.8.1 Pin-On-Disk Test

A coated specimen (the disk) is rotated against a stationary pin under controlled load and speed conditions. The amount of material worn away from the disk is measured, and wear resistance is quantitatively assessed. This test is commonly used to evaluate adhesive and abrasive wear.

Sivaprakasam et al. investigated the wear behaviour of high-strength titanium alloys (Grade 23) under different conditions using a pin-on-disc tribometer in accordance with ASTM G99-95a [51]. Key variables such as load (30–50 N), sliding speed (1–2 m/s), and sliding distance (1000–2000 m) were analysed to determine their effects on wear mass loss (WML) and coefficient of friction (CoF). Using Response Surface Methodology (RSM) and statistical analysis (ANOVA), the study found that sliding distance had the most significant impact on WML and CoF, contributing 60% and 51%, respectively. Additionally, the interaction between sliding speed and distance, and the quadratic effects of sliding speed also influenced wear outcomes. The optimal conditions for minimising WML and CoF were identified as a 30 N load, 1.0 m/s sliding speed, and 1000 m sliding distance, resulting in the lowest WML of 0.0262 g and CoF of 0.3308. Conversely, the highest wear mass loss of 110.6 mg occurred at a 50 N load, 2 m/s sliding speed, and 2000 m sliding distance.

2.8.2 Erosion Testing

Erosion testing exposes the coating to a stream of abrasive particles or fluid jets, simulating erosive wear conditions. The material loss is measured to evaluate the coating's resistance to erosive forces.

2.8.3 Sliding Wear Testing

This method involves sliding a coated surface against a counterface under controlled conditions. The wear track is analysed to assess the extent of material removal and the wear mechanism involved.

2.8.4 Reciprocating Wear Testing

In this test, a coated specimen is subjected to back-and-forth sliding motion against a counter material. It simulates wear conditions experienced in components like pistons or seals, where the direction of movement changes periodically.

Wear resistance is a vital property for coating materials, directly influencing their durability and performance in various applications. A combination of material composition, microstructure, coating thickness, deposition techniques, and environmental factors determines it. Understanding and optimising wear resistance through careful selection of materials and processing methods are crucial for developing coatings that can withstand the mechanical demands of their intended use, ultimately enhancing the longevity and reliability of coated components across multiple industries.

2.9 *Fatigue Resistance*

Fatigue resistance is a critical property of coating materials that determines their ability to withstand cyclic or repetitive mechanical loading without experiencing failure [4]. In many practical applications, coatings are subjected to fluctuating stresses, such as those caused by vibration, thermal expansion, or repeated mechanical contact, which can lead to the initiation and growth of cracks over time. The ability of a coating to resist such fatigue-induced damage is crucial for ensuring its long-term durability and performance. Fatigue failure in coatings typically occurs through the initiation and propagation of cracks under cyclic loading. Even if the applied stress is below the coating's tensile strength, repeated loading can gradually accumulate microscopic damage, ultimately resulting in crack formation. Once a crack is initiated, continued cyclic loading can cause it to grow, eventually leading to the coating's failure [60]. This process is particularly concerning when coatings are exposed to mechanical vibrations, thermal cycling, or repetitive stress.

3 Factors Influencing Mechanical Properties of Coating Materials

The mechanical properties of coating materials are crucial determinants of their performance, durability, and suitability for specific applications. These properties are not inherent to the material alone but are influenced by various factors that affect the overall behavior of the coating when subjected to mechanical stresses. Understanding these factors is essential for designing coatings that meet the stringent demands of industrial applications. The different factors influencing the mechanical properties of coating materials are composition, microstructure, thickness, deposition process, substrate-coating interaction, and environmental conditions [5, 22, 37, 61–65].

3.1 *Coating Composition*

The chemical composition of a coating material is one of the most fundamental factors influencing its mechanical properties. The choice of base materials, alloying elements, and any additional components like fillers or reinforcements can significantly alter properties such as hardness, elasticity, and toughness.

- **Base Material**: The primary material of the coating defines its essential mechanical characteristics. For instance, metallic coatings (e.g., nickel, chromium) exhibit higher hardness and wear resistance than polymeric coatings, which may be more flexible but less durable [39].
- **Alloying Elements**: The inclusion of alloying elements can enhance specific properties. For example, adding tungsten or molybdenum to a metallic coating can

increase its hardness and temperature resistance. In polymer coatings, additives like plasticisers can improve flexibility but may reduce hardness [66].

- **Reinforcements**: In composite coatings, particulate or fibrous reinforcements can drastically improve mechanical properties. For example, adding ceramic particles to a polymer matrix can enhance wear resistance and hardness [22].

3.2 Microstructure

The microstructure of a coating, which includes grain size, phase distribution, and defect structure, plays a critical role in determining its mechanical properties.

- **Grain Size**: The grain size of crystalline coatings is a significant factor governed by the Hall–Petch relationship, where smaller grains generally increase hardness and strength due to the grain boundary strengthening mechanism. However, extremely small grains can lead to embrittlement, affecting toughness [21, 67].
- **Phase Distribution**: The presence and distribution of different phases within the coating can affect its mechanical properties. For example, in multi-phase alloys or ceramic coatings, the distribution of hard and soft phases can influence the overall hardness, wear resistance, and fracture toughness [68].
- **Defects and Porosity**: Defects such as dislocations, voids, or porosity can weaken a coating by providing sites for crack initiation and propagation. High porosity levels, often resulting from certain deposition techniques, can reduce the density and mechanical integrity of the coating [59, 69].

3.3 Coating Thickness

The thickness of a coating is another crucial factor that influences its mechanical properties. Both excessively thick and thin coatings can pose challenges.

- **Thick Coatings**: While thicker coatings may provide enhanced protection and durability, they are also more susceptible to cracking, delamination, and residual stresses due to their inability to conform to substrate deformations flexibly. Additionally, thicker coatings may experience differential thermal expansion, leading to stress accumulation and potential failure [70].
- **Thin Coatings**: Thin coatings, while often more flexible and less prone to cracking, may offer limited mechanical protection, particularly in abrasive or high-load environments. Thin coatings may also wear away quickly, reducing their effective lifespan [70, 71].

3.4 Deposition Process

The method used to apply the coating significantly affects its microstructure and, consequently, its mechanical properties.

- **Physical Vapor Deposition (PVD)**: PVD techniques, such as sputtering or evaporation, generally produce coatings with high density and strong adhesion, resulting in improved hardness and wear resistance. However, the process can induce residual stresses that may affect the coating's toughness [6, 72].
- **Chemical Vapor Deposition (CVD)**: CVD coatings are known for their high purity and strong bonding to substrates, which enhances their mechanical properties. However, the high temperatures required for CVD can lead to thermal stresses [5].
- **Thermal Spray Coating**: Thermal spraying techniques, like plasma spraying or flame spraying, produce coatings with high thickness and variable microstructures, which can lead to anisotropy in mechanical properties. The high porosity typical of thermal spray coatings can reduce their mechanical strength unless post-deposition treatments like sealing are applied [22, 69].
- **Electrodeposition**: This method can produce coatings with very fine microstructures and good adhesion, enhancing hardness and toughness. However, electrodeposition can sometimes result in coatings with internal stresses and grain boundary defects [73–75].

3.5 Substrate-Coating Interaction

The interaction between the coating and the substrate is a critical factor influencing the mechanical behaviour of the coated system.

- **Adhesion**: Good adhesion between the coating and substrate is essential for effective load transfer and mechanical stability. Poor adhesion can lead to delamination and failure under mechanical stress [76].
- **Substrate Properties**: The mechanical properties of the substrate, such as hardness, elasticity, and thermal expansion coefficient, affect the overall performance of the coating. A mismatch in thermal expansion between the substrate and coating can lead to residual stresses and cracking during thermal cycling [77].
- **Interfacial Layer**: An interfacial layer or bonding agent is sometimes used to improve adhesion. This layer can influence the stress distribution and energy dissipation at the interface, affecting the coating's mechanical properties [78, 79].

3.6 Environmental Conditions

Environmental factors, including temperature, humidity, and exposure to chemicals, can significantly impact the mechanical properties of coatings.

- **Temperature**: Thermal exposure can lead to thermal expansion or contraction, affecting the stress state within the coating and between the coating and substrate. High temperatures can also cause phase transformations, oxidation, or diffusion of elements at the interface, altering the coating's mechanical properties [80–82].
- **Humidity and Corrosive Environments**: Exposure to moisture or corrosive chemicals can degrade coatings through processes like hydrolysis, oxidation, or corrosion, leading to embrittlement, reduced adhesion, and accelerated wear [22, 66].
- **Mechanical Load Conditions**: The mechanical load applied during service, such as static loads, cyclic loading, or impact stresses, can influence the coating's response, potentially leading to fatigue, cracking, or delamination [83–87].

A complex interplay of factors, including composition, microstructure, coating thickness, deposition process, substrate interaction, and environmental conditions, influences the mechanical properties of coating materials. Understanding and controlling these factors is essential for optimising the performance of coatings for specific applications. By carefully considering each of these factors during the design and application of coatings, it is possible to tailor the mechanical properties to meet the demands of various industrial environments.

4 Challenges and Future Directions

Evaluating the mechanical properties of coating materials is essential for their industrial applications, but it presents several challenges as coatings become increasingly complex. Modern coatings often consist of multi-layered structures, nanocomposites, or functionally graded materials, each with distinct mechanical behaviours that interact under stress, making accurate predictions and measurements difficult. Additionally, mechanical properties such as hardness and elastic modulus can vary depending on the measurement scale, complicating the correlation between laboratory tests and real-world performance. Environmental factors further complicate evaluations, as coatings are exposed to harsh conditions that are difficult to replicate in controlled settings, and they may degrade over time, altering their mechanical properties. The lack of standardised testing protocols for advanced coatings contributes to inconsistencies in data, while ensuring reproducibility remains a challenge, especially at the nanoscale. Moreover, advanced characterisation techniques generate large datasets that require sophisticated analysis to extract meaningful insights, and misinterpretation of data can lead to incorrect conclusions about coating performance.

To address these challenges, future research should develop multi-scale testing methods that bridge the gap between nano- and macroscale measurements, providing a more comprehensive understanding of coating behaviour. Advanced in-situ and operando testing techniques will be crucial for observing coatings in real-time under realistic service conditions, offering valuable data for developing durable and reliable coatings. Integrating machine learning and data analytics into the evaluation process holds promise for identifying patterns, optimising materials, and developing predictive models, ultimately improving efficiency and accuracy. As environmental regulations tighten, there is also a growing need for sustainable coatings made from non-toxic, renewable materials, with evaluation methods that consider both mechanical performance and environmental impact. High-throughput testing techniques will play a key role in rapidly evaluating a wide range of coating compositions and accelerating the discovery of new materials. Finally, understanding and optimising interfacial properties, such as adhesion strength and fracture toughness, will be essential for developing coatings that are not only mechanically robust but also durable and reliable in service.

5 Conclusions

Evaluating the mechanical properties of coating materials is crucial for understanding their performance under different operational conditions. This chapter has highlighted various methodologies, including tensile, compressive, and bending tests, that are used to assess coatings. Additionally, the analysis of hardness, adhesion, and wear resistance provides insight into the durability and longevity of coatings in practical applications. The complex interactions between coating materials and substrates significantly influence these properties, necessitating a tailored approach for each application. Advanced techniques such as nanoindentation and scratch testing have further refined the accuracy of these evaluations, offering detailed insights at micro and nano scales. Integrating these methods with modern computational tools enhances the predictive capabilities for coating performance, leading to more efficient and reliable material selection and design. A comprehensive understanding of mechanical properties is indispensable for optimising the functionality and reliability of coatings across various industries, from aerospace to biomedical engineering. Future research should focus on developing more robust, standardised testing protocols and exploring the synergy between coating materials to achieve superior mechanical performance. Additionally, the ongoing advancements in testing technology and computational modelling are expected to play an essential role in the evolution of coating evaluation techniques, ultimately contributing to developing next-generation materials with unprecedented mechanical properties.

References

1. Patra N, Shukla M, Salerno M (2023) Thermoplastic polymer-based nanocoatings. https://doi.org/10.1016/B978-0-32-390778-1.00010-4
2. Patra N, Mameda N, Salerno M (2023) Thermoset polymer-based nanocoatings. https://doi.org/10.1016/B978-0-32-390778-1.00011-6
3. Wen W, Becker AA, Sun W (2017) Determination of material properties of thin films and coatings using indentation tests: a review. J Mater Sci 52(21):12553–12573. https://doi.org/10.1007/s10853-017-1348-3
4. McCrea J, Palumbo G (2011) Nanocoatings for commercial and industrial applications. Nanostructured Met Alloy Process Microstruct Mech Prop Appl 663–686. https://doi.org/10.1533/9780857091123.4.663
5. Nguyen-Tri P, Nguyen TA, Carriere P, Ngo Xuan C (2018) Nanocomposite coatings: preparation, characterisation, properties, and applications. Int J Corros 2018 (ii). https://doi.org/10.1155/2018/4749501
6. Yilmaz O, Yorgancioglu A (2018) Nanocoatings: preparation, properties, and biomedical applications. Elsevier Inc. https://doi.org/10.1016/B978-0-12-813932-5.00008-X
7. Benabdi M, Roche A (2012) Mechanical properties of thin and thick coatings applied to various substrates. Part I. An elastic analysis of residual stresses within coating materials. 12:16–20. https://doi.org/10.1163/156856197X00363
8. Merlin M, Fortini A (2021) Mechanical properties and structural analysis of coatings and engineered surfaces. Coatings 11(8):10–11. https://doi.org/10.3390/coatings11080875
9. Koleske JV (2000) Mechanical properties of solid coatings. Encycl Anal Chem 1–15. https://doi.org/10.1002/9780470027318.a0608
10. Demir B, Abdelaal AF, Baig MMA, Mohammed AS (2023) Mechanical, thermal, and surface properties of fusion-bonded epoxy nanocomposite coatings. J Coatings Technol Res 20(4):1489–1498. https://doi.org/10.1007/s11998-023-00760-z
11. Holmberg K Matthews A (2009) Coatings tribology: Properties, mechanisms, techniques and applications in surface engineering (2nd ed.). Amsterdam: Elsevier. ISBN 978-0-444-52750-9.
12. Mohan S, Mohan A (2014) Wear, friction and prevention of tribo-surfaces by coatings/nanocoatings. Elsevier Ltd. https://doi.org/10.1016/B978-0-85709-211-3.00001-7
13. Yang Z, Peng H, Wang W, Liu T (2010) Abrasion resistance of thermoplastic polyurethane materials blended with ethylene–propylene–diene monomer rubber. J Appl Polym Sci 116(5):2658–2667. https://doi.org/10.1002/app
14. Guo WC, Rauchs G, Papeleux L, Zhang WH, Gitzhofer F, Ponthot JP (2012) Evaluation of the mechanical properties of plasma-sprayed coating by nanoindentation technology. Mech Ind 13(3):151–162. https://doi.org/10.1051/meca/2012014
15. Calibration Issues for Nanoindentation Experiments (2010) Direct atomic force microscopy measurements and indirect methods. Microsc Res Tech. https://doi.org/10.1002/jemt.20850
16. Shen S, Zuo Y (2014) The improved performance of Mg-rich epoxy primer on AZ91D magnesium alloy by addition of ZnO. Corros Sci 87:167–178. https://doi.org/10.1016/j.corsci.2014.06.020
17. Wang DZ, Hu QW, Zeng XY (2015) Influences of parameters on microstructures and mechanical properties of Cr13Ni5Si2 based composite coating by laser-induction hybrid cladding. Surf Coatings Technol 280:359–369. https://doi.org/10.1016/j.surfcoat.2015.09.029
18. Podgorski L, de Meijer M, Lanvin JD (2017) Influence of coating formulation on its mechanical properties and cracking resistance. Coatings 7(10):1–11. https://doi.org/10.3390/coatings7100163
19. Vijaya Kumar B, Madhusudhan Reddy G, Mohandas T (2015) Influence of process parameters on physical dimensions of AA6063 aluminium alloy coating on mild steel in friction surfacing. Def Technol 11(3):275–281. https://doi.org/10.1016/j.dt.2015.04.001
20. Ammar S, Ramesh K, Vengadaesvaran B, Ramesh S, Arof AK (2016) Amelioration of anticorrosion and hydrophobic properties of epoxy/PDMS composite coatings containing nano ZnO particles. Prog Org Coatings 92:54–65. https://doi.org/10.1016/j.porgcoat.2015.12.007

21. Errichello R, Eckert R, Milburn A (2023) Influence of grain size on metallurgical properties. www.geartechnology.com
22. Kim GE, Brochu M (2014) 19—thermal spray nanostructured ceramic and metal-matrix composite coatings. Elsevier Ltd. https://doi.org/10.1016/B978-0-85709-211-3.00019-4
23. Engels HW, Pirkl HG, Albers R, Albach RW, Krause J, Hoffmann A, Casselmann H, Dormish J (2013) Polyurethanes: versatile materials and sustainable problem solvers for today's challenges. Angew Chemie Int Ed 52(36):9422–9441. https://doi.org/10.1002/anie.201302766
24. Cheong Z, Sorce FS, Ngo S, Lowe C, Taylor AC (2020) The effect of substrate material properties on the failure behaviour of coatings in the erichsen cupping test. Prog Org Coatings 2021(151):106087. https://doi.org/10.1016/j.porgcoat.2020.106087
25. Bhushan T, Chandrashekhar A, Venkat Prasat S, Rajasri Reddy I (2020) Effect of substrate surface roughness on adhesion of titanium nitride coatings deposited by physical vapour deposition technique. IOP Conf Ser Mater Sci Eng 981(4). https://doi.org/10.1088/1757-899X/981/4/042022
26. Caio F, Moreau C (2019) Influence of substrate shape and roughness on coating microstructure in suspension plasma spray. Coatings 9(11). https://doi.org/10.3390/coatings9110746
27. Zhan X, Liu Y, Yi P, Feng W, Feng Z, Jin Y (2021) Effect of substrate surface texture shapes on the adhesion of plasma-sprayed Ni-based coatings. J Therm Spray Technol 30(1–2):270–284. https://doi.org/10.1007/s11666-020-01126-2
28. Tan AWY, Sun W, Bhowmik A, Lek JY, Song X, Zhai W, Zheng H, Li F, Marinescu I, Dong Z, Liu E (2019) Effect of substrate surface roughness on microstructure and mechanical properties of cold-sprayed Ti6Al4V coatings on Ti6Al4V substrates. J Therm Spray Technol 28(8):1959–1973. https://doi.org/10.1007/s11666-019-00926-5
29. Wang Z, Rong X, Zhao L, Xing X, Ma H (2022) Effects of substrate surface characteristics on the adhesion properties of geopolymer coatings. ACS Omega 7(14):11988–11994. https://doi.org/10.1021/acsomega.2c00170
30. Jiang J, Arnell RD (2000) The effect of substrate surface roughness on the wear of DLC coatings. Wear 239(1):1–9. https://doi.org/10.1016/S0043-1648(99)00351-8
31. Miszczyk A, Darowicki K (2003) Effect of environmental temperature variations on protective properties of organic coatings. Prog Org Coatings 46(1):49–54. https://doi.org/10.1016/S0300-9440(02)00188-1
32. Stojanović I, Škrlec B, Kurtela M, Alar V, Odeljan M (2023) Physical and chemical properties of high-temperature silicone-based polymer coatings applied on different surface roughnesses. Coatings 13(12). https://doi.org/10.3390/coatings13122100
33. Evans AG, Hutchinson JW (2007) The mechanics of coating delamination in thermal gradients. Surf Coatings Technol 201(18):7905–7916. https://doi.org/10.1016/j.surfcoat.2007.03.029
34. Zhou YC, Hashida T (2002) Thermal fatigue failure induced by delamination in thermal barrier coating. Int J Fatigue 24(2–4):407–417. https://doi.org/10.1016/S0142-1123(01)00096-2
35. Dehri I, Erbil M (2000) The effect of relative humidity on the atmospheric corrosion of defective organic coating materials: an EIS study with a new approach. Corros Sci 42(6):969–978. https://doi.org/10.1016/S0010-938X(99)00126-2
36. Mishra A, Mishra PK, Bansal G (2024) Recent advances in coating characterisation techniques, INC. https://doi.org/10.1016/B978-0-443-19009-4.00024-2
37. Fernando RH (2009) Nanocomposite and nanostructured coatings: recent advancements. ACS Symp Ser 1008:2–21. https://doi.org/10.1021/bk-2009-1008.ch001
38. Tri PN, Nguyen TA, Rtimi S, Plamondon CMO (2019) Nanomaterials-based coatings: an introduction. In: Tri PN, Rtimi S, Plamondon CMO (eds) Micro and nano technologies, nanomaterials-based coatings. Elsevier, pp. 1–7, ISBN9780128158845. https://doi.org/10.1016/B978-0-12-815884-5.00001-6
39. Callister WD, Rethwisch DG (2010) Materials science and engineering: An introduction (8th ed., Vol. 1). Hoboken, NJ: John Wiley & Sons.
40. Naseer A, Evaristo M, Oliveira J, Kalin M, Cavaleiro A (2024) A sustainable deposition method for diamond-like nanocomposite coatings—insights into the evolution of atomic structure and properties. Surf Coatings Technol 489:131082. https://doi.org/10.1016/j.surfcoat.2024.131082

41. Ma Z, Zhang W, Du P, Zhu X, Krishnaswamy S, Lin L, Lei M (2017) Non-destructive measurement of elastic modulus for thermally sprayed WC-Ni coatings based on acoustic wave mode conversion by small angle incidence. NDT E Int 2018(94):38–46. https://doi.org/10.1016/j.ndteint.2017.11.003
42. Chai Y, Zou Y, Liu S, Wang D, Li Y (2022) Characterization of elastic modulus of laser cladding coatings using laser ultrasonic method. Laser Optoelectron Prog 59(1):107567. https://doi.org/10.3788/LOP202259.0114014
43. Zhu W, Wu Q, Yang L, Zhou YC (2020) In situ characterization of high temperature elastic modulus and fracture toughness in air plasma sprayed thermal barrier coatings under bending by using digital image correlation. Ceram Int 46(11):18526–18533. https://doi.org/10.1016/j.ceramint.2020.04.158
44. Luka S, Josse C, Balbo A, Monticelli C, Erika S, Kosec T, Gartner N, Martini C, Masi G, Chiavari C, Bernardi E, Bignozzi MC (2019) Practical adhesion measurements of protective coatings on bronze by three-point bending test. 16(5):1465–1477. https://doi.org/10.1007/s11998-019-00230-5
45. Yilbas BS, Arif AFM, Gondal MA (2005) HVOF coating and laser treatment: three-point bending tests. J Mater Process Technol 164–165:954–957. https://doi.org/10.1016/j.jmatprotec.2005.02.091
46. Alfano M, Pagnotta L (2006) Determining the elastic constants of isotropic materials by modal vibration testing of rectangular thin plates. J Sound Vib 293(1–2):426–439. https://doi.org/10.1016/j.jsv.2005.10.021
47. Tarancón S, Tejado E, Richou M, Pastor JY (2021) Evaluation of tensile and elastic properties of W/Cu cold-spray coatings for application to the FGM DEMO divertor concept. Fusion Eng Des 171. https://doi.org/10.1016/j.fusengdes.2021.112719
48. Varis T, Mäkelä A, Suhonen T, Laurila J, Vuoristo P (2022) Integrity of APS, HVOF and HVAF sprayed NiCr and NiCrBSi coatings based on the tensile stress-strain response. Surf Coatings Technol 2023:452. https://doi.org/10.1016/j.surfcoat.2022.129068
49. Kuroda PAB, Rossi MC, Grandini CR, Afonso CRM (2023) Assessment of applied voltage on the structure, pore size, hardness, elastic modulus, and adhesion of anodic coatings in Ca-, P-, and Mg-rich produced by MAO in Ti–25Ta–Zr alloys. J Mater Res Technol 26:4656–4669. https://doi.org/10.1016/j.jmrt.2023.08.165
50. Bull SJ, Berasetegui EG (2006) An overview of the potential of quantitative coating adhesion measurement by scratch testing. Tribol Int 39(2):99–114. https://doi.org/10.1016/j.triboint.2005.04.013
51. Jackson GA, Bai M, Pala Z, Hussain T, Sun W (2019) Small punch creep testing of thermally sprayed stellite 6 coating : a comparative study of as-received vs post-heat treatment. Mater Sci Eng A 749:137–147. https://doi.org/10.1016/j.msea.2019.02.030
52. Duda M, Pach J, Lesiuk G (2019) Influence of polyurea composite coating on selected mechanical properties of AISI 304 steel. Materials (Basel) 12(19). https://doi.org/10.3390/ma12193137
53. van Dam JPB, Tiringer U, Abrahami ST, Milošev I, Terryn H, Kovač J, Mol JMC (2024) Surface engineering of aerospace aluminium alloys: understanding alloying effects on chemical pretreatment and sol-gel coating adhesion. Surf Coatings Technol 485:0–14. https://doi.org/10.1016/j.surfcoat.2024.130901
54. Milionis A, Loth E, Bayer IS (2016) Recent advances in the mechanical durability of superhydrophobic materials. Adv Colloid Interface Sci 229:57–79. https://doi.org/10.1016/j.cis.2015.12.007
55. Verma A, Banik D, Tiwary CS, Bhattacharya J (2024) Co-pyrolysis of waste wood and plastic to produce quasi graphitic carbon: enhanced biochar properties for metal coatings through safe sequestration and green technology. Carbon Trends 15:100369. https://doi.org/10.1016/j.cartre.2024.100369
56. Haritha AH, Faturíková K, Duran A, Galusek D, Castro Y, Velázquez JJ (2024) High performance double layer hydrophobic antireflective coatings on glass prepared by sol–gel method. Open Ceram 18. https://doi.org/10.1016/j.oceram.2024.100607

57. Lei F, Hamdi M, Liu P, Li P, Mullins M, Wang H, Li J, Krishnamoorti R, Guo S, Sue HJ (2017) Scratch behavior of epoxy coating containing self-assembled zirconium phosphate smectic layers. Polymer (Guildf. 112:252–263. https://doi.org/10.1016/j.polymer.2017.02.020
58. Kalidindi RSR, Subasri R (2014) Sol-gel nanocomposite hard coatings. Elsevier Ltd. https://doi.org/10.1016/B978-0-85709-211-3.00005-4
59. Odhiambo JG, Li WG, Zhao YT, Li CL (2019) Porosity and its significance in plasma-sprayed coatings. Coatings 9(7):1–19. https://doi.org/10.3390/coatings9070460
60. Bouzakis KD, Skordaris G, Bouzakis E, Makrimallakis S, Kombogiannis S, Lemmer O (2013) Fatigue strength of diamond coatings' interface assessed by inclined impact test. Surf Coatings Technol 237:135–141. https://doi.org/10.1016/j.surfcoat.2013.09.030
61. Pawlowski L (2008) The science and engineering of thermal spray coatings (2nd ed.). Chichester, UK: John Wiley amp; Sons. https://doi.org/10.1002/9780470754085
62. Wicks ZW, Jones FN, Pappas SP, Wicks DA (2006) Organic coatings: science and technology. Org Coatings Sci Technol 1–722. https://doi.org/10.1002/9780470079072
63. Knudsen OØ, Forsgren A (2017) Corrosion control through organic coatings, 2nd edn. CRC press, Taylor and Francis group. https://doi.org/10.1201/9781315153186
64. Tri PN, Nguyen TA, Rtimi S, Plamondon CMO (2019) Nanomaterials-based coatings: an introduction, Elsevier Inc. https://doi.org/10.1016/B978-0-12-815884-5.00001-6
65. Ammar S, Ma IAW, Ramesh K, Ramesh S (2019) Polymers-based nanocomposite coatings. Elsevier Inc. https://doi.org/10.1016/B978-0-12-815884-5.00002-8
66. Ahmad Z (2006) selection of materials for corrosive. Princ Corros Eng Corros Control 1:1–646. https://doi.org/10.1016/B978-075065924-6/50010-6
67. Shi X, Nguyen TA, Suo Z, Liu Y, Avci R (2009) Effect of nanoparticles on the anticorrosion and mechanical properties of epoxy coating. Surf Coatings Technol 204(3):237–245. https://doi.org/10.1016/j.surfcoat.2009.06.048
68. Bian Y, Cao L, Zeng D, Cui J, Li W, Yu Z, Zhang P (2023) The tribological properties of two-phase hard and soft composite wear-resistant coatings on titanium alloys. Surf Coatings Technol 456:129256. https://doi.org/10.1016/j.surfcoat.2023.129256
69. Iqbal AA, Lim MJ (2023) A relationship of porosity and mechanical properties of spark plasma sintered Scandia stabilized zirconia thermal barrier coating. Results Eng 19:101263. https://doi.org/10.1016/j.rineng.2023.101263
70. Bogdan M, Peter I (2024) A comprehensive understanding of thermal barrier coatings (TBCs): applications, materials, coating design and failure mechanisms. Metals (Basel) 14(575):2–33. https://doi.org/10.3390/met14050575
71. Benabdi M, Roche AA (1997) Mechanical properties of thin and thick coatings applied to various substrates. Part i. an elastic analysis of residual stresses within coating materials. J Adhes Sci Technol 11(2):281–299. https://doi.org/10.1163/156856197X00363
72. Wendel JF, Lenz Leite M, Furtat P, Tangermann-Gerk K, Schafföner S, Motz G (2022) New coatings systems with improved mechanical properties by combining polymer derived ceramic and physical vapor deposition coating methods. Adv Mater Interfaces 9(34). https://doi.org/10.1002/admi.202201654
73. Nguyen VC, Lee CY, Chen FJ, Lin CS, Liu TY (2012) Study on the internal stress of nickel coating electrodeposited in an electrolyte mixed with supercritical carbon dioxide. Surf Coatings Technol 206(14):3201–3207. https://doi.org/10.1016/j.surfcoat.2012.01.005
74. Kovenskiy IM, Kulemina AA (2016) Influence of the conditions of electrode position and heat treatment on the structure and properties of metallic coatings. IOP Conf Ser Mater Sci Eng 127(1). https://doi.org/10.1088/1757-899X/127/1/012041
75. Godon A, Creus J, Cohendoz S, Conforto E, Feaugas X, Girault P, Savall C (2010) Effects of grain orientation on the hall-petch relationship in electrodeposited nickel with nanocrystalline grains. Scr Mater 62(6):403–406. https://doi.org/10.1016/j.scriptamat.2009.11.038
76. Bhattacharyya AS, Bhattacharyya K (2024) Effect of loading rate and coating thickness on wear and adhesion during sliding indents of Si–C–N/glass coatings useful for automotive applications. Results Surf Interfaces 14:100196. https://doi.org/10.1016/j.rsurfi.2024.100196

77. Seidl WM, Bartosik M, Kolozsvári S, Bolvardi H, Mayrhofer PH (2018) Influence of coating thickness and substrate on stresses and mechanical properties of (Ti, Al, Ta)N/(Al, Cr)N multilayers. Surf Coatings Technol 347:92–98. https://doi.org/10.1016/j.surfcoat.2018.04.060
78. Fu Q, Zhuang L, Ren Q, Feng L, Li H, Guo Y (2015) Carbon nanotube-toughened interlocking buffer layer to improve the adhesion strength and thermal shock resistance of SiC coating for C/C–ZrC–SiC composites. J Mater 1(3):245–252. https://doi.org/10.1016/j.jmat.2015.07.006
79. Yang J, Yue Y, Lv H, Ren B, Zhang Y (2023) Effect of adding intermediate layers on the interface bonding performance of WC-Co diamond-coated cemented carbide tool materials. Molecules 28(16). https://doi.org/10.3390/molecules28165958
80. Tien CL, Chiang CY, Wang CC, Lin SC (2024) Temperature-dependent residual stresses and thermal expansion coefficient of VO_2 thin films. Inventions 9(3):1–10. https://doi.org/10.3390/inventions9030061
81. Levi CG, Hutchinson JW, Vidal-Sétif MH, Johnson CA (2012) Environmental degradation of thermal-barrier coatings by molten deposits. MRS Bull 37(10):932–941. https://doi.org/10.1557/mrs.2012.230
82. Song Y, Zhuan X, Wang TJ, Chen X (2014) Thermal stress in fabrication of thermal barrier coatings. J Therm Stress 37(12):1390–1415. https://doi.org/10.1080/01495739.2014.937249
83. Han L, Zhang Z, Dai J, Li X, Bai J, Huang Z, Guo C, Xue F, Chu C (2021) The influence of alternating cyclic dynamic loads with different low frequencies on the bio-corrosion behaviors of AZ31B magnesium alloy in vitro. Bioact Mater 2022(7):263–274. https://doi.org/10.1016/j.bioactmat.2021.05.049
84. Liu H, Wang J, Yu H, Zhang Y (2023) The influence of cyclic load amplitude on mechanical response and acoustic emission characteristics of granite. Sustainability 15(12). https://doi.org/10.3390/su15129228
85. Yang HQ, Zhang Q, Li YM, Liu G, Huang Y (2020) Effects of mechanical stress on protective properties of a marine coating on mild steel substrate. Corros Sci 177:108986. https://doi.org/10.1016/j.corsci.2020.108986
86. Kesler OE, Hobbs LW (1999) Residual stresses and properties of layered and graded coatings. Thesis (Sc.D.)--Massachusetts Institute of Technology, Dept. of Materials Science and Engineering, 1999.
87. Yáñez-Contreras P, León-Rodríguez M, Medina-Flores JM, Jiménez-García JA, Santander-Bastida FJ, Yáñez-Rodríguez J (2020) Determination of residual stresses in thermal barrier coating due to the amount of CMAS infiltration. DYNA 87(215):76–83. https://doi.org/10.15446/dyna.v87n215.86941

Chapter 3
Engineering Coating Materials

Sumitra Dutta, Aishwarya Madhuri, and Bibhu Prasad Swain

Abstract Engineering coating materials play a crucial role in enhancing the performance and longevity of components across numerous industries. This abstract explores the application of diamond-like carbon (DLC) and its alloys, metal carbides and metal nitrides, titanium nitride (TiN), titanium carbide (TiC) and its alloys, molybdenum disulphide (MoS_2), aluminium oxide (Al_2O_3) and carbo-nitride (C_3N_4) coatings in engineering materials, focusing on their composition, deposition methods and properties. Among these, C_3N_4 coatings stand out for their unique combination of hardness, wear resistance and thermal stability. The review examines the impact of the above coatings on the performance of cutting tools, bearings, and other high-stress components, highlighting their potential to improve efficiency and durability in diverse applications. A comprehensive exploration of engineering coating materials through case studies and practical examples provides insights into the critical considerations for selecting and engineering coatings for specific applications, ensuring optimal performance and durability in demanding environments. Essential topics of conversation include the connection between the performance and microstructure of coatings, how surface engineering can prolong the life of industrial components, and the developments in nanostructured coatings that provide better mechanical and thermal qualities. The chapter also looks at adhesion, thickness control, and developing eco-friendly processes in coating technology.

Keywords DLC · Metal carbides and nitrides · TiN and TiC · MoS_2 · Al_2O_3 · C_3N_4 · Mechanical properties · Surface engineering

1 Introduction

Engineering coating materials are substances applied to surfaces to enhance their performance, durability and functionality. These coatings serve various purposes, including protection against corrosion, wear resistance, thermal insulation and

S. Dutta (✉) · A. Madhuri · B. P. Swain
Department of Physics, National Institute of Technology Manipur, Langol, Imphal, Manipur, India
e-mail: sumidutt09@gmail.com

B. P. Swain (ed.), *Advances in Mechanical Coating*, Materials Horizons: From Nature to Nanomaterials, https://doi.org/10.1007/978-981-96-7484-8_3

aesthetic improvement. They are used across multiple industries, from aerospace and automotive to electronics and construction. This chapter explores the multifaceted world of engineering coating materials, delving into the principles of coating science, the variety of materials used, and the technologies employed to apply these coatings. The field encompasses various materials, from traditional metallic and ceramic coatings to advanced polymeric and composite coatings. Each type of coating material offers unique properties and advantages tailored to meet the specific demands of its application.

1.1 Types of Engineering Coating Materials

Metal coatings such as zinc, aluminium and chromium protect against corrosion and wear. They are often used in environments where mechanical strength and durability are critical. Ceramics like alumina and zirconia are chosen for their hardness, chemical resistance, and thermal stability, which are ideal for high-temperature applications and environments with abrasive wear. Polymers, including organic coatings, epoxies, polyurethanes and polyesters, offer versatility and protection against corrosion and chemicals. They are also used for decorative purposes and can provide impact resistance. Composite coatings combine polymers with other materials like ceramics or metals to enhance thermal stability, electrical conductivity, and mechanical strength. For example, polymer-matrix composites offer improved strength and thermal stability. Silica-based coatings are used for their optical clarity, chemical resistance and durability. They are applied in optical devices and high-temperature environments. Incorporating nanoparticles into coatings can enhance properties such as self-cleaning, anti-microbial effects and increased strength, which are used in high-tech applications.

2 Engineering Coating Technology

Engineering coating technology involves applying thin layers of material to surfaces to enhance their properties such as durability, corrosion resistance and appearance. It encompasses various methods and materials, including:

2.1 Physical Vapour Deposition (PVD)

PVD is a process used to deposit thin films or coatings onto a substrate. In this process, a material (usually a metal or semiconductor) is heated in a vacuum chamber until it evaporates or is sputtered off, turning into a vapour. The vapourized material travels through the vacuum chamber. The vapour condenses onto the substrate, forming a thin, solid film. PVD is used in various applications, including coating surfaces for wear resistance, improving optical properties and enhancing electronic devices. Standard techniques within PVD include evaporation and sputtering.

Important PVD techniques include:

2.1.1 Evaporation

Evaporation-PVD is a specific PVD process used to deposit thin films or coatings. The material (often metal or semiconductor) is placed in a source or evaporation boat during this process. This material is heated to its boiling point in a vacuum chamber, causing it to evaporate. The evaporated material forms a vapour that travels through the chamber's vacuum environment. The vapour then condenses onto the substrate, usually held at a lower temperature. As the vapour cools, it solidifies and forms a thin, uniform coating on the substrate. The process is conducted under a high vacuum to prevent material contamination and oxidation, ensuring the deposited film's high purity and quality. Evaporation PVD is commonly used in industries to coat optical components, semiconductors, and decorative finishes.

2.1.2 Ion Plating

Ion plating is a variation of PVD that enhances film quality and adhesion. A target material is vaporized in a vacuum chamber using thermal evaporation or sputtering methods. The vaporized material is then ionized by bombarding it with high-energy electrons or an electric field. This process generates positively charged ions from the vaporized material. The ions are accelerated towards the substrate, condensing to form a thin film. Simultaneously, they bombard the substrate surface, which helps clean and improve the adhesion of the coating. The ion bombardment causes the vaporized material to be deposited more densely and uniformly on the substrate, enhancing film properties such as hardness and adhesion. Ion plating is used for high-performance coatings in various applications, including hard coatings for tools, decorative finishes and wear-resistant layers.

2.1.3 Magnetron Sputtering

Magnetron sputtering is a PVD technique that deposits thin films onto surfaces. It involves using a magnetron, a type of magnet that creates a magnetic field in a vacuum chamber. In this process, the target of the material to be deposited is placed in the chamber. A gas, often argon, is introduced into the chamber and ionized to form a plasma by applying a high voltage. The magnetron creates a magnetic field that confines and enhances the plasma near the target. Ions from the plasma bombard the target material, causing it to eject atoms or molecules. These ejected atoms or molecules travel through the vacuum and deposit onto a substrate, forming a thin film. This technique is widely used in various industries for coating applications because it produces high-quality, uniform coatings.

2.1.4 Cathodic Arc Deposition

Cathodic arc deposition is a PVD technique used to deposit thin films and coatings with high performance and durability. A high-voltage electrical arc is struck between a cathode (the target material) and an anode inside a vacuum chamber. This arc generates intense heat and ionizes the surrounding gas (often argon). The heat from the arc causes the target material to evaporate or sputter away, forming a plasma of ions and neutral atoms. The plasma created in the arc region expands and the ions and neutral atoms are accelerated towards the substrate. These particles then condense on the substrate surface, forming a thin film. As the material condenses, it forms a high-density, hard coating with good adhesion and mechanical solid properties. Cathodic arc deposition is known for producing high-quality, adherent coatings that are useful in various applications, including wear-resistant coatings, decorative finishes and functional coatings in electronics. Figure 1 shows a schematic for some of the important PVD techniques: (a) Evaporation, (b) Ion Plating, (c) Magnetron Sputtering and (d) Cathodic Arc Deposition.

2.2 Chemical Vapour Deposition (CVD)

CVD is a process used to produce thin films, coatings or materials on a substrate through chemical reactions of vapourized precursors. In CVD, gaseous chemicals react or decompose on the surface of a heated substrate, forming a solid film. In this process, vapourized chemical precursors are introduced into a reaction chamber. The precursors undergo a chemical reaction or decomposition on the heated substrate surface. The reaction produces a solid material that deposits onto the substrate, forming a thin film or coating. Any gaseous by-products are evacuated from the chamber. CVD is widely used in various industries, including electronics, semiconductor fabrication, solar panel production, and coatings in materials science. It allows precise control over the film's thickness, composition and uniformity. Essential CVD techniques include:

2.2.1 Low-Pressure CVD (LPCVD)

LPCVD is a variant of CVD that operates at reduced pressure, typically in the range of 0.1–1 Torr. The reduced pressure improves the uniformity and quality of the deposited films. This process is conducted in a vacuum or near-vacuum environment, which lowers the pressure inside the deposition chamber. Gaseous chemical precursors are introduced into the chamber. At elevated temperatures, usually between 300

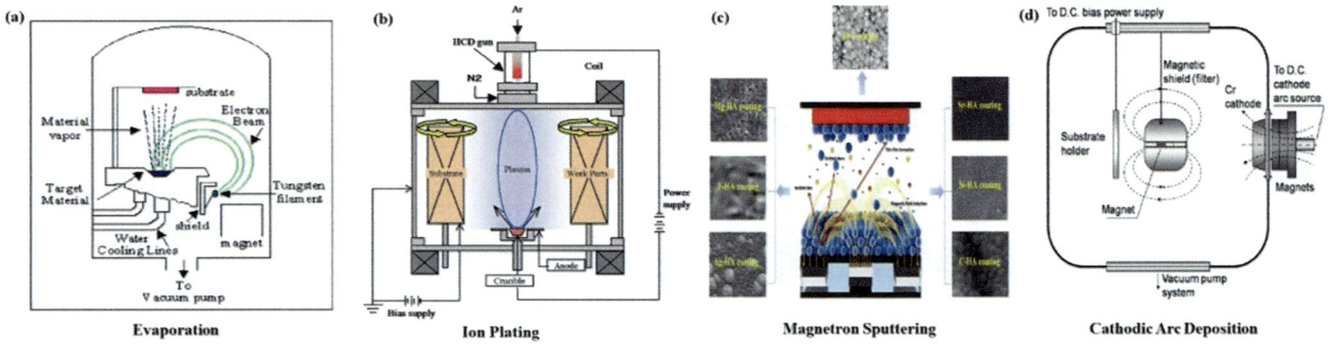

Fig. 1 Some important PVD techniques: **a** evaporation, **b** ion plating, **c** magnetron sputtering and **d** cathodic arc deposition [1–4]

and 800 °C, the precursors decompose or react on the substrate surface, forming a solid film. The solid material from the chemical reaction deposits onto the substrate, creating a thin film. Gaseous by-products from the reaction are removed from the chamber through vacuum pumps. LPCVD is favoured for its ability to produce high-quality films with good conformality and uniformity over large areas, making it especially useful in the semiconductor industry for insulating layers, protective coatings and diffusion barriers.

2.2.2 Atmospheric Pressure CVD (APCVD)

APCVD is a CVD process conducted at atmospheric pressure. In APCVD, the deposition of thin films occurs at the same pressure as the surrounding environment, typically around 1 atmosphere. This process occurs at normal atmospheric pressure, simplifying equipment requirements compared to low-pressure methods. Gaseous precursors are introduced into the deposition chamber. These gases are typically mixed with carrier gases to ensure proper transport and dispersion. The precursor gases are heated, often using a flame or electrical heating elements, causing them to react or decompose on the substrate surface. This reaction forms a solid film or coating. The solid material deposits onto the substrate, forming a uniform film. The gas flow carries away any by-products of the reaction and is typically vented out of the chamber. APCVD is widely used for depositing films and producing glass, ceramics, and semiconducting coatings. It is known for its simplicity, cost-effectiveness and ability to cover large areas, though it may offer less control over film uniformity and thickness compared to low-pressure techniques.

2.2.3 Metal–Organic CVD (MOCVD)

MOCVD is a technique used to deposit thin films and coatings of metal–organic compounds onto a substrate. This process is precious for creating high-quality semiconductor layers and other advanced materials. MOCVD uses metal–organic compounds as precursors. These compounds contain metal atoms bonded to organic ligands. The precursors are vaporized and transported into the deposition chamber. In the chamber, the precursors are heated to high temperatures, typically between 200 and 1000 °C. The metal–organic compounds decompose or react on the substrate surface. The metal atoms from the precursors deposit onto the substrate and form a solid film. The organic components are often removed as gaseous by-products. The substrate, a semiconductor wafer or another material, is usually heated to facilitate the deposition process and ensure high-quality film growth. The gaseous by-products from the reaction are extracted from the chamber by a flow of carrier gas and/or vacuum systems. MOCVD is widely used in the fabrication of semiconductor devices, including LEDs, solar cells and high-electron-mobility transistors (HEMTs), due to its ability to deposit precise and uniform layers of materials such as gallium nitride (GaN), indium phosphide (InP) and various other compound semiconductors.

2.2.4 Plasma-Enhanced CVD (PECVD)

PECVD is a variation of CVD that utilises plasma to enhance the chemical reactions of vaporized precursors, allowing for deposition at lower temperatures than conventional CVD. A plasma is generated in the deposition chamber using an electric field. This is typically achieved by applying radio frequency (RF) or microwave energy to a gas, creating a highly ionized and reactive environment. Gaseous chemical precursors are introduced into the chamber. These gases are often less reactive at lower temperatures but become highly reactive when exposed to the plasma. The plasma ionizes the precursors, breaking them into reactive species such as ions, radicals and electrons. These reactive species facilitate the chemical reactions required for film formation at lower temperatures. The reactive species from the plasma interact with the substrate, producing a thin film or coating. This process can be controlled to achieve specific film properties. A vacuum or gas flow system removes gaseous by-products from the chamber. PECVD is particularly useful for depositing thin films of materials such as silicon dioxide, silicon nitride and various polymers. It is favoured for applications requiring low deposition temperatures, such as the deposition of films on temperature-sensitive substrates or in the fabrication of microelectronic devices (Fig. 2).

3 Coating Materials

3.1 Diamond-Like Carbon Coatings

Diamond-Like Carbon (DLC) coatings are a class of amorphous carbon materials that exhibit some of the desirable properties of diamond, such as high hardness, low friction, and chemical inertness, which makes them resistant to scratching, wear, corrosion and chemical attack, and improves the efficiency of mechanical systems. Some forms of DLC can be optically transparent, which is useful for optical applications. DLC coatings vary based on the sp^2 (graphite-like) ratio to sp^3 (diamond-like) carbon bonds. These variations lead to different properties: (a) Hydrogenated DLC (a-C) contains hydrogen, which can help reduce internal stress and improve adhesion; (b) Non-hydrogenated DLC (ta-C) contains a higher fraction of sp^3 bonds, offering greater hardness and wear resistance. CVD is a common method where a carbon-containing gas (like methane) is decomposed on a substrate to form the DLC coating. DLC coatings are used on engine components like piston rings and fuel injectors to reduce friction and wear, improving fuel efficiency and engine life. Due to their biocompatibility, DLC coatings are used on medical devices, implants and surgical tools to improve their wear resistance and reduce the risk of infection. These optical lenses and windows use these coatings for their protective and anti-reflective properties. They are used in hard disk drives to protect the disks and read/write heads from wear and damage. The optimization process to produce DLC films employing

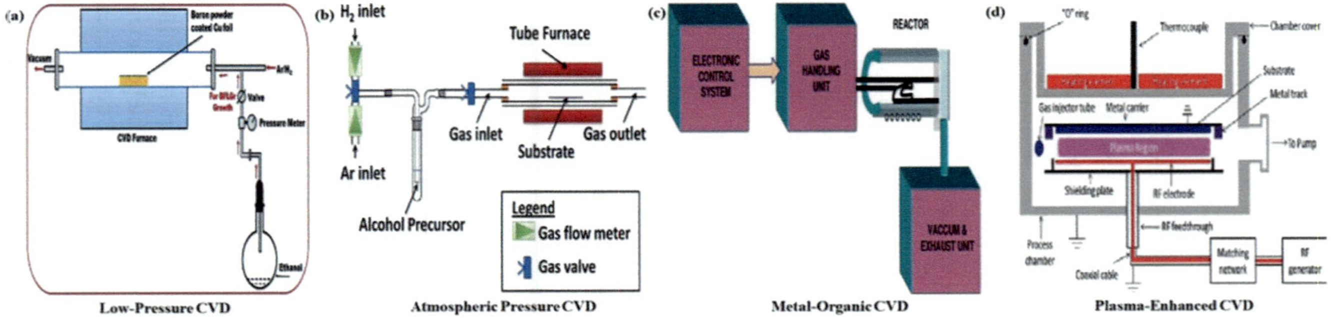

Fig. 2 Some important CVD techniques: **a** low-pressure CVD, **b** atmospheric pressure CVD, **c** metal–organic CVD and **d** plasma-enhanced CVD [5–8]

genetic algorithms (GAs) by Ghadai et al. involves encoding the deposition parameters as chromosomes, which evolved over successive generations [9]. The optimized coatings exhibit superior hardness, reduced wear, and lower friction coefficients. Another study optimized the PECVD parameters to maximize the hardness of the resulting DLC films [10]. The highest values of Young's modulus (E) and hardness (H) for the DLC coating deposited using APCVD by Ghadai et al. are observed to be 193.97 GPa and 22.46 GPa, respectively, as the H_2 flow rate increases [11]. In another study, Ghadai et al. explore the incorporation of various metals, such as titanium (Ti), chromium (Cr), or tungsten (W), into the DLC matrix. These metals aim to improve the mechanical and tribological properties of the DLC films, such as hardness, toughness, and wear resistance [12]. Sikora et al. find that femtosecond laser ablation tends to produce films with a higher fraction of sp^3 (diamond-like) bonds, leading to a more diamond-like structure with potentially enhanced mechanical properties [13].

In contrast, nanosecond laser ablation results in a higher fraction of sp^2 (graphite-like) bonds, which can influence the hardness and wear resistance of the film. By applying an external magnetic field, Modabberasl et al. report increased sp^3 (diamond-like) bond content within the DLC films, improving hardness, reduced friction, and better wear resistance [14]. Tallant et al. identify the temperature ranges above 300 °C, where the transition from sp^3 to sp^2 bonding occurs [15]. As the temperature increases, DLC films lose their diamond-like character above 260 °C, reducing hardness and wear resistance. The main issues with DLC coatings are their significant internal stress and the potential deterioration of mechanical characteristics caused by heat treatment.

3.2 Influence of Synthesis Method and Process Parameters on DLC Properties

Various synthesis methods, such as CVD or PVD, can affect the structure, composition and bonding of DLC films. Process parameters like temperature, pressure, gas composition, and substrate material are crucial in determining the properties of DLC coatings, such as hardness, adhesion, wear resistance, and friction coefficient. The initial research in the early 1970s that paved the way for the development of DLC coatings involved the utilization of ion beam deposition techniques by Aisenberg and Chatbot et al. [16]. The properties of DLC coatings, including their mechanical and physicochemical characteristics, are influenced by both the process parameters and the method of production [10]. Trava-Airoldia et al. [17] demonstrated that there are notable variations in hardness, tribological properties, internal stress and adherence of DLC coatings produced through different methods like radio frequency PACVD, IBAD and EP-DC PECVD even when the coatings are of equal thickness. The stress values varied between approximately 0.5 to 3 GPa. The choice of carbon-containing gas during the CVD process for synthesizing DLC was crucial as it could impact the speed at which the coating was formed [18]. Libardi et al. [19] presented data indicating that coatings have the highest hardness at around 10% hydrocarbon content. Research in the field has shown a correlation between the roughness of DLC surfaces and factors such as surface temperature [20] or the energy of ion impingement [21]. Chowdhury et al. [22] studied the synthesis of DLC coatings using RF magnetron sputtering and found that finds that films deposited at lower substrate temperatures tend to have higher hardness due to a more significant fraction of sp^3 bonding, while higher temperatures lead to softer films with more sp^2 bonding.

3.3 *Doping of DLC*

Silicon doping is shown to potentially reduce the internal stress in DLC films, which can otherwise lead to cracking and delamination. The maximum hardness of 17.95 GPa and Young's modulus of 186.65 GPa were observed for Si-DLC films with 70 sccm and 90 sccm H_2 flow rate by Ghadai et al. [23]. Titanium (Ti) enhances the toughness and adhesion of DLC coatings, reducing internal stresses and improving wear resistance. Ma et al. examined how Ti doping influences the microstructure of the DLC films, particularly the ratio of sp^2 to sp^3 carbon bonding, with a mean water contact angle of $104.2 \pm 1°$ [24]. Tungsten (W) increases the hardness and wear resistance while maintaining a low friction co-efficient. Song et al. demonstrate that the sp^3-rich DLC is incorporated in the W_2C or WC_{1-x} composite structure, and the smooth surface is encompassed by W-DLC films [25]. W doping significantly increased the adhesion and hardness from 6.59 to 22.44 GPa of the DLC film. Cr-doped DLC coatings exhibit enhanced hardness and wear resistance compared to undoped DLC coatings. Adding Cr forms a more robust and durable coating, making it suitable for applications requiring high performance under severe conditions. The Mo addition significantly improved the decrease of residual stress in the DLC film.

Additionally, the wettability changed from a hydrophilic to a hydrophobic state with the maximum water contact angle of 94.2° at 20.1% Mo doping before significantly decreasing to 92.7° as the Mo content increased [26]. Boron increases the hardness and chemical stability, which is useful for applications requiring high wear resistance. Phosphorus improves the electrical conductivity and wear resistance of the DLC films. Hydrogen (H) is commonly present in a-C DLC, influencing the sp^2/sp^3 bonding ratio, hardness, and elasticity. PECVD is a standard method for incorporating dopants into DLC during deposition in CVD. Reactive gases containing the dopant elements are introduced into the plasma, allowing them to be incorporated into the growing film. In PVD, targets containing the dopant materials are sputtered along with carbon to create a doped DLC film. The dopant material is introduced into the carbon arc stream to achieve doping. Dopant ions are accelerated and implanted into the surface of the DLC coating, modifying its properties locally. Doped DLC coatings, particularly those doped with metals like W and Ti, are used in engine components to reduce friction and wear, improving fuel efficiency and component lifespan. Si-DLC and N-DLC coatings are employed in medical implants and are

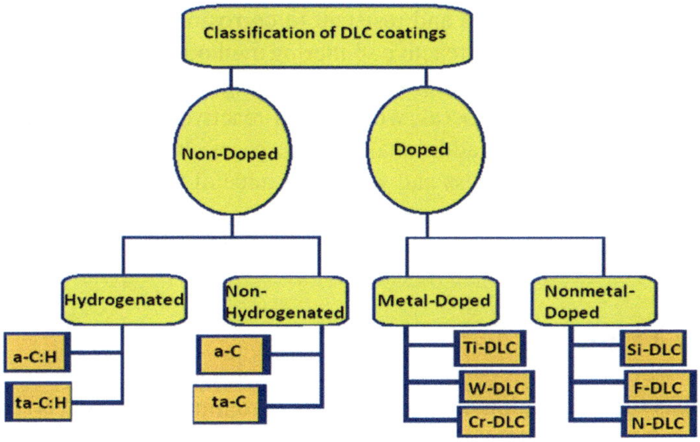

Fig. 3 Key factors that influence the mechanical properties of DLC coatings [30]

tools to enhance biocompatibility, minimise wear, and resist bacterial adhesion. P-doped DLC can enhance the electrical conductivity of coatings, making them suitable for use in electronic devices and components. F-doped DLC coatings reduce surface energy and improve anti-reflective properties, which is helpful for optical lenses and windows. One standard method for enhancing DLC adhesion involves using single and multilayers and introducing gradient layers [27]. Various metallic and non-metallic dopants such as Si, F, O or N have changed DLC's hydrophobic/hydrophilic characteristics.

Wan et al. find that phosphorus doping enhances the field emission performance of DLC films, reducing the turn-on electric field and increasing the emission current density [28]. Oxygen doping reduced the refractive index and increased the optical bandgap of the DLC films, which is attributed to changes in the sp^2/sp^3 bonding ratio and the introduction of oxygen-related defects in the carbon matrix [29]. Oxygen doping enhanced the optical transparency of DLC films in the visible and near-infrared regions. This makes oxygen-doped DLC films particularly attractive for applications requiring high optical clarity. Figure 3 shows the key factors that influence the mechanical properties of DLC coatings.

3.4 Metal Carbides and Metal Nitrides

3.4.1 Metal Carbides

Metal carbides such as tungsten carbide (WC) and titanium carbide (TiC) are known for their extreme hardness, making them ideal for wear-resistant applications. They possess high melting points suitable for high-temperature applications. Metal carbides often have good thermal conductivity, aiding in heat dissipation. They are generally chemically stable and resistant to corrosion and oxidation. Glechner et al. compare reactive and non-reactive sputtering methods for depositing WC-based coatings [31]. In reactive sputtering, a reactive gas (e.g., nitrogen or oxygen) is introduced during the deposition process, while in non-reactive sputtering, only the target material is sputtered without additional reactive gases. It is observed that reactive sputtering enhances the hardness and wear resistance of the WC-N or WC-O coatings due to the formation of more complex phases or compounds. This also leads to increased brittleness compared to non-reactive sputtered WC coatings.

3.4.2 Metal Nitrides

Metal nitrides such as titanium nitride (TiN) and chromium nitride (CrN) offer excellent hardness and wear resistance. Das et al. focus on how different nitrogen flow rates influence the growth, composition, and properties of TiN films deposited by CVD [32]. Higher N_2 flow rates typically result in more nitrogen-rich TiN films with less surface roughness than films deposited with lower N_2 flow and enhance the hardness and wear resistance of the TiN coatings. Nitrides are resistant to corrosion, particularly in harsh chemical environments. Protective coatings protect against wear, corrosion, and high temperatures, typical in aerospace, automotive, and manufacturing industries. TiN is biocompatible and used in medical implants and surgical instruments to enhance wear resistance and reduce friction. Nitrides are chosen for environments requiring a balance of hardness, corrosion resistance and electrical conductivity. Madhuri et al. examine the effect of N_2 flow rate on the atomic coordination in AlN thin films [33]. Higher N_2 flow rates lead to a nitrogen-rich environment, affecting the stoichiometry and coordination of Al and N atoms in the AlN lattice.

3.4.3 Ceramic Protective Coatings

Ceramic coatings are typically made from silicon carbide, aluminium oxide, zirconium oxide and titanium dioxide. These materials provide excellent hardness and thermal stability. Common application methods include thermal spraying, PVD, CVD and sol–gel processes. Ceramic coatings can withstand high temperatures, making them ideal for components exposed to extreme heat, such as engine parts and turbine blades. They provide a protective barrier against corrosive environments, crucial for marine and chemical processing applications. The hardness of ceramic coatings protects surfaces from abrasion and mechanical wear, extending the lifespan of components. Some ceramic coatings have excellent electrical insulating properties that benefit electronic components. Ceramic coatings can be more expensive than other types, but their long-term benefits often justify the investment. Microarc oxidation ceramic coatings formed on aluminium alloy substrates typically exhibit a porous outer layer and a dense inner layer, with phases such as alumina (Al_2O_3) contributing to their hardness and wear resistance [34]. The dense inner layer of the MAO coatings provides excellent corrosion protection by acting as a barrier to corrosive agents, such as chlorides and acids, preventing the penetration of corrosive species.

3.4.4 Toughness Enhancement in Transition Metal Nitride (TMN) Alloys

Introducing additional elements can improve the toughness of TMN alloys. Adding aluminium (Al) can form complex nitrides like TiAlN, enhancing toughness while maintaining hardness. Das et al. found that the TiAlN films deposited by CVD exhibit the highest hardness of 22 GPa at 1200 °C [35]. The Al content in the TiAlN films forms a protective aluminium oxide layer at high temperatures, which prevents further oxidation and maintains the integrity of the coating. Incorporating chromium (Cr) can enhance toughness and oxidation resistance.

Small additions of silicon (Si) can lead to the formation of amorphous phases that enhance toughness. Nanocomposite structures are created, where nanocrystalline TMN grains are embedded in an amorphous or nanocrystalline matrix, and these structures can absorb more energy during deformation, leading to improved toughness. The interfaces between multilayer coatings with alternating layers of different TMNs can deflect cracks and prevent their propagation, thus increasing the overall toughness of the coating. Creating coatings with a compositional gradient, where the composition changes gradually from the substrate to the surface, can reduce stress concentrations and enhance toughness. This approach can help manage thermal expansion differences and improve adhesion. Some TMN alloys can undergo phase transformations that absorb energy and enhance toughness. For example, certain alloying additions can stabilize metastable phases that transform under stress. Modifying the valence electron concentration by alloying W in TiN alloys has been demonstrated to enable adjustment of the Cauchy pressure and G/B ratio [36], leading to the creation of ductile materials. Experimental confirmation of the theoretical projections of enhanced toughness in ternary TMN alloys was achieved through the examination of $V_{0.5}Mo_{0.5}N(001)$ thin films deposited on MgO(001) substrates [37].

3.4.5 The Effect of Atomic Ordering on Mechanical Properties of TMN Alloys

In TMN alloys, atomic ordering can increase hardness by reducing the mobility of dislocations. This occurs because an ordered atomic structure introduces more obstacles to dislocation motion, making it harder for the material to deform plastically. For instance, in the case of CrN alloys, atomic ordering in the crystal lattice can enhance the hardness by creating a more stable and rigid structure that resists deformation. Polcra et al. highlight that CrN coatings exhibit excellent wear resistance and maintain a low coefficient of friction at elevated temperatures, typically up to 700 °C [38]. This is attributed to forming a protective oxide layer (chromium oxide) on the surface, which reduces wear and friction. A more disordered atomic arrangement, such as in amorphous or partially ordered TMN phases, may result in lower hardness due to the increased ease of dislocation movement. However, in some cases, disordered phases can provide better toughness or resistance to cracking. Madhuri et al. aimed at optimizing the mechanical stability of AlWN alloys by adjusting the Al to W ratio and showed that WN tends to have more ionic character, whereas AlN has a strong covalent bonding nature [39]. This bonding character difference influenced the alloys' hardness and wear resistance. TMN alloys with high degrees of atomic ordering tend to have reduced ductility due to the stiffening effect of the ordered structure. However, this can also increase brittleness, making these materials prone to cracking under stress. To improve toughness, partial disorder or alloying can be introduced to disrupt the strict atomic ordering, allowing for more plastic deformation and energy absorption. For example, in Ti–Al–N alloys, adding aluminium can create a partially ordered structure that balances hardness with improved toughness. Atomic ordering is crucial in phase stability, directly affecting mechanical properties. In TMN alloys, changes in atomic ordering can lead to phase transformations (e.g., from cubic to hexagonal phases), which can either enhance or weaken mechanical properties depending on the phase formed. For example, in TiN alloys, a high degree of atomic ordering stabilizes the cubic structure, which is associated with high hardness. However, introducing disorder may lead to phase transformations that can improve ductility while sacrificing some hardness. Introducing additional elements into TMN alloys can affect the degree of atomic ordering and thus influence mechanical properties. Elements like aluminium (Al), silicon (Si), and boron (B) are often used to modify the atomic structure of TMN alloys. Atomic ordering can also affect the thermal stability of TMN alloys. More ordered structures are thermally stable, maintaining their hardness at elevated temperatures. This is important for applications such as cutting tools or protective coatings in high-temperature environments.

3.4.6 Elastic Properties of Ternary Transition Metal Carbide Alloys

The specific transition metals and their ratios significantly affect the alloy's properties. Common transition metals in these alloys include titanium (Ti), vanadium (V), chromium (Cr) and tungsten (W). The arrangement and size of grains, as well as the presence of any secondary phases or precipitates, can impact the elastic behaviour of the material. The nature of the chemical bonds between the metal and carbon atoms affects the rigidity and strength of the material. Elastic properties can vary with temperature, especially for applications in high-temperature environments. Ti–Al–C alloys are known for their high hardness and moderate ductility, useful in lightweight structural components. Cr–Mo–C alloys exhibit excellent wear resistance and are used in coatings and surface treatments. W-Ti-C alloys are known for their high melting points and excellent mechanical properties and are suitable for extreme conditions. Hugosson et al. [40] utilized valence electron concentration (VEC) tuning to align the energies of the cubic and hexagonal phases, encouraging the creation of stacking faults that impede dislocation movement and lead to higher hardness. Theoretical studies suggest that maximum hardness is achieved with a VEC of 8.4 in cubic transition metal carbo-nitrides [41] as this fully occupies shear-resistant metal-N bonds.

4 Synthesis and Properties of Titanium and Titanium-Based Alloys

The Kroll process extracts Titanium from its ore, mainly ilmenite ($FeTiO_3$) and rutile (TiO_2). Titanium dioxide is reduced with chlorine gas in the presence of carbon to form titanium tetrachloride ($TiCl_4$).

$$TiO_2 + 2Cl_2 + C \rightarrow TiCl_4 + CO_2 \quad (1)$$

The $TiCl_4$ is purified by fractional distillation. The purified $TiCl_4$ is reduced with magnesium in an inert atmosphere or under a vacuum.

$$TiCl_4 + 2Mg \rightarrow Ti + 2MgCl_2 \quad (2)$$

A titanium sponge is produced and magnesium chloride is removed by vacuum distillation. Titanium alloys are created by melting titanium with other metals such as aluminium, vanadium, molybdenum and iron. In vacuum arc re-melting (VAR), titanium and alloying elements are melted using an electric arc in a vacuum. This method helps to remove impurities. Electron beam melting (EBM) uses an electron beam in a vacuum to melt the titanium and alloying elements. The materials are melted using an induction furnace, often under an inert gas atmosphere to prevent contamination. Using forge and subsequent heat treatments is common to alter the microstructure and properties of titanium alloys [42]. Different nucleation temperatures can impact the microstructures of TiB_2 and FeB alloys in the RF thermal plasma method [43]. Wang et al. developed a novel method for creating porous titanium by following the calciothermic reaction of porous TiO precursors [44]. Research by Baehre et al. suggests that titanium prevents the material from dissolving electrochemically [45].

One of the most commonly used titanium alloys with a good balance of strength, corrosion resistance and workability is Ti–6Al–4V (Grade 5). It is used in aerospace, medical devices and marine applications. Ti–3Al–2.5V (Grade 9) is known for its excellent weldability and formability and is often used in tubing for bicycle frames and aerospace hydraulic systems. Ti–6Al–7Nb is developed explicitly for medical implants, providing excellent biocompatibility and mechanical properties. By adjusting the β stabilizer levels, it is possible to produce implants that are both strong and have a low modulus [46].

4.1 *Synthesis and Properties of Titanium-Based Ceramics Material*

Titanium-based ceramics encompass materials with exceptional properties suitable for various high-performance applications. Here is an overview of their synthesis and properties.

4.1.1 Titanium Carbide (TiC)

The carbothermal reduction method involves reducing titanium dioxide (TiO_2) with carbon at high temperatures (above 1800 °C).

$$TiO_2 + 3C \rightarrow TiC + 2CO \tag{3}$$

It is a simple process, cost-effective for large-scale production, requires high temperatures and long reaction times, and produces CO gas. In self-propagating high-temperature synthesis (SHS), titanium and carbon powders are mixed and ignited to initiate a highly exothermic reaction.

$$Ti + C \rightarrow TiC \tag{4}$$

It is fast, energy-efficient, can produce highly pure TiC and requires precise control of reactant ratios and conditions. In chemical vapor deposition (CVD), titanium tetrachloride ($TiCl_4$) reacts with methane (CH_4) and hydrogen (H_2) at high temperatures to deposit TiC on substrates.

$$TiCl_4 + CH_4 + H_2 \rightarrow TiC + 4HCl \tag{5}$$

It produces high-purity, uniform coatings. It is a complex setup, expensive, and limited to thin films. Titanium and carbon powders are mechanically milled to induce solid-state reactions, forming TiC. It has a low temperature, suitable for producing fine powders. It has long processing times and potential contamination from milling media. In Spark Plasma Sintering (SPS), titanium and carbon powders are compacted and subjected to pulsed electric current, rapidly sintering into dense TiC ceramics. It has rapid densification, lower sintering temperatures, and high-density products. It requires specialized equipment.

TiC is extremely hard with a Vickers hardness of up to 3200 HV. This makes it ideal for cutting tools and wear-resistant coatings. TiC has a melting point of around 3160 °C, enabling its use in high-temperature environments. It has excellent thermal conductivity (up to 21 W/m–K), which is beneficial for thermal management applications. TiC is chemically inert and exhibits excellent corrosion resistance, particularly in oxidizing and reducing environments. Unlike many ceramics, TiC is electrically conductive with a conductivity of approximately 2.5×10^6 S/m. This property is useful in electronic and electrical applications. The combination of hardness and chemical stability provides outstanding wear resistance, essential for industrial tooling and protective coatings. At temperatures exceeding 1000 °C, TiO and TiN are produced when the substance is exposed to air [47]. Using smaller carbide particles to create fine-grained composites is too costly. Paraye et al. employed a TIG (inert tungsten gas) arcing process using a high-energy arc to melt the surface of the ferrous substrate [48]. Titanium and carbon are introduced into the molten pool, where they react to form TiC directly on the surface. This results in the formation of a hard, wear-resistant TiC layer. Sivasankaran et al. [49] explored an innovative sonochemical method to synthesize TiC using high-frequency ultrasonic waves, which generate acoustic cavitation. This process led to high-temperature, high-pressure microbubbles that collapse violently, creating localized conditions favourable for chemical reactions. Zhang et al. employed a thermal reactive diffusion process that involved immersing low-carbon steel in a molten salt bath containing titanium and carbon sources [50]. At high temperatures (around 800–1100 °C), the titanium and carbon atoms diffuse into the steel surface, reacting to form a TiC layer. Nyanor et al. discovered that nano-TiC particles improved the ductility of a carbon aluminium nanotube composite produced through Spark Plasma Sintering (SPS) [51]. Additionally, TiC can be created by carbothermal reduction of TiO_2 in an inert atmosphere at a higher temperature of 1450 °C [52].

4.1.2 Titanium Nitride (TiN)

In CVD, titanium tetrachloride ($TiCl_4$) and nitrogen (N_2) or ammonia (NH_3) gases are introduced into a reaction chamber.

$$TiCl_4 + N_2 \rightarrow TiN + 2Cl_2 \tag{6}$$

In PVD, a titanium target is bombarded with high-energy ions in a nitrogen atmosphere, causing titanium atoms to be ejected and react with nitrogen to form TiN on the substrate. Titanium is evaporated in the presence of nitrogen gas, leading to TiN deposition on a substrate. Titanium metal is directly reacted with nitrogen gas.

$$Ti + N_2 \rightarrow TiN \tag{7}$$

Titanium dioxide (TiO_2) is mixed with carbon and heated in a nitrogen atmosphere.

$$TiO_2 + 3C + N_2 \rightarrow TiN + 2CO \tag{8}$$

TiN is golden yellow in colour. It has a density of approximately 5.22 g/cm^3 and a melting point of around 2950 °C. It has a face-centred cubic (FCC) crystal structure. It is very hard (Vickers hardness of about 1800–2100 HV). It has Young's Modulus of around 450–590 GPa and a low friction coefficient (0.4–0.9), which is beneficial for wear resistance. It is an excellent electrical conductor with low resistivity (20–25 μΩ cm). It has excellent resistance to corrosion and oxidation and is stable in most environments but can react with strong acids at high temperatures. Das et al. deposited TiN thin films via CVD, where titanium precursor and nitrogen gas are used in the reaction chamber at different substrate temperatures to study their influence on film growth, morphology, and structural integrity [53]. Guha et al. discussed various deposition techniques, such as RF magnetron sputtering, ion beam-assisted deposition, and chemical vapour deposition, which have been widely used to deposit TiSiN films and showed how the incorporation of silicon into titanium nitride enhanced the mechanical properties of the films [54]. Das et al. deposited TiN, TiAlN and TiAlSiN films by CVD in a nitrogen atmosphere and observed TiAlSiN with a slightly higher elastic modulus compared to TiAlN and TiN, indicating its superior mechanical rigidity [55]. Yu et al. employed a laser and plasma mixing technique to develop TiN coatings with varying laser power levels [56]. Tacikowski et al. combined to produce titanium layers resistant to corrosion [57]. Wu et al. discovered a method to generate nanocrystalline TiN from a unique nitrogen and titanium precursor in an autoclave at 650 °C [58].

4.2 *Synthesis and Properties of Titanium-Based Composite Material*

Titanium-based composite materials are widely used in various industries, including aerospace, automotive and biomedical, due to their excellent mechanical properties, corrosion resistance and biocompatibility. The synthesis and properties of these composites depend on the matrix material, reinforcement phase and processing methods.

4.2.1 TiN-TiC

Titanium, nitrogen and carbon powders are mixed to achieve a homogeneous blend. The powder mixture is compacted to form a green body using cold isostatic or uniaxial pressing (CIP). The compacted mixture is sintered at high temperatures, typically in a nitrogen atmosphere, to promote the formation of TiN and TiC phases. This process may involve solid-state reactions between titanium, nitrogen and carbon to form the desired composite structure. Elemental powders of titanium, carbon and nitrogen or a mixture of TiC and TiN powders are subjected to high temperatures. During sintering, reactions occur to form the composite material. Self-propagating high-temperature synthesis (SHS) involves a highly exothermic reaction between titanium, carbon and nitrogen powders, which propagates through the mixture, forming TiN-TiC in-situ. This method is energy-efficient and produces a highly homogeneous composite. In the CVD process, $TiCl_4$, CH_4 and N_2 gases are used in a high-temperature reactor. The gases react on the substrate surface, depositing a thin film of TiN-TiC composite material. This method is commonly used for coatings on cutting tools and wear-resistant surfaces. The ratio of TiN to TiC can be controlled by adjusting the gas flow rates and reaction conditions. In SPS process, a mixture of Ti, C and N powders or pre-synthesized TiN and TiC powders is placed in a die and subjected to a pulsed electric current and pressure. The rapid heating and cooling rates lead to dense composites with fine microstructures. SPS offers lower sintering temperatures and shorter processing times than conventional sintering, resulting in improved mechanical properties. Matsuda et al. examine a series of TiCN alloys where Mo and W are introduced as alloying elements to modify their properties [59]. These elements influence the hardness, thermal conductivity, and elastic moduli of transition metal carbonitrides. Cardinal et al. explored the synthesis of TiC–TiN-based cermets, which typically consist of hard ceramic particles (TiC and TiN) embedded within a metallic binder matrix using powder metallurgy techniques, which involve blending, pressing, and sintering of powders to achieve a dense composite [60]. One of the important aspects of cermet design is the ratio of ceramic to metal phases. Liu et al. synthesized TiC-based cermets with varying nano- and micro-sized TiN [61]. The materials are typically prepared using powder metallurgy techniques, where TiC, TiN, and metallic binders (such as nickel or cobalt) are mixed, pressed, and sintered. The study likely explored the influence of the ratio between nano- and micro-TiN particles on the resulting cermet properties. This includes examining the effect of different amounts of TiN additions and how the combination of nano- and micro-scale particles affects the sintering process and densification. Yang et al. systematically studied how the carbon-to-nitrogen ratio affects the elastic properties of $Ti(C_xN_{1-x})$ [62]. Gordienko et al. observed that the thermal conductivities of Ti(C,N) were lower than those of Ti(C,N)-Ni cermets, as expected [63]. Pazhouhanfar et al. found that the TiN additive wholly dissolved into the TiC matrix in the composite sample, forming an in-situ phase (C_3N_4) [64]. The TiN-doped sample exhibited a flexural strength of 450 MPa and Vickers hardness of 2750 $HV_{0.1}$. Zhao et al. utilized reactive RF magnetron sputtering to develop a novel TiC-TiN/Al_2O_3 tandem solar selective absorption coating [65]. Mura et al. observed that various stages of mutual dissolution occur during the mechanical processing of TiC-TiN powder mixes through ball milling for alloying [66].

TiN-TiC composites exhibit very high hardness, often exceeding that of individual TiN or TiC, making them ideal for cutting tools and wear-resistant applications. The composite nature provides a good balance of hardness and toughness, reducing brittleness compared to pure ceramics. They have high compressive and flexural strength, contributing to their performance under mechanical stress. They have excellent resistance to high temperatures, making them suitable for high-temperature applications and moderate thermal conductivity, which can be advantageous in applications requiring heat dissipation. They have exceptional wear resistance due to the

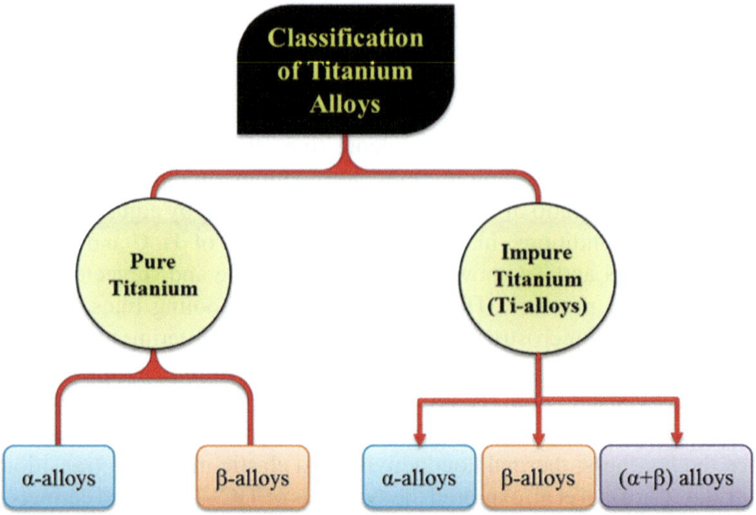

Fig. 4 Different categories of materials made from titanium [67]

hardness of TiN and TiC phases, making these composites suitable for abrasive environments and having good resistance to oxidation and chemical attack, particularly in harsh industrial environments. TiN and TiC are good conductors of electricity, allowing the composites to be used in electrical applications where wear resistance is required. They are reflective and can be used in coatings for decorative purposes or for their thermal radiation properties. The interest in an exceptional combination of thermo-mechanical properties extends to titanium-based alloys, ceramics and matrix composites, as depicted in Fig. 4.

4.3 *Applications of Titanium-Based Materials*

Titanium alloys are critical in the aerospace industry due to their high strength-to-weight ratio, resistance to high temperatures, and corrosion resistance. The biocompatibility of Ti-based materials makes it an ideal choice for medical applications, such as dental implants, hip and knee replacements, and spinal fusion devices. The review by Saurabh et al. likely highlights the use of titanium in aircraft frames, turbine engines, and landing gear components and also covers the importance of titanium in reducing rejection rates and enhancing patient outcomes in surgical implants [68]. In the automotive sector, titanium alloys are increasingly used in high-performance vehicles to reduce weight, improve fuel efficiency, and enhance performance. Exhaust systems, engine components, and suspension parts made from titanium alloys provide excellent durability and strength. Due to their superior corrosion resistance, titanium and its alloys are used in the chemical processing industry, especially for handling corrosive substances such as acids and chlorides. Heat exchangers, reactors, and piping systems made from titanium are critical in extending the lifespan of equipment exposed to harsh chemical environments. Titanium is used in architectural applications for its aesthetic appeal, durability and resistance to corrosion. It is often used in building facades, roofing and structural components. Titanium composites allow a weight reduction of up to 50% compared to solid superalloys.

5 Structure and Polymorphs of Molybdenum Disulphide (MoS_2)

MoS_2 typically has a layered structure similar to graphite, where layers of molybdenum atoms are sandwiched between layers of sulphur atoms. These layers are held together by weak van der Waals forces, which allow them to slide over each other easily, contributing to their lubricating properties. It has unique mechanical properties, making it ideal for solid lubricant use. It displays polytypism with three different configurations (1T, 2H and 3R), falling under different point groups (D_{6d}, D_{6h} and C_{3V}). 2H-MoS_2 (hexagonal MoS_2) is the most common and stable polymorph with a semiconducting phase. In this structure, each molybdenum atom is co-ordinated to six sulphur atoms, forming trigonal prismatic units. The layers are arranged in an ABAB stacking sequence. It has a space group of $P6_3/mmc$ and lattice parameters of $a = 3.16$ Å and $c = 12.3$ Å. This 3R-MoS_2 (rhombohedral MoS_2) less common polymorph has a stacking sequence of ABCABC, differing from the hexagonal form. It has a space group of R_3m and lattice parameters of $a = 3.16$ Å and $c = 18.6$ Å. In 1T-MoS_2 (trigonal MoS_2), six sulphur atoms co-ordinated each molybdenum atom octahedrally. The layers have an AAAAA stacking sequence, distinct from the prismatic coordination in the 2H phase. It has a space group of P3m1 and lattice parameters of $a = 3.16$ Å and $c = 6.14$ Å. The 1T phase is the metallic phase of MoS_2, characterized by octahedral coordination around the Mo atoms. This phase is of interest for catalysis, particularly for hydrogen evolution reactions (HER), due to its enhanced conductivity and increased catalytic activity compared to the 2H phase. Gan et al. focused on inducing the transition from the semiconducting 2H phase to the metallic 1T phase by introducing sulfur vacancies through an electrochemical process [69]. The vacancies are introduced into the MoS_2 lattice during electrochemical cycling, disrupting the 2H structure and promoting the 1T phase formation. 2H MoS_2 in Fig. 5a is known to be the most stable form of MoS_2. This material is an n-type semiconductor with a charge carrier mobility of 100 cm^2 $(Vs)^{-1}$. It remains thermodynamically stable until around 1203 °C after which it transforms into Mo_2S_3. Due to its 4d anti-parallel spin electrons, 2H MoS_2 displays non-magnetic semiconductor properties. Heating it above 800 °C under high pressure causes a transition from a semiconductor to a metallic state. Doping with rubidium (Rb) can transform the semiconducting properties of a material into superconducting properties and this conversion can be achieved through intercalation at low temperatures of 6.9 K.

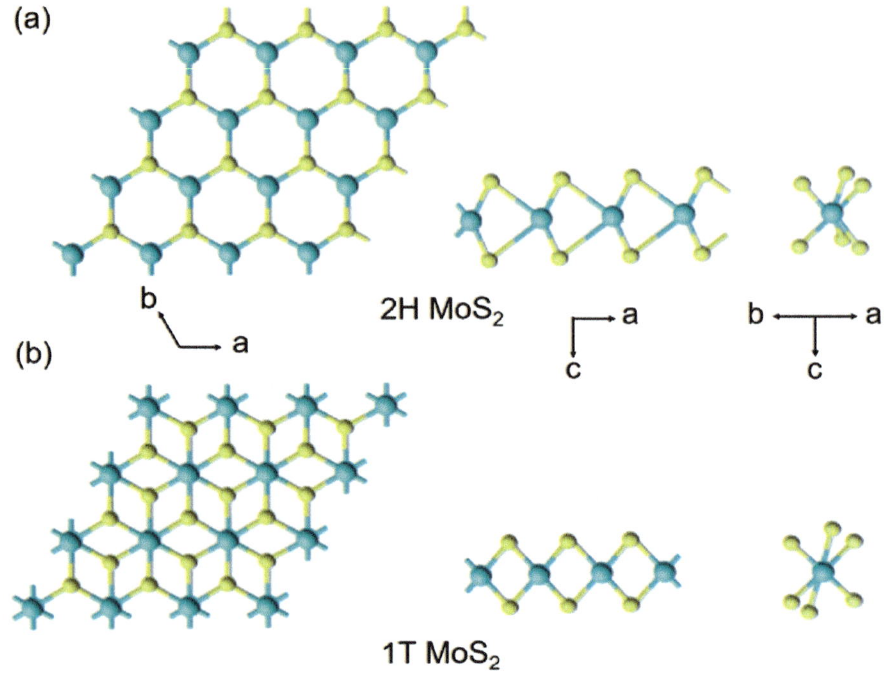

Fig. 5 The top and side views of the **a** 2H and **b** 1T configurations of MoS2 are shown. In the 2H-MoS_2 arrangement, the octahedral coordination around the Mo atom is illustrated with S atoms in yellow and Mo atoms in cyan [74]

2H MoS_2 is commonly used in hydrodesulfurization processes to remove sulphur from fuel and organic compounds, producing cleaner fuels. UV–Vis spectrum analysis exhibits significant peaks at around 580 nm (2.1 eV) and 640 nm (1.9 eV). Raman scattering reveals that MoS_2 exhibits four distinct vibration modes: $E^1{}_{1g}$ (at 383 cm^{-1}), A_{1g} (at 409 cm^{-1}), E_{1g} (at 287 cm^{-1}) and $E^2{}_{2g}$ (at 33 cm^{-1}) [70].

A different metastable state of MoS_2 is the tetragonal (1T) phase, illustrated in Fig. 5b with lattice parameters, $a = 5.6$ Å and $c = 5.99$ Å. It processes both paramagnetic and metallic properties. The process of 1T formation includes three intermediate polymorphs. Oxidation for 15 min forms a first polymorph with an a × 2a superstructure; after 60 min, a second polymorph forms with a 2a × 2a superstructure. Further intense oxidation leads to the creation of a superstructure of $a\sqrt{3} \times a\sqrt{3}$ [71]. 1T-MoS_2, created through CVD, is a viable option for the source and drain electrodes in FET devices based on 2H-MoS_2 [72]. This is because of its electrical and structural properties' similarity to the primary material, resulting in decreased contact resistance (~200 Ω μm) and enhanced performance. Additionally, the less common 3R (rhombohedral) MoS_2 in small quantities in molybdenite ore is suitable for non-linear optical devices due to its unique non-linear properties resulting from its broken inversion symmetry [73].

5.1 Various Aspects and Properties of MoS_2

5.1.1 Mechanical Properties of MoS_2

Single-layer MoS_2 has a Young's Modulus (elastic modulus) of approximately 270 GPa. Bulk MoS_2 has a Young's Modulus lower than single-layer, typically around 200 GPa. This indicates the stiffness and resistance to deformation. Single-layer MoS_2 has a tensile strength of around 15–23 GPa, highlighting its ability to withstand significant tensile stress before breaking. MoS_2 is relatively soft with a Mohs hardness of about 1–2. Despite this, it is widely used as a solid lubricant due to its low shear strength. MoS_2 has low fracture toughness, which can be brittle and susceptible to cracking under stress. The mechanical properties vary significantly between single-layer and bulk forms. Single-layer MoS_2 is much stronger and stiffer than its bulk counterpart due to fewer defects and the intrinsic strength of the 2D structure. The in-plane stiffness and Poisson's ratio of a single layer of MoS_2, analyzed in the harmonic elastic strain range, were found to be 123 Nm^{-1} and 0.25, respectively, showcasing MoS_2 as softer than graphene [75]. The low pre-tension and high elasticity of MoS_2 make it a prospective material for semiconductor composite films and flexible optoelectronic devices.

Johari et al. focused on mechanical strain as an effective method to tune the electronic band structure of MoS_2 without introducing defects or changing the material composition [76]. The 2H-phase of MoS_2 and WS_2 exhibits a transition from an indirect bandgap in their bulk form to a direct bandgap in the monolayer limit, enhancing their light absorption and emission properties [77]. A lattice mismatch between MoS_2 and HfO_2 layers in advanced transistors acts like tensile strain, improving carrier mobility.

5.2 Preparation Methods Based on the Top-Down Approach

5.2.1 Mechanical Exfoliation Method

Bulk MoS_2 is repeatedly peeled using adhesive tape (Scotch tape method) until monolayers or few-layer MoS_2 are obtained. The exfoliated layers are then transferred onto a substrate such as silicon with a silicon dioxide layer. It produces high-quality monolayers and is simple and cost-effective. It has low yield and scalability, and has difficulty controlling layer thickness. Radisavljevic et al. [78] successfully created MoS_2 monolayers mechanically through a micro-exfoliation technique from bulk MoS_2 crystals that is effective for analogue and digital circuits as well as highly sensitive photodetectors for mobility engineering analysis. Fabrication of a field-effect transistor (FET) based on single-layer MoS_2, with a channel length of 1 μm was also carried out. The device leverages the direct bandgap of MoS_2 (approximately 1.8 eV), which allows for efficient current modulation and switching. Magda et al. [79] successfully created monolayers of MoS_2 with lateral sizes of typically <10 μm using mechanical exfoliation. They observed increased adhesion of MoS_2 to gold substrates, allowing for further exploration of the material at a fundamental level.

5.2.2 Exfoliation in Liquid Phase

Bulk MoS_2 is dispersed in a solvent with ultrasonication. The dispersed layers are then separated by centrifugation to obtain monolayers or a few layers of nanosheets. It is scalable, relatively simple, and can produce large quantities of exfoliated MoS_2. It has possible structural defects and has difficulty in achieving a uniform monolayer thickness. One method for exfoliating MoS_2 involves mechanical actions such as bubbling, shearing, sonication, grinding and stirring. This process mainly relies on physical techniques but also incorporates some chemistry, such as using surfactants like SDC (sodium deoxycholate), bile salt, chitosan to prevent recombination of the exfoliated flakes. The bubbles are pushed out from the material interface, leading to the exfoliation process. Forsberg et al. demonstrate the successful exfoliation of MoS_2 in water, achieving stable suspensions without using any chemical additives [80]. The exfoliation process involved ultrasonication of bulk MoS_2 in deionized water, producing few and single MoS_2 flakes. Ultrasonication generates cavitation, which breaks the weak van der Waals forces between layers, separating monolayers. Fan et al. This study combines lithium intercalation with sonication and infrared laser treatment to efficiently produce high-quality 2H-phase MoS_2 from bulk material [81]. Lithium ions expand the interlayer spacing of MoS_2, making it easier to exfoliate into monolayers or few-layer nanosheets. After lithium intercalation, the ultrasonic waves generate cavitation forces that help separate the MoS_2 layers, leading to exfoliation. Then, the infrared laser treatment induces phase reversion from 1T to 2H. Walia et al. [82] suggested using the drop-casting method, followed by drying the suspension to create thin films of MoS_2 for examining its electrical and optical characteristics.

As shown in Fig. 6, Liu et al. showcased a straightforward approach to producing MoS_2 nanosheets through liquid exfoliation with the assistance of salts [83]. They utilized isopropanol as the exfoliation medium and selected salts such as potassium sodium tartrate and potassium ferrocyanide as assistants, which proved effective in aiding the exfoliation process. Specifically, adding potassium ferrocyanide $K_4Fe(CN)_6$ significantly improved the exfoliation efficiency by 73 times and increased the concentration of dispersed MoS_2 nanosheets to 0.24 mg mL^{-1}. Different combinations of MoS_2 and salts were mixed, centrifuged and sonicated to achieve the desired nanosheets. Thick or unexfoliated flakes were removed through centrifugation, washing and resuspension steps, ensuring a stable dispersion of deep brown MoS_2 nanosheets for up to a month without aggregation, as shown in Fig. 6. This method shows great potential for producing high-quality MoS_2 materials for energy storage applications.

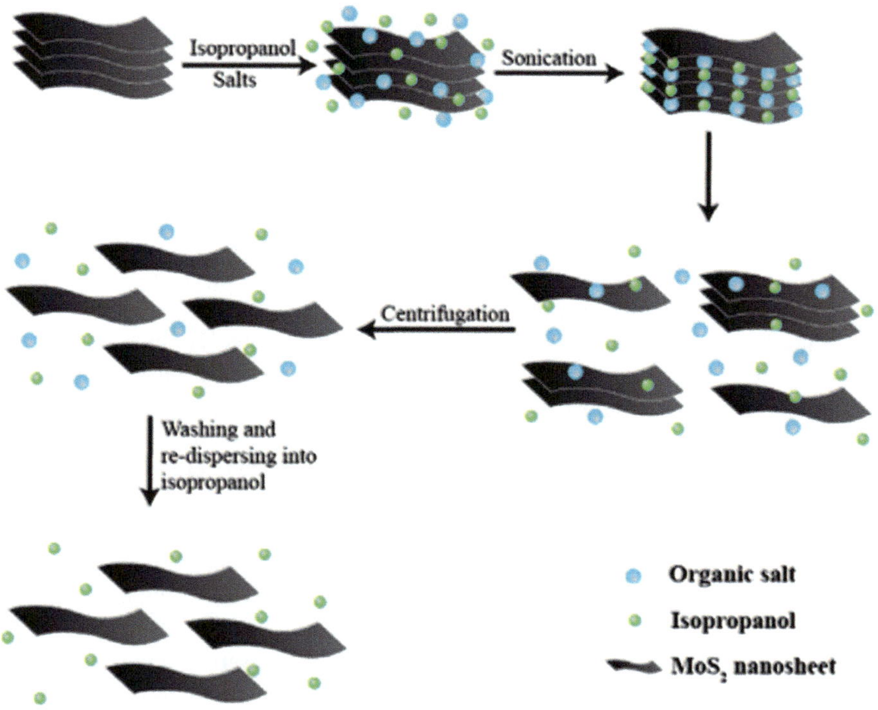

Fig. 6 MoS_2 nanosheets were fabricated in isopropanol with the help of salt [83]

5.3 Preparation Methods Based on Bottom-Up Approach

5.3.1 Chemical Vapour Deposition (CVD)

A precursor containing molybdenum (e.g., MoO_3) and a sulphur source (e.g., H_2S or S powder) are vapourized and reacted in a high-temperature furnace. The reaction takes place on a substrate, leading to the formation of MoS_2 layers. It has high-quality and uniform monolayer scalability, potential for large-area synthesis, and good layer thickness and morphology control. It requires high temperatures, a controlled environment, a complex setup and precise control. CVD primarily uses three types of precursors: molybdenum trioxide powder (MoO_3), ammonium thiomolybdate solution $(NH_4)_2MoS_4$ and elemental Mo. Pondick et al. conducted sulfurization in a CVD furnace, where MoO_3 is exposed to sulfur vapour at elevated temperatures and progressively transformed into MoS_2 through intermediate phases, such as oxygen-deficient molybdenum oxysulfide (MoO_xS_γ) [84]. The sulfurization process follows a stepwise mechanism: $MoO_3 \rightarrow MoO_xS_\gamma \rightarrow MoS_2$. The main innovation of the work by Ahn et al. is the ability to synthesize MoS_2 thin films at the low substrate temperature of 150 °C, making it suitable for deposition on flexible plastic substrates, which have lower thermal tolerances compared to rigid substrates like silicon [85].

Kumar et al. discuss optimising crucial CVD factors such as temperature, pressure, precursor concentration, gas flow rates, and substrate choice, directly influencing the quality of MoS_2 crystals [86].

5.3.2 Atomic Layer Deposition (ALD)

Molybdenum and sulphur precursors are alternately introduced to the substrate cyclically. The precursors react on the substrate to form MoS_2 in a layer-by-layer fashion. It has atomic-level control over thickness, uniform and conformal coating on various substrates, and high-quality films with precise stoichiometry. It has a slow deposition rate and requires sophisticated equipment and accurate control of precursor flow. ALD was used to deposit MoS_2 thin films through alternating exposures of the molybdenum precursor [$(NtBu)_2(NMe_2)_2Mo$] and 1-propane thiol by Kalanyan et al. [87]. Tan et al. used H_2S and $MoCl_5$ in vapour form in ALD to cultivate monolayer and a few-layer MoS_2 on c-plane single crystal sapphire substrates at 300 ºC [88]. The resulting samples, containing one or more layers of MoS_2, were annealed at 800 °C under saturated sulfur vapour in which triangular-shaped crystals exhibited strong PL in the visible region, which is evidence of the crystallinity of the film. Figure 7a and b show the images captured from a bright field view of MoS_2 film using TEM, revealing polycrystalline grains with dimensions ranging from 15 to 20 nm. The electron diffraction pattern shown in Fig. 7b inset matches the 2H-MoS_2 structure, displaying a sixfold coordination symmetry and honeycomb arrangement. Fast Fourier transformation and reciprocal space analysis confirmed the presence of single crystal structures measuring around 20 nm. MoS_2 nanosheets fabricated using the ALD method are efficient for applications like catalysts, sensors, and electronic devices.

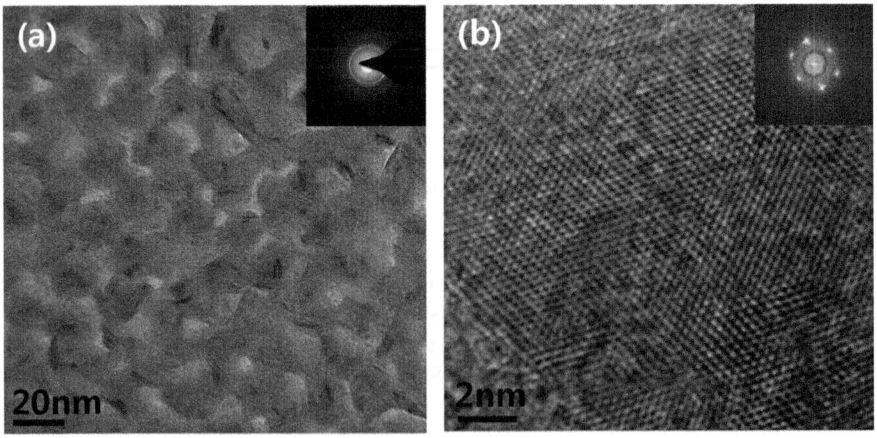

Fig. 7 Plan-view of ALD-MoS_2 films TEM image: **a** bright-field image of an ALD-MoS_2 film, deposited on SiO_2 and **b** high resolution (HR) image [89]

5.3.3 Solution Chemical Process

Various techniques for producing MoS_2 through solution chemistry, including solvothermal and hydrothermal synthesis [85]. These processes typically start at room temperature and atmospheric pressure but often require post-annealing. In hydrothermal synthesis, molybdenum and sulphur precursors are dissolved in a solvent and heated in a sealed autoclave. The reaction takes place at elevated temperatures and pressures, forming MoS_2 nanostructures. It is relatively simple and low-cost, can produce various morphologies (e.g., nanosheets, nanoflowers), and is suitable for large-scale production. It requires long reaction times and has limited control over layer thickness and uniformity. Feng et al. focused on the ability to control the shape, size, and layer number of the MoS_2 nanosheets by mixed–solvothermal technique [90]. Zhou et al. synthesized MoS_2 nanospheres with a flower-like morphology, characterized by ultrathin, layered nanosheets radiating from the centre using the hydrothermal method, which involves dissolving molybdenum and sulfur precursors in water, followed by heating the solution in an autoclave at elevated temperatures [91]. This structure provides a high surface area, which is advantageous for electrochemical applications like supercapacitors. The mixed-solvothermal method involves a combination of solvents under high temperature and pressure to promote the growth of MoS_2 nanosheets. Solvothermal synthesis, which uses organic solvents instead of water, is similar to hydrothermal synthesis. Molybdenum and sulphur precursors are dissolved in the solvent and heated in a sealed vessel. Additionally, methods such as photo-assisted deposition and electrochemical deposition are sometimes employed.

5.3.4 Physical Vapour Deposition (PVD)

In Molecular Beam Epitaxy, molecular or atomic beams of precursors such as molybdenum and sulphur are directed onto a heated substrate under ultra-high vacuum conditions, resulting in epitaxial growth. The atoms condense and react on the substrate to form MoS_2 layers. Zhan et al. deposited MoS_2 layer-by-layer onto a SiO_2/Si substrate [92]. Monolayer MoS_2 has a direct bandgap, while multilayer MoS_2 has an indirect bandgap.

5.4 MoS_2 Applications

5.4.1 Electronic Applications

MoS_2 can fabricate field-effect transistors (FETs) with high on/off ratios, making it suitable for low-power and high-performance electronic devices. Due to its high surface-to-volume ratio and sensitivity to adsorbed molecules, MoS_2 is used in gas, biosensors and chemical sensors. MoS_2's direct bandgap in monolayer form allows it to efficiently absorb light, making it suitable for photodetectors and optical sensors.

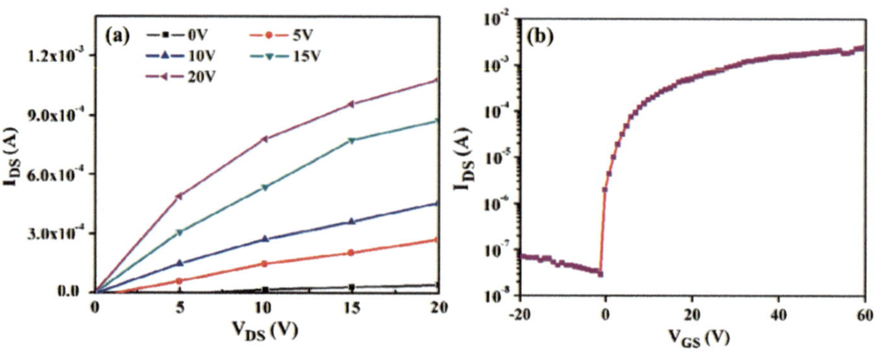

Fig. 8 **a** I_{DS} versus V_{DS} curve and **b** I_{DS} versus V_{GS} curves for MoS_2 TFTs in the saturated region [95]

The mechanical flexibility and strength of MoS_2 enable the development of flexible and wearable electronic devices. Their 2D structure enables control over its electrostatic properties, allowing for the creation of short-channel FETs and low-power electronic devices.

Radisavljevic et al. showcased a transistor that can be switched based on a single layer of MoS_2, indicating its potential as a viable option for low-power integrated circuits and in the realm of transparent and flexible electronics [93]. Ye et al. created a heterojunction using black phosphorus and MoS_2, held together by van der Waals forces for use in photodetectors, spanning the visible to infrared spectrum. This heterojunction displayed strong detectivity at specific wavelengths, reaching 3.1×10^{11} Jones at 532 nm and 2.13×10^{9} Jones at 1550 nm at room temperature [94]. Furthermore, Tyagi et al. utilized magnetron sputtering to investigate the manufacture of a Thin Film Transistor (TFT) with a hexagonal, vertically aligned crystal structure [95]. Figure 8a and b display the output and transfer characteristics of a TFT utilizing thin films of MoS_2. The TFT operates in n-channel enhancement mode, exhibiting distinct current saturation and pinch-off regions. The transfer characteristics were analysed by adjusting the gate voltage (V_{GS}) from −20 to 60 V with a constant drain to source voltage (V_{DS}) of 10 V, as illustrated in Fig. 8b. Kwon et al. explored pulsed laser-annealed Ti/Au contacts to enhance the performance of multilayer FET based on MoS_2 on a bendable plastic substrate, resulting in an increase in field effect mobility from 24.84 to 44.84 cm^2/Vs [96].

5.4.2 Solid Lubricants

MoS_2 is commonly used in coatings as a dry film lubricant to reduce friction and wear in mechanical systems. These coatings are beneficial in environments with impractical liquid lubricants, such as vacuum or space applications. Coatings containing MoS_2 can significantly reduce wear in components subjected to sliding or rolling contact, such as gears, bearings and piston rings. Solid lubricants such as MoS_2

are a cost-effective and lightweight alternative to grease lubrication systems. Unlike graphite, MoS_2 is particularly promising in harsh environmental conditions because it does not need water vapour pressure for lubrication. When sputtered with Ti, MoS_2 improves its mechanical properties and shields it from moisture. Kong et al. deposited alternating MoS_2, Ti, and Si layers, forming a multilayer nanocomposite using magnetron sputtering [97]. This design aimed to combine the lubricating properties of MoS_2 with the hardness and toughness of Ti and Si layers.

6 Aluminium Oxide (Al_2O_3) Coatings

Al_2O_3 is a versatile ceramic material that is extensively used in various industries. It is known for its toughness, non-reactive nature and ability to withstand high temperatures. It exhibits excellent resistance to wear, erosion and other damage forms, making it suitable for various applications. Moreover, its high-purity versions boast impressive electrical insulation and thermal conductivity properties. Due to its cost-effectiveness and diverse usability, it is a popular choice across different sectors. Shen et al. used the microarc oxidation process, where high-voltage electrical discharges are applied to the surface of the aluminium or aluminium-containing alloy in an electrolytic bath for the deposition of Al_2O_3 Coatings [98]. This creates a plasma discharge at the surface, forming a ceramic oxide layer (Al_2O_3) due to the intense localized heat and chemical reactions between the electrolyte and the metal surface.

6.1 Al_2O_3 Coating Benefits

Al_2O_3 exhibits superior wear and abrasion resistance properties up to 3000 F and a high service temperature that acts as a thermal barrier. Its dielectric strength is approximately 300–400 V per 0.001 inch thickness at room temperature. Furthermore, it remains insoluble in common acidic and alkaline aqueous solutions. Figure 9 shows the SEM image of aluminium oxide.

6.2 Applications of Al_2O_3 Coatings

Al_2O_3 coatings are highly recommended for use in the chemical and textile industries due to their excellent wear properties and chemical resistance. Additionally, the dielectric properties of Al_2O_3 make it ideal for industrial components, requiring the prevention of conductivity or electrical shorts. This coating has been successfully employed in motor housing, electric railways, and other elements. Its ability to withstand high service temperatures has also made it a preferred choice for applications such as rocket engine thrust chambers and jet engines. Common examples of

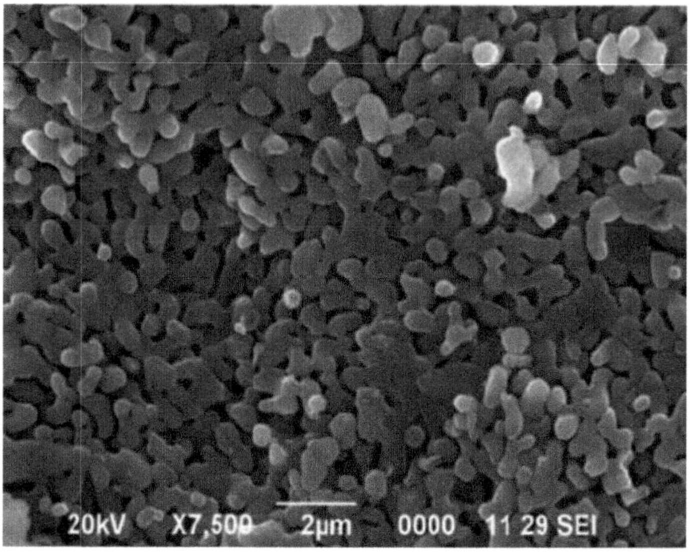

Fig. 9 SEM image of aluminium oxide [98]

Al_2O_3 applications include armature shafts, electrically insulative housings, liners, non-conductive rollers, pump seals, substrates for electrical components, textile yarn guides, draw rolls and wear surfaces. Dhawale et al. describe the common synthesis method used to prepare aluminium oxide (Al_2O_3) nanoparticles, including sol–gel, chemical precipitation, and hydrothermal methods [99]. Al_2O_3 nanoparticles are tested for their catalytic ability to break down azo dyes in aqueous solutions. This process typically involves the adsorption of dye molecules onto the surface of the nanoparticles, followed by catalytic degradation, often in the presence of an oxidizing agent like hydrogen peroxide (H_2O_2).

7 Carbon-Based Coatings

Graphene coatings are known for their exceptional strength, electrical and thermal conductivity. They are utilized in the field of electronics, sensors and protective coatings. They are a two-dimensional honeycomb structure made of a single layer of carbon atoms. Carbon Nanotube (CNT) coatings utilize the unique properties of carbon nanotubes, which include high mechanical strength, electrical conductivity and thermal stability. CNT coatings are used in applications such as field emission devices, composite materials and electronics. Amorphous carbon coatings do not have a crystalline structure and can be tailored to exhibit a range of properties, from those similar to graphite to those closer to diamond. They are often used in wear-resistant and corrosion-resistant applications. Fullerenes are carbon molecules found in the shape of a hollow sphere, ellipsoid or tube. Coatings based on fullerenes are used in various advanced materials and nanotechnology applications. Depending on the specific application and desired properties, these coatings are typically applied using techniques such as PVD, CVD or sputtering. Thin films based on C have shown to be highly appealing because of their mechanical, tribological properties and structural versatility. They have low friction, μ and high-wear resistance, high hardness in tetrahedral amorphous carbon (ta-C) films and high elasticity in fullerene-like carbon nitride ($FL\text{-}CN_x$) films. These films are already being used as protective coatings in technology like hard disk drives and applications in biomedicine. Brotman et al. emphasize the use of a high ionization deposition technique, such as high power impulse magnetron sputtering (HiPIMS) or filtered cathodic vacuum arc (FCVA), to enhance the adhesion of carbon-based coatings to the steel substrates [100]. These techniques involve creating a high degree of ionization in the deposition process, allowing for better control over the film structure and improving the coating-substrate interface.

7.1 Carbon Nitride (C_3N_4)

C_3N_4 materials display outstanding thermal stability, making them ideal for high-pressure situations. Graphitic carbon nitride ($g\text{-}C_3N_4$) is a nitrogen-rich polymeric material consisting of carbon and nitrogen atoms. It has a layered structure similar to graphite, where van der Waals forces hold layers together. $g\text{-}C_3N_4$ has a bandgap that allows it to absorb visible light, making it suitable for photocatalytic applications, hydrogen generation, and CO_2 reduction. It has semiconducting properties, which are useful in electronic and optoelectronic applications. Nitrogen shares hybridization states with carbon and can form covalent bonds with an electronic configuration of $[He]2s^22p^3$. It can have sp^3- and sp^2-hybridized states. However, the additional electron leads to adjustments in bond angles in the sp^3 configuration and results in two configurations when bonding with sp^2 states, as shown in Fig. 10. The incorporation of these elements affects the structure of the film with sp^3 formation, resembling that of ammonia, which results in unpaired electrons. In contrast, sp^2 states are comparable to either pyridine or a threefold coordination, replacing a carbon atom in a graphitic location.

Thermal polymerization is the most common method, and it involves heating precursors like melamine, dicyandiamide, or cyanamide to form $g\text{-}C_3N_4$. Hydrothermal and solvothermal methods involve reactions in a sealed vessel at high temperatures and pressures, allowing for the synthesis of various forms of carbon nitride. Carbon nitride, a type of carbon material, gained attention following the prediction of the metastable β phase of C_3N_4 in 1990 by Liu et al. [101]. This phase was thought to have a bulk modulus very similar to that of diamond. To evaluate their environmental impact, Yuan et al. studied the life cycle assessment (LCA) of PVD-fabricated $g\text{-}C_3N_4$ over biomass-based substrate, which is typically less common than

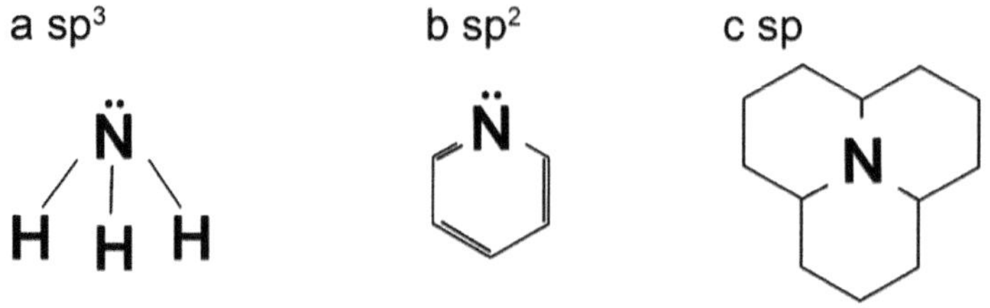

Fig. 10 N atoms exhibit different types of bonding configurations, including **a** sp^3-hybridized bonds seen in ammonia, **b** sp^2-hybridized bonds found in pyridine and **c** sp-hybridized bonds with threefold coordination with C atoms [104]

other fabrication techniques like thermal polymerization [102]. Neidhardt et al. highlight the differences between fullerene-like (FL) carbon nitride and amorphous (a) carbon nitride coatings [103]. While both are carbon-based materials with nitrogen incorporated into the structure, they exhibit different bonding characteristics, leading to variations in performance. FL-C_3N_4 features a curved, cage-like structure resembling fullerenes, offering bond angle flexibility and enhancing toughness. In contrast, a-C_3N_4 has a disordered, amorphous structure, resulting in different mechanical properties. The curved structure FL-C_3N_4 can absorb stress more effectively, resulting in superior fracture resistance compared to the amorphous variant.

7.2 *Synthesis of Carbon Nitride Coatings*

7.2.1 Sputter Deposition

Mid-frequency magnetron sputtering (MFMS) uses mid-frequency to deposit insulating materials, leading to unstable discharges with arcing. Fernández-Ramos et al. assessed the thermal stability of the C_3N_4 thin films deposited using reactive magnetron sputtering [105]. The films are subjected to annealing at different temperatures and above 1023 K, nitrogen begins to escape from the films, leading to a decrease in nitrogen content and, consequently, a reduction in hardness and mechanical stability. Hellgren et al. showed how plasma parameters such as power, pressure, gas composition, and magnetic field strength play critical roles in determining the characteristics of the deposited CN_x films produced by magnetron sputtering [106]. Kusano et al. used a magnetron sputtering source combined with an additional ionization stage to deposit carbon nitride films [107]. The ionized magnetron sputtering process enhances the ionization of the sputtered material, leading to improved film quality. This method can produce films with enhanced density and better adhesion properties than conventional sputtering techniques.

7.2.2 C_3N_4 Coatings Deposited by Magnetron Sputtering

For C_3N_4 deposition, the target material in magnetron sputtering can be a carbon target, which is sputtered in the presence of a nitrogen-containing gas (such as nitrogen or ammonia) to form carbon nitride films. Tejasvi et al. ensured that the g-C_3N_4 thin films retain the stoichiometry and structural characteristics of the bulk g-C_3N_4 using RF magnetron sputtering, which involved optimizing the sputtering parameters to transfer the exact composition and structure from the bulk target to the film [108]. Schmidt et al. discovered that in high-power impulse magnetron sputtering (HiPIMS), the ion fluxes are significantly greater than in direct-current magnetron sputtering (DCMS), leading to higher plasma densities [109]. HiPIMS often provides higher deposition rates and better film composition control due to the sputtered material's high ionisation.

7.3 Coating Characterization Methods

7.3.1 Mechanical Characterization

Profilometry

Profilometry is a crucial technique for characterizing the surface properties of carbon nitride (CN_x) coatings. It provides detailed information about surface roughness, thickness and topographical features, which are essential for evaluating the quality and performance of the coatings. In contact profilometry, a stylus physically contacts the surface and moves across it to measure variations in height. It has high resolution and accuracy in measuring surface roughness and step heights. It can potentially damage delicate surfaces and may not be suitable for very soft or extremely hard coatings. In non-contact profilometry, optical profilometry uses light to scan the surface, typically employing interferometry or confocal microscopy techniques; interferometric profilometry measures surface variations using the interference patterns of light and confocal profilometry provides high-resolution, 3D images by focusing light at different depths. They are non-destructive and capable of measuring soft or delicate surfaces. Non-contact methods are generally less accurate than contact methods for measuring very fine details.

In sample preparation, it is ensured that the sample surface is clear and devoid of any contaminants. If using contact profilometry, it is ensured that the surface is sufficiently robust to withstand stylus contact. The profilometer is calibrated using standard reference samples to provide accurate measurements. Scans are performed across the coating surface, following pre-defined paths to capture representative data. For thickness measurements, a step is created by partially masking the substrate during deposition or by etching away a section of the coating. The acquired data is analyzed to calculate roughness parameters thickness and generate 3D surface maps. Specialized software is used to process and visualize the data.

Nanoindentation

Nanoindentation is widely used to determine the hardness and elastic modulus of thin films and coatings such as CNx. This method involves pressing a hard, sharp indenter (typically made of diamond) into the material's surface with controlled force and then measuring the depth of the indentation. The formula determines hardness:

$$H = \frac{P}{A} \tag{9}$$

where 'P' represents contact pressure and 'A' represents the contact area between the tip and the film surface. Calculating the contact area is essential to accurately assess the true contact between the tip and the surface, which is achieved through a calibration process known as 'tip area function calibration' before conducting nanoindentation measurements. Oliver and Pharr's [110] method of utilizing a Berkovich tip for nanoindentation on fused silica is employed to calculate the tip area function through a fitting process:

$$A(h_c) = 24.5h_c^2 + C_1h_c^1 + C_2h_c^{1/2} + C_3h_c^{1/4} + \cdots + C_8h_c^{1/128} \tag{10}$$

where C_1 to C_8 are constants.

Elastic modulus (E) is determined using the unloading curve, often through the Oliver-Pharr method. This involves fitting a power-law relation to the unloading data to extract the stiffness (S), which is the slope of the unloading curve at maximum depth.

$$S = \frac{dP}{dh} \tag{11}$$

It is ensured that the indentation depth is shallow compared to the thickness of the coating to avoid substrate effects. A smoother surface yields more accurate results. Temperature and humidity can influence the measurements and should be controlled. The reduced elastic modulus of thin films can be determined by knowing the contact area. This can be represented as

$$E_r = \frac{\sqrt{\pi}}{2} \frac{S}{\sqrt{A}} \tag{12}$$

where E_r is the reduced elastic modulus and S is the contact stiffness. Elastic recovery, ER, is calculated as:

$$ER = \frac{h_t - h_c}{h_t} 100\% \tag{13}$$

Adhesive force is the attraction between the nanoindenter tip and the coating surface. Smooth surfaces generally exhibit higher adhesive forces due to increased contact area. The chemical and physical properties of both the coating and the indenter material affect adhesion. Temperature and humidity can also impact the adhesive forces. Work of adhesion quantifies the energy required to separate two surfaces that are in contact. For nanoindentation, it represents the energy needed to detach the indenter from the coating surface. Stronger chemical bonds at the interface increase the work of adhesion. Larger contact areas generally require more energy to separate. The high surface energies of the materials involved lead to greater adhesion work. To determine the adhesive forces, a meniscus must be created between the tip and the steel surface as they come into contact during the approach of the tip to the sample surface.

Nanoindentation measures mechanical properties at the nanoscale by pressing a hard, sharp tip into the material. Load–displacement curves show how the material responds to applied force. Adhesive forces can cause hysteresis in these curves. The maximum load and the corresponding indentation depth help determine hardness and modulus. Controlled rates are crucial for accurate measurements of adhesive forces and adhesion work. Precise calibration of the indenter tip is necessary for reliable data. There are multiple methods to calculate the work of adhesion W_{adh} of a surface, with the most commonly used theories being the Derjaguin-Miller-Toporov (DMT) and Johnson-Kendall-Roberts (JKR) models. The DMT approach is the most suitable theory for determining the W_{adh} of the low adhesive metallic surfaces. In DMT terminology, W_{adh} is determined by using Eq. (14):

$$F_{adh} = 2\pi W_{adh} R \tag{14}$$

where F_{adh} represents the adhesive force while R indicates the radius of the tip. The actual contact area is calculated considering any elastic recovery upon unloading. The work of adhesion is determined from the area under the unloading curve, representing the energy needed to detach the indenter. The peak adhesive force is measured during unloading, often seen as a negative force before complete detachment.

8 Conclusion

Engineering coating materials enhance various substrates' performance, durability and functionality across many industries. These materials, which include polymers, ceramics, metals and composites, are engineered to provide specific properties such as corrosion resistance, wear resistance, thermal insulation and aesthetic enhancement. The development of advanced coating technologies such as nanocoatings, smart coatings and environmentally friendly coatings reflects the ongoing innovation in the field. These advancements address contemporary challenges, such as environmental sustainability and the need for higher performance in extreme conditions.

The successful application of coating materials requires a comprehensive understanding of the substrate properties, environmental conditions and desired performance outcomes. Through continued research and development, the field of engineering coating materials will undoubtedly evolve, offering even more sophisticated solutions to meet the ever-growing demands of modern engineering and industry. Ultimately, the future of engineering coatings lies in integrating novel materials and technologies, fostering enhanced capabilities and paving the way for new applications and improved efficiency in a wide array of sectors.

References

1. Savale PA (2016) Physical vapor deposition (PVD) methods for synthesis of thin films: a comparative study. Arch Appl Sci Res 8(5):1–8
2. Yasuoka M, Wang P, Murakami R (2012) Comparison of the mechanical performance of cutting tools coated by either a TiC_xN_{1-x} single-layer or a $TiC/TiC_{0.5}N_{0.5}/TiN$ multilayer using the hollow cathode discharge ion plating method. Surf Coat Technol 206(8–9):2168–2172
3. Qadir M, Li Y, Wen C (2019) Ion-substituted calcium phosphate coatings by physical vapor deposition magnetron sputtering for biomedical applications: a review. Acta Biomater 89:14–32
4. Yate L, Martínez-de-Olcoz L, Esteve J, Lousa A (2012) Effect of the bias voltage on the structure of nc-CrC/a-C: H coatings with high carbon content. Surf Coat Technol 206(11–12):2877–2883
5. Srivastava S, Jain SK, Gupta G, Senguttuvan TD, Gupta BK (2020) Boron-doped few-layer graphene nanosheet gas sensor for enhanced ammonia sensing at room temperature. RSC Adv 10(2):1007–1014
6. Huang J, Fam D, He Q, Chen H, Zhan D, Faulkner SH, Nimmo MA, AlY T (2014) The mechanism of graphene oxide as a growth template for complete reduced graphene oxide coverage on an SiO_2 substrate. J Mater Chem C 2:109
7. Jogade S, Joshi PS, Jamadar B, Sutrave DS, Dayanand ODBF (2011) MOCVD of cobalt oxide using Co-actylacetonate As precursor: thin film deposition and study of physical properties. J Nano- Electron Phys 3(1):203–211
8. Long J, Yin Y, Sian SYR, Ren Z, Wang J, Vayalakkara P, Venkataraj S, Aberle AG (2012) Doped microcrystalline silicon layers for solar cells by 13.56 MHz plasma-enhanced chemical vapour deposition. Energy Procedia 15:240–247
9. Ghadai RK, Kalita K, Mondal SC, Swain BP (2019) Genetically optimized diamond-like carbon thin film coatings. Mater Manuf Process 34(13):1476–1487
10. Ghadai RK, Kalita K, Mondal SC, Swain BP (2018) PECVD process parameter optimization: towards increased hardness of diamond-like carbon thin films. Mater Manuf Process 33(16):1905–1913
11. Ghadai RK, Das S, Kalita K, Swain BP, Davim JP (2021) Structural and mechanical analysis of APCVD deposited diamond-like carbon thin films. SILICON 13(12):4453–4462
12. Ghadai RK, Singh K, Sharma A, Roy MK, Swain BP (2021) Mechanical and tribological properties of metal incorporated DLC thin film. Nanostruct Mater Appl 229–263
13. Sikora A, Garrelie F, Donnet C, Loir A, Fontaine J, Sanchez-Lopez JC (2010) Structure of diamond-like carbon films deposited by femtosecond and nanosecond pulsed laser ablation. J Appl Phys 108:113516
14. Modabberasl A, Kameli P, Ranjbar M, Salamati H, Ashiri R (2015) Fabrication of DLC thin films with improved diamond-like carbon character by the application of external magnetic field. Carbon 94:485–493

15. Tallant DR, Parmeter JE, Siegal MP, Simpson RL (1995) The thermal stability of Diamond like carbon. Diam Relat Mater 4:191–199
16. Aisenberg S, Chabot R (1971) Ion-beam deposition of thin films of diamond-like carbon. J Appl Phys 42:2953
17. Trava-Airoldia VJ, Bonettia LF, Capotea G, Santosa LV, Corata EJ (2007) A comparison of DLC film properties obtained by PACVD, IBAD and enhanced pulsed-DC PACVD. Surf Coat Technol 202:549–554
18. Robertson J (2002) Diamond-like amorphous carbon. Mater Sci Eng R 37:129–281
19. Libardi J, Grigorov K, Massi M, Otani C, Ravagnani SP, Maciel HS, Guerino M, Ocampo JMJ (2004) Comparative studies of the feed gas composition effects on the characteristics of DLC films deposited by magnetron sputtering. Thin Solid Films 459:282–285
20. Sattel S, Robertson J, Ehrhardt H (1997) Effect of deposition temperature on the properties of hydrogenated tetrahedral amorphous carbon. J Appl Phys 82:4566–5476
21. Peng XL, Barber ZH, Clyne TW (2001) Surface roughness of diamond-like carbon films prepared using various techniques. Surf Coat Technol 138:23–32
22. Chowdhury S, Laugier MT, Rahman IZ (2004) Effects of substrate temperature on bonding structure and mechanical properties of amorphous carbon films. Thin Solid Films 447–448:174–180
23. Ghadai RK, Das S, Kumar D, Mondal SC, Swain BP (2018) Correlation between structural and mechanical properties of silicon doped DLC thin films. Diam Relat Mater 82:25–32
24. Ma G, Gong S, Lin G, Zhang L, Sun G (2012) A study of structure and properties of Ti-doped DLC film by reactive magnetron sputtering with ion implantation. Appl Surf Sci 258(7):3045–3050
25. Song R, Chen S, Liu Z, Huo C, Chen Q (2023) Effect of W-doping on the structure and properties of DLC films prepared by combining physical and chemical vapor deposition. Diam Relat Mater 132:109687
26. Zhao D, Mei H, Ding JC, Cheng Y, Zhang L, Zhang TF, Kwang HK, Zheng J (2023) Microstructure and properties of Mo doped DLC nanocomposite films deposited by a hybrid sputtering system. Vacuum 208:111732
27. Schwarz C, Heeg J, Rosenberg M, Wienecke M (2008) Investigation on wear and adhesion of graded Si/SiC/DLC coatings deposited by plasma-enhanced-CVD. Diam Relat Mater 17:1685–1688
28. Wan S, Wang L, Zhang J, Xue Q (2009) Field emission properties of DLC and phosphorus-doped DLC films prepared by electrochemical deposition process. Appl Surf Sci 255:3817–3821
29. Safaie P, Eshaghi A, Bakhshi SR (2016) Optical properties of oxygen doped diamond-like carbon thin films. J Alloys Compd 672:426–432
30. Rajak DK, Kumar A, Behera A, Menezes PL (2021) Diamond-like carbon (DLC) coatings: classification, properties, and applications. Appl Sci 11(10):4445
31. Glechner T, Hahn R, Zauner L, Rißlegger S, Kirnbauer A, Polcik P, Riedl H (2021) Structure and mechanical properties of reactive and non-reactive sputter deposited WC based coatings. J Alloys Compd 885:161129
32. Das S, Guha S, Ghadai R, Sharma A (2021) Influence of nitrogen gas over microstructural, vibrational and mechanical properties of CVD Titanium nitride (TiN) thin film coating. Ceram Int 47(12):16809–16819
33. Madhuri A, Jena S, Gupta M, Swain BP (2023) Nitrogen flow rate dependent atomic coordination, phonon vibration and surface analysis of DC magnetron sputtered nitrogen rich-AlN thin films. Physica B 666:415141
34. Wei T, Yan F, Tian J (2005) Characterization and wear-and corrosion-resistance of microarc oxidation ceramic coatings on aluminum alloy. J Alloys Compd 389(1–2):169–176
35. Das S, Guha S, Ghadai R, Kumar D, Swain BP (2017) Structural and mechanical properties of CVD deposited titanium aluminium nitride (TiAlN) thin films. Appl Phys A 123:1–5
36. Buchinger J, Koutná N, Kirnbauer A, Holec D, Mayrhofer PH (2022) Heavy-element-alloying for toughness enhancement of hard nitrides on the example Ti-W-N. Acta Mater 231:117897

37. Kindlund H, Sangiovanni DG, Martínez-de-Olcoz L, Lu J, Jensen J, Birch J, Petrov I, Greene JE, Chirita V, Hultman L (2013) Toughness enhancement in hard ceramic thin films by alloy design. APL Mater 1(4):042104
38. Polcar T, Martinez R, Vítů T, Kopecký L, Rodriguez R, Cavaleiro A (2009) High temperature tribology of CrN and multilayered Cr/CrN coatings. Surf Coat Technol 203(20–21):3254–3259
39. Madhuri A, Jena S, Gupta M, Swain BP (2023) Tweaking the mechanical stability of $Al_{1-x}W_xN$ ternary nitride alloys for hard coating applications, and study on their structural, photoluminescence and surface chemical analysis. J Alloys Compd 968:17185
40. Hugosson HW, Jansson U, Johansson B, Eriksson O (2001) Restricting dislocation movement in transition metal carbides by phase stability tuning. Sci 293:2434–2437
41. Jhi S-H, Ihm J, Louie SG, Cohen ML (1999) Electronic mechanism of hardness enhancement in transition-metal carbonitrides. Nature 399:132–134
42. Zheng XQ, Liu Y, Li J, He L, Zhang LC (2015) Boride precipitation and mechanical behaviour of high boron stainless steel with boron and titanium additions. Int J Mater Prod Technol 51(4):332
43. Cheng Y, Choi S, Watanabe T (2013) Effect of nucleation temperature and heat transfer on synthesis of Ti and Fe boride nanoparticles in RF thermal plasmas. Powder Technol 246:210–217
44. Wang Z, Xu B, Yang J, Yang B, Xiong H, Yang G, Wang F (2020) Investigation on preparation of porous titanium through calciothermic reduction of porous TiO precursors. J Mater Res Technol 9(6):13137–13146
45. Baehre D, Ernst A, WeiBhaar K, Natter H, Stolpe M, Busch R (2016) Electrochemical dissolution behaviour of titanium and titanium-based alloys in different electrolytes. Procedia CIRP 42:137–142
46. Pitchi CS, Priyadarshini A, Sana G, Narala SKR (2020) A review on alloy composition and synthesis of β-Titanium alloys for biomedical applications. Mater Today: Proc 263:297–3304
47. Ahmed YMZ, Zaki ZI, Bordia RK, Besisa DHA, Amin AMM (2016) Simultaneous synthesis and sintering of TiC/Al_2O_3 composite via self propagating synthesis with direct consolidation technique. Ceram Int 42(15):16589–16597
48. Paraye NK, Ghosh PK, Das S (2021) Surface modification via in situ formation of titanium carbide in ferrous matrix through TIG arcing. Mater Lett 283:128723
49. Sivasankaran S, Kishor Kumar MJ (2015) A novel sonochemical synthesis of nanosize silicon nitride and titanium carbide. Ceram Int 41(9):11301–11305
50. Zhang J, Li S, Lu C, Sun C, Pu S, Qi X, Lin Y, Huang M (2019) Anti-wear titanium carbide coating on low-carbon steel by thermo-reactive diffusion. Surf Coat Technol 364:265–272
51. Nyanor P, Bahador A, El-Kady OA, Umeda J, Kondoh K, Hassan MA (2020) Improved ductility of spark plasma sintered aluminium-carbon nanotube composite through the addition of titanium carbide microparticles. Mater Sci Eng A 795:139959
52. Wu KH, Jiang Y, Jiao SQ, Chou KC, Zhang GH (2019) Preparations of titanium nitride, titanium carbonitride and titanium carbide via a two-step carbothermic reduction method. J Solid State Chem 277:793–803
53. Das S, Guha S, Ghadai R, Sharma A, Chatterjee S (2022) Morphological, mechanical property analysis and comparative study over structural properties of CVD TiN film grown under different substrate temperature in nitrogen gas atmosphere. SILICON 14(1):183–199
54. Guha S, Bandyopadhyay A, Das S, Swain BP (2018) Synthesis and characterization of Titanium Silicon Nitride (TiSiN) thin film: a review. IOP Conf Ser: Mater Sci Eng 377:012181
55. Das S, Guha S, Ghadai R, Swain BP (2021) A comparative analysis over different properties of TiN, TiAlN and TiAlSiN thin film coatings grown in nitrogen gas atmosphere. Mater Chem Phy 258:123866
56. Yu H, Tan T, Wu W, Tian C, An Y, Sun F (2012) Thermal stability of titanium nitride coatings prepared by the mixing technology with laser and plasma. Curr Appl Phys 12(1):152–154
57. Tacikowski M, Banaszek M, Smolik J (2014) Corrosion-resistant composite titanium nitride layers produced on the AZ91D magnesium alloy by a hybrid method. Vacuum 99:298–302

58. Wu M (2009) Low temperature synthesis of nanocrystalline titanium nitride from a single-source precursor of titanium and nitrogen. J Alloy Compd 486(1–2):223–226
59. Matsuda T, Matsubara H (2013) Thermophysical and elastic properties of titanium carbonitrides containing molybdenum and tungsten. J Alloy Compd 562:90–94
60. Cardinal S, Malchere A, Garnier V, Fantozzi G (2009) Microstructure and mechanical properties of TiC–TiN based cermets for tools application. Int J Refract Met Hard Mater 27(3):521–527
61. Liu N, Xu YD, Li H, Li GH, Zhang LD (2002) Effect of nano-micro TiN addition on the microstructure and mechanical properties of TiC based cermets. J Eur Ceram Soc 22(13):2409–2414
62. Yang Q, Lengauer W, Koch T, Scheerer M, Smid I (2000) Hardness and elastic properties of $Ti(C_xN_{1-x})$, $Zr(C_xN_{1-x})$ and $Hf(C_xN_{1-x})$. J Alloy Compd 309(1–2):L5–L9
63. Gordienko SP (2001) Reaction of titanium with boron nitride under self-propagating high-temperature synthesis conditions. Powder Metall Metal Ceram 40(1):58–60
64. Pazhouhanfar Y, Sabahi Namini A, Delbari SA, Nguyen TP, Le QV, Shaddel S, Pazhouhanfar M, Shokouhimehr M, Shahedi Asl M (2020) Microstructural and mechanical characterization of spark plasma sintered TiC ceramics with TiN additive. Ceram Int 46(11):18924–18932
65. Zhao S-S, Gao X-H, Qiu X-L, Yu D-M, Tian G-K (2020) A novel TiC-TiN based spectrally selective absorbing coating: structure, optical properties and thermal stability. Infrared Phys Technol 110:103471
66. Mura G, Musu E, Delogu F (2013) Early stages of the mechanical alloying of TiC-TiN powder mixtures. Mater Chem Phys 137(3):1039–1045
67. Asad M, Sana M (2023) Potential of titanium based alloys in the biomedical sector and their surface modification techniques: a review. Proc Inst Mech Eng C: J Mech Eng Sci 237(23):5503–5532
68. Saurabh A, Meghana CM, Singh PK, Verma PC (2022) Titanium-based materials: synthesis, properties, and applications. Mater Today Proc 56:412–419
69. Gan X, Zhao H, Lo TW, Ho KHW, Lee LYS, Lei DY, Wong K (2018) 2H/1T phase transition of multilayer MoS_2 by electrochemical incorporation of S vacancies. ACS Appl Energy Mater 1(9):4754–4765
70. Krishnan U, Kaur M, Singh K, Kumar M, Kumar A (2019) A synoptic review of MoS_2: Synthesis to applications. Superlattices Microstruct 128:274–297
71. Das S, Swain G, Parida K (2021) One step towards the 1T/2H-MoS 2 mixed phase: a journey from synthesis to application. Mater Chem Front 5(5):2143–2172
72. Kappera R, Voiry D, Yalcin SE, Branch B, Gupta G, Mohite AD, Chhowalla M (2014) Phase-engineered low-resistance contacts for ultrathin MoS_2 transistors. Nat Mater 13:1128–1134
73. Shi J, Yu P, Liu F, He P, Wang R, Qin L, Zhou J, Li X, Zhou J, Sui X, Zhang S, Zhang Y, Zhang Q, Sum TC, Qiu X, Liu Z, Liu X (2017) 3R MoS_2 with broken inversion symmetry: a promising ultrathin nonlinear optical device. Adv Mater 29:1–9
74. Gupta D, Chauhan V, Kumar R (2020) A comprehensive review on synthesis and applications of molybdenum disulfide (MoS_2) material: past and recent developments. Inorg Chem Comm 121:108200
75. Yue Q, Kang J, Shao Z, Zhang X, Chang S, Wang G, Qin S, Li J (2012) Mechanical and electronic properties of monolayer MoS_2 under elastic strain. Phys Lett A 376:1166–1170
76. Johari P, Shenoy VB (2012) Tuning the electronic properties of semi-conducting transition metal dichalcogenides by applying mechanical strains. ACS Nano 6:5449–5456
77. Yun WS, Han SW, Hong SC, Kim IG, Lee JD (2012) Thickness and strain effects on electronic structures of transition metal dichalcogenides: 2H-MX_2 semiconductors (M=Mo, W; X=S, Se, Te). Phys Rev B 85(3):033305
78. Radisavljevic B, Whitwick MB, Kis A (2012) Small-signal amplifier based on single-layer MoS_2. Appl Phys Lett 101:043103
79. Magda GZ, Pető J, Dobrik G, Hwang C, Biró LP, Tapasztó L (2015) Exfoliation of large-area transition metal chalcogenide single layers. Sci Rep 5:14714

80. Forsberg V, Zhang R, Bäckström J, Dahlström C, Andres B, Norgren M, Andersson M, Hummelgård M, Olin H (2016) Exfoliated MoS_2 in water without additives. PLoS ONE 11:1–12
81. Fan X, Xu P, Zhou D, Sun Y, Li YC, Nguyen MAT, Terrones M, Mallouk TE (2015) Fast and efficient preparation of exfoliated 2H-MoS_2 nanosheets by sonication-assisted lithium intercalation and infrared laser-induced 1T to 2H phase reversion. Nano Lett 15:5956–5960
82. Walia S, Balendhran S, Wang Y, Kadir RA, Zoolfakar AS (2014) Characterization of metal contacts for two-dimensional MoS_2 nanoflakes. Appl Phys Lett 103:232105
83. Liu H, Xu L, Liu W, Zhou B, Zhu Y, Zhu L, Jiang X (2018) Production of mono- to few-layer MoS_2 nanosheets in isopropanol by a salt-assisted direct liquid-phase exfoliation method. J Colloid Interface Sci 515:27–31
84. Pondick JV, Woods JM, Xing J, Zhou Y, Cha JJ (2018) Stepwise sulphurization from MoO_3 to MoS_2 via chemical vapour deposition. ACS Appl Nano Mater 10:5655–5661
85. Ahn C, Lee J, Kim HU, Bark H, Jeon M, Ryu GH, Lee Z, Yeom GY, Kim K, Jung J, Kim Y, Lee C, Kim T (2015) Low-temperature synthesis of large-scale molybdenum disulphide thin films directly on a plastic substrate using plasma-enhanced chemical vapour deposition. Adv Mater 27:5223–5229
86. Kumar VK, Dhar S, Choudhury TH, Shivashankar SA, Raghavan S (2015) A predictive approach to CVD of crystalline layers of TMDs: the case of MoS_2. Nanoscale 7:7802–7810
87. Kalanyan B, Beams R, Katz MB, Davydov AV, Maslar JE, Kanjolia RK (2019) MoS_2 thin films from a $(N^tBu)_2(NMe_2)_2Mo$ and 1-propanethiol atomic layer deposition process. J Vac Sci Technol A 37:010901
88. Tan LK, Liu B, Teng JH, Guo S, Low HY, Loh KP (2014) Atomic layer deposition of a MoS_2 film. Nanoscale 6:10584–10588
89. Jang Y, Yeo S, Kim H, Kim SH (2016) Wafer-scale, conformal and direct growth of MoS2 thin films by atomic layer deposition. Appl Surf Sci 365:160–165
90. Feng X, Tang Q, Zhou J, Fang J, Ding P, Sun L, Shi L (2013) Novel mixed-solvothermal synthesis of MoS_2 nanosheets with controllable morphologies. Cryst Res Technol 6:1–6
91. Zhou X, Xu B, Lin Z, Shu D, Ma L (2014) Hydrothermal synthesis of flower-like MoS_2 nanospheres for electrochemical supercapacitors. J Nanosci Nanotechnol 14:7250–7254
92. Zhan L, Wan W, Zhu Z, Shih TM, Cai W (2017) MoS2 materials synthesized on SiO_2/Si substrates via MBE. J Phys: Conf Ser IOP Publ 864(1):012037
93. Radisavljevic B, Radenovic A, Brivio J, Giacometti V, Kis A (2011) Single-layer MoS_2 transistors. Nat Nanotechnol 6:147–150
94. Ye L, Li H, Chen Z, Xu J (2016) Near-infrared photodetector based on MoS_2/black phosphorus heterojunction. ACS Photon 3:692–699
95. Tyagi S, Kumar A, Kumar M, Singh BP (2019) Large area vertical aligned MoS_2 layers towards the application of thin film transistor. Mater Lett 250:64–67
96. Kwon H, Choi W, Lee D, Lee Y, Kwon J, Yoo B, Grigoropoulos P, Kim S (2014) Selective and localized laser annealing effect for high-performance flexible multilayer MoS_2 thin-film transistors. Nano Res 7:1137–1145
97. Kong N, Wei B, Li D, Zhuang Y, Sun G, Wang B (2020) A study on the tribological property of MoS2/Ti–MoS2/Si multilayer nanocomposite coating deposited by magnetron sputtering. RSC Adv 10(16):9633–9642
98. Shen YZ, Guo XZ, Lin YB, Tao J (2014) Al_2O_3 coatings fabricated on stainless steel/ aluminium composites by microarc oxidation. Surf Eng 30(10):735–740
99. Dhawale VP, Khobragade V, Kulkarni SD (2018) Synthesis and characterization of aluminium oxide (Al_2O_3) nanoparticles and its application in azodye decolourisation. Int J Environ Anal Chem 2(1):10–17
100. Broitman E, Hultman L (2012) Adhesion improvement of carbon-based coatings through a high ionization deposition technique. J Phys: Conf Ser IOP Publ 370(1):012009
101. Liu A, Cohen M (1990) Structural properties and electronic structure of low compressibility materials; β-Si_3N_4 and hypothetical β-C_3N_4. Phys Rev B 41(15):10727–10734

102. Yuan B, Wang Y, Elnaggar AY, Azab IHE, Huang M, Mahmoud MHH, El-Bahy SM, Guo M (2022) Physical vapor deposition of graphitic carbon nitride (g-C_3N_4) films on biomass substrate: optoelectronic performance evaluation and life cycle assessment. Adv Compos Hybrid Mater 5(2):813–822
103. Neidhardt J, Hultman L, Broitman E, Scharf T, Singer I (2004) Structural, mechanical and tribological behaviour of fullerene-like and amorphous carbon nitride coatings. Diam Relat Mater 13:1882–1888
104. Bakoglidis KD (2016) Low-friction and wear-resistant carbon nitride coatings for bearing components grown by magnetron sputtering. Linkopings Universitet (Sweden)
105. Fernández-Ramos C, Sayagués MJ, Rojas TC, Alcalá MD, Real C, Fernández A (2000) Study of the thermal stability of carbon nitride thin films prepared by reactive magnetron sputtering. Diam Rel Mater 9(2):212–218
106. Hellgren N, Macák K, Broitman E, Johansson M, Hultman L, Sundgren J-E (2003) Influence of plasma parameters on the growth and properties of magnetron sputtered CN_x thin films. J Appl Phys 524(1):524–532
107. Kusano Y, Barber ZH, Evetts JE, Hutchings IM (2000) Tribological and mechanical properties of carbon nitride films deposited by ionised magnetron sputtering. Surf Coat Technol 124(2–3):104–109
108. Tejasvi R, Basu S (2020) Formation of C_3N_4 thin films through the stoichiometric transfer of the bulk synthesized gC_3N_4 using RFM sputtering. Vacuum 171:108937
109. Schmidt S, Czigány Z, Wissting J, Greczynski G, Janzén E, Jensen J, Ivanov I, Hultman L (2016) A comparative study of direct current magnetron sputtering and high power impulse magnetron sputtering processes for CN_x thin film growth with different inert gases. Diam Relat Mater 64:13–26
110. Oliver W, Pharr G (1992) An improved technique for determining hardness and elastic modulus using load and displacement sensing indentation experiments. J Mater Res 7(6):1564–1583

Chapter 4
Adhesion and Surface Modification of Coating Materials

Siddharth Parthasarathy, Aishwarya Madhuri, and Bibhu Prasad Swain

Abstract This book chapter explores the intricate relationship between adhesion and surface modification in coating materials, emphasising the pivotal role that surface treatments play in enhancing the performance and durability of coatings. The chapter delves into various surface modification techniques, including mechanical, chemical, and physical treatments, and their impact on adhesion properties. Through a comprehensive analysis of how surface roughness, energy, and chemical composition influence adhesion, the chapter provides insights into optimising coating processes for diverse industrial applications. The discussion also highlights recent advancements in nanotechnology and their contributions to improving adhesion through innovative surface modification strategies. This chapter serves as a valuable resource for researchers and engineers aiming to develop more robust and efficient coatings for various environments. Surface treatment encompasses mechanical, chemical, and physical procedures to prepare substrates for coating application. Mechanical treatments like abrasive blasting and grinding enhance surface roughness for improved mechanical interlocking. Chemical treatments such as acid etching and alkaline cleaning boost surface energy and cleanliness, while physical methods like plasma treatment, ion implantation, and laser surface treatment alter surface properties to enhance coating adhesion and performance. The chapter investigates how these processes heighten coating adhesion, prevent corrosion, and strengthen wear resistance. It also explores characterisation and evaluation methods like surface energy measurement, chemical analysis, topographical analysis, and adhesion testing to gauge treatment effectiveness. Real-world examples from the aerospace, automotive, and electronics sectors demonstrate the practical applications and advantages

S. Parthasarathy · B. P. Swain (✉)
Department of Physics, Centurion University of Technology and Management, Bhubaneswar, Odisha, India

S. Parthasarathy
e-mail: siddharthnayak999@gmail.com

A. Madhuri · B. P. Swain
Department of Physics, National Institute of Technology Manipur, Langol, Imphal West, Manipur, India
e-mail: aishwaryamadhuri99@gmail.com

B. P. Swain (ed.), *Advances in Mechanical Coating*, Materials Horizons: From Nature to Nanomaterials, https://doi.org/10.1007/978-981-96-7484-8_4

of surface passivation and treatment techniques. The chapter concludes by examining current obstacles and prospects, emphasising the necessity for eco-friendly and sustainable practices and technological advancements to meet the evolving demands of diverse industries. The comprehensive overview underscores the critical role of surface modification and treatment in ensuring optimal performance and longevity of coating materials.

Keywords Surface treatment · Surface modification · Oxide layer · Corrosion resistance · Plasma implantation

1 Introduction

Surface modification and treatment are pivotal in enhancing coating materials' performance, durability, and adhesion across various industries. These techniques involve creating protective layers and applying mechanical, chemical, and physical procedures to prepare surfaces for coatings. Surface passivation aims to prevent corrosion and environmental degradation by forming a shield on material surfaces, while surface treatment methods such as abrasive blasting, acid etching, and plasma treatment are utilised to optimise adhesion and performance. In this introduction, we delve into the mechanisms, methods, and applications of surface passivation and treatment, highlighting their importance in improving coating properties and ensuring longevity in industrial applications. These treatments improve surface roughness, enhance cleanliness and surface energy, and modify surface properties to promote better adhesion and coating performance. Adhesion is a fundamental aspect of coating technologies, determining the effectiveness and longevity of coatings across a wide range of applications. In industrial settings, the bond strength between a coating and its substrate is crucial for ensuring the coated material's durability, protection, and functionality. However, achieving optimal adhesion is not a straightforward process; the properties of the substrate surface heavily influence it. Surface modification techniques have become indispensable in enhancing the adhesion of coatings to various materials. Altogether, surface passivation and treatment are critical for increasing coating adhesion, preventing corrosion, and improving wear resistance, ensuring that coatings maintain their optimal properties and functionality over time.

Figure 1 shows the schematic presentation of different surface treatment techniques. The roughness of the coating surface can significantly impact how other materials interact with it, which can change things like adhesion, wear resistance, and general functionality. The wettability, mechanical interlocking, and chemical bonding at the interface can all be impacted by roughness at the micro- and nanoscale, which in turn influences the strength and longevity of the adhesive bonds. It is possible to precisely regulate the qualities of a coating by engineering surface roughness using a variety of approaches, including mechanical texturing, chemical treatments, and physical vapour deposition. Surface modification involves altering a substrate's physical, chemical, or mechanical properties to improve its interaction with a coating

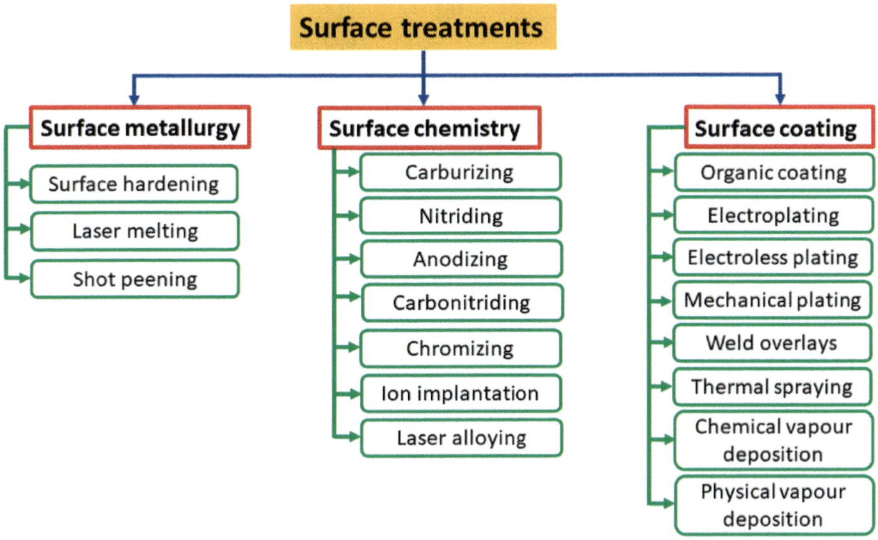

Fig. 1 Schematic classification of different types of surface treatment techniques

material. This chapter explores the critical role of surface modification in adhesion, examining how different treatments, such as mechanical roughening, chemical etching, and the application of thin films, affect the bonding performance of coatings. The discussion includes a review of traditional methods and recent advancements in nanotechnology that offer new possibilities for enhancing adhesion through precise surface engineering.

2 Mechanisms of Surface Passivation

Chemical passivation involves the formation of a stable, protective layer on the material's surface through chemical reactions. This method is widely used for metals like stainless steel and aluminium. The formation of oxide layers is a fundamental method of surface passivation used to protect metals from corrosion and environmental degradation. Oxide layers are typically created through chemical reactions between the metal surface and oxygen from the air or an oxidising agent. These oxide layers serve as protective barriers that prevent further oxidation and corrosion of the underlying metal. This section explores the mechanisms, methods, and benefits of oxide layer formation for surface passivation.

A range of studies have explored the mechanisms of surface passivation in different materials. Chen et al. found that water vapor induces the formation of a bilayer oxide film on aluminium surfaces, consisting of a crystalline-like $Al(OH)_3$ top layer and an inner layer of amorphous Al_2O_3 [1]. Xia et al. reviewed various

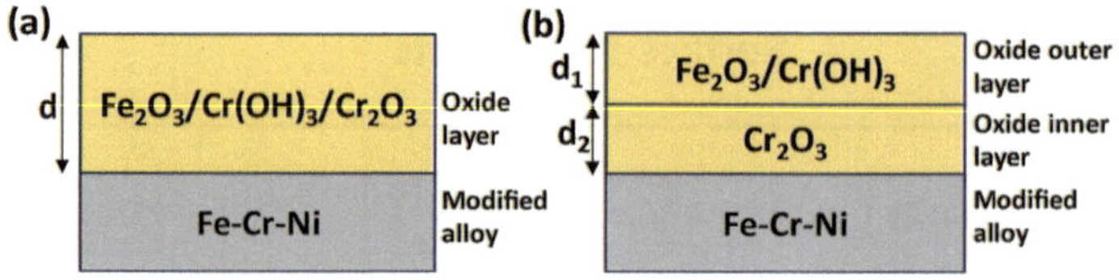

Fig. 2 Models of surface oxide film formation on FeCrNi stainless steel **a** monolayer and **b** bilayer

surface passivation strategies for perovskite solar cells, including Lewis acid–base interactions, low-dimensional perovskites, inorganic molecules, and polymers [2]. Ma et al. investigated the passivation mechanisms and pre-oxidation effects on model surfaces of FeCrNi austenitic stainless steel, finding that pre-oxidation promotes the formation of a protective oxide film [3]. Figure 2a and b show the passive layer chemical structure models for monolayer and bilayer over the Fe–Cr-Ni modified alloy. According to the single-layer model, there is just one surface oxide layer from which all oxide and hydroxide components originate. In the bilayer model, atomic transport is slowed by an inner barrier between the electrolyte and the outside layer of exchange. Bae et al. discussed surface passivation's challenges and potential benefits in nanoscale zerovalent iron, highlighting the importance of understanding passivation chemistry and its related mechanisms [4].

2.1 *Mechanisms of Oxide Layer Formation*

2.1.1 Oxidation Reaction

The basic chemical reaction for oxide layer formation involves the metal reacting with oxygen to form a metal oxide, as shown in Eqs. (1) and (2) below. This initial reaction forms a thin, protective oxide layer.

$$2Al + 3O_2 \rightarrow Al_2O_3 \tag{1}$$

$$4Fe + 3O_2 \rightarrow 2Fe_2O_3 \tag{2}$$

In the case of aluminium coatings, the passivation process involves the formation of an aluminium oxide layer (Al_2O_3), known for its excellent adhesion and protective properties. The growth process of the oxide layer can occur through two primary mechanisms: (a) outward diffusion, where the metal ions from the underlying coating diffuse outward through the oxide layer to react with oxygen at the oxide-gas interface and (b) inward diffusion, where the oxygen molecules or ions diffuse inward through the oxide layer to react with the metal at the metal-oxide interface. The oxidation reaction mechanisms of surface passivation have been extensively studied. Yuan et al. used molecular dynamic simulations to show that oxidant species diffuse into

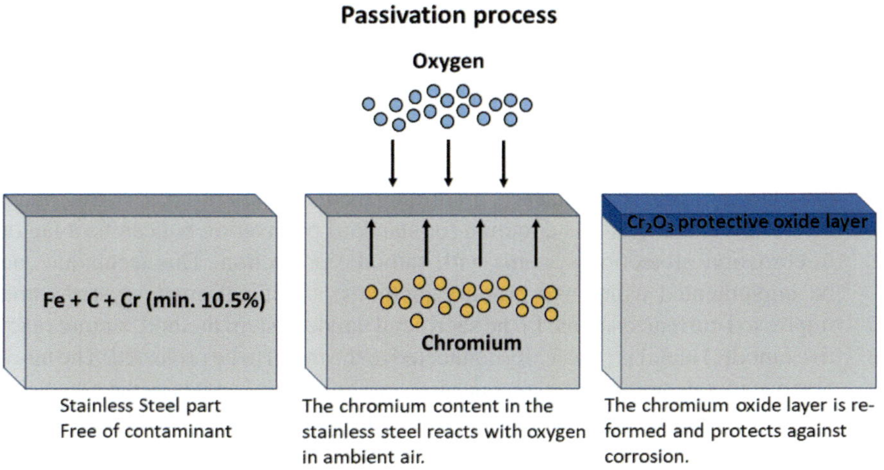

Fig. 3 Model of different steps included in the passivation process

the Si substrate through peroxy-like structures during the oxidation of chemically passivated silicon surfaces [5]. Choi et al. identified stable initial surface products for O adsorption and proposed two channels for SiO desorption during the oxidation of Si (100) by atomic oxygen [6]. Chen et al. observed a two-stage oxidation process during the passivation of aluminium surfaces, with the first stage dominated by intralayer atomic shuffling and the second stage by interlayer atomic disordering [7]. Figure 3 shows the models of different steps in the passivation process. The passivating procedure consists of three methods that produce an entirely homogeneous, thin oxide coating on the metallic surface. Loo et al. obtained a high level of surface passivation of c-Si surfaces using atomic layer deposited transparent conductive ZnO films [8]. The field effect passivation caused by a decrease in the surface hole concentration is responsible for the increasing surface passivation levels in ZnO with increasing doping levels.

2.1.2 Chemical Treatment (Cathodic and Anodic)

While oxidation plays a significant role in passivation, techniques also involve reduction (electron gain) achieved through chemical treatments. Chemical treatment, specifically passivation, has been shown to improve the corrosion resistance of various materials significantly. Chembath et al. found that nitric acid passivation enhanced NiTi alloy's corrosion resistance, while Baron et al. demonstrated that chemical passivation of austenitic steel improved its resistance to corrosion [9, 10]. Anodic treatment with the specific group via anions has also been shown to enhance the passivation process for zinc [11]. Furthermore, Mansfeld et al. reported that chemical passivation in a $CeCl_3$ solution resulted in highly corrosion-resistant surfaces for aluminium alloys and metal matrix composites [12]. These collectively highlight the effectiveness of chemical treatment, particularly passivation, in improving the corrosion resistance of various materials. These treatments can be categorised into two main types based on the applied electrical potential: Cathodic and Anodic.

(a) **Cathodic Passivation** involves applying a negative potential (becoming a cathode) to the surface. This attracts positively charged oxidising agents (like dissolved oxygen) to the surface, promoting the formation of a passivating oxide layer. This method is less common for standard passivation but can be a factor in corrosion protection systems with cathodic protection. This technique can be implemented using two primary methods: sacrificial anode systems and impressed current systems. In the sacrificial anode system method, a more reactive (anodic) metal is physically connected to the metal to be protected. The more reactive metal (such as zinc, magnesium, or aluminium) serves as the sacrificial anode, corroding preferentially and protecting the less reactive metal (such as steel). The sacrificial anode corrodes over time, releasing electrons that flow to the protected metal, making it a cathode and thereby preventing its oxidation. The impressed current system method uses an external power source (a DC power supply) to provide a continuous current flow. An inert anode material (such as titanium or graphite) is placed in the electrolyte and connected to the positive terminal of the power supply. The metal to be protected is connected to the negative terminal, making it the cathode. The external current ensures the protected metal remains in a cathodic (reduced) state, preventing oxidation. Applying a negative potential to the metal increases the electron density at the metal surface, reducing the metal cations and preventing their dissolution.

(b) **Anodic Passivation**: This is a more widely used method for chemical treatment passivation. A positive potential is applied to the surface (making it an anode). This drives the oxidation of the surface atoms, directly forming a passivating oxide layer. For aluminium, the anodising process is represented in Eq. (3) below;

$$2Al + 3H_2O \rightarrow Al_2O_3 + 6H + +6e^{-2} \tag{3}$$

The resulting aluminium oxide (Al_2O_3) layer is hard, corrosion-resistant, and can be dyed to different colours for aesthetic purposes.

Common Anodic Passivation Techniques:

- **Nitric Acid Passivation:** A popular technique for stainless steel uses a nitric acid (HNO_3) bath. The iron layer on the surface is dissolved by the acid, increasing the amount of chromium present and promoting the formation of a chromium oxide (Cr_2O_3) passivating layer. Adding sodium dichromate to nitric acid accelerates the passivation process for some particular alloys. This method is less popular because soaking in nitric acid increases the risks due to the presence of sodium dichromate.

Research on the passivation of various nitric acid materials has yielded mixed results. Callen et al. found that nitric acid passivation reduced the oxide thickness on Ti6Al4V but increased trace element release [13]. O'Laoire et al. observed increased Cr: Fe ratios at the surface of 304 and 316 stainless steels following nitric acid passivation, with citric acid-based passivating agents resulting in more coherent oxide surfaces [14]. Yue et al. identified the critical factor determining the corrosion resistance of the passive film on 304 stainless steel during nitric acid passivation and determined the optimised passivating time [15]. Grant et al. investigated the passivation of a Si (100) surface by silicon dioxide grown in nitric acid, finding that the process did not passivate the silicon surface before annealing but did so after nitrogen annealing [16]. Padhy et al. observed a rise in the transmissive potential, a drop in passive current density, and an overall resistance to corrosion of the sputter-deposited TiO_2 on austenitic type 304L stainless steel assessed in the oxidising medium of 1 and 8 M nitric acid [17].

- **Citric Acid Passivation**: A milder alternative to nitric acid, particularly for applications where avoiding harsh chemicals is preferred. Citric acid removes surface contaminants and promotes natural oxide layer formation through air oxidation.

Citric acid passivation has been shown to improve the corrosion resistance of various materials, including titanium dental implants and AISI 316L stainless steel [18, 19]. Using citric acid in passivation methods has also been explored and optimised for common stainless-steel alloys, with promising results [20]. Compared to nitric acid passivation, citric acid has resulted in more coherent oxide surfaces that are more corrosion-resistant [14].

2.1.3 Plasma Implantation

Plasma implantation in surface passivation involves using a plasma source to modify the surface of materials, such as metals, plastics, and ceramics [21]. This technique has improved industrial applications' microhardness, wear properties and tool life. In ZnO nanocombs, plasma immersion ion implantation (PIII) with low bias voltages has quenched surface-originated exciton emission, leading to surface state passivation [22]. In crystalline silicon solar cells, the passivation of both cell surfaces is crucial for achieving high conversion efficiencies, with plasma silicon nitride identified as a promising method [23]. Furthermore, plasma-assisted atomic layer deposited hafnium oxide films have been investigated for silicon surface passivation, demonstrating effective surface recombination velocity and interface defect density [24]. Leng et al. used direct current plasma oxidation to deposit titanium oxide layers on commercially pure titanium and TiN and found that following plasma oxidation, the micro-hardness of the pure Ti sample increased by about 70%, while that of the TiN sample increased by around 30% [25]. Tang et al. demonstrated that the energy band structure of the surface and other FTO features can be effectively modulated by oxygen plasma treatment, which offers an excellent deal for interface and surface modification and, ultimately, the enhancement of photoelectric device performance [26]. Figure 4 demonstrates the effect of oxygen plasma treatment on the surface of FTO. Since the oxyanion hole was the primary cause of FTO conduction, the carrier

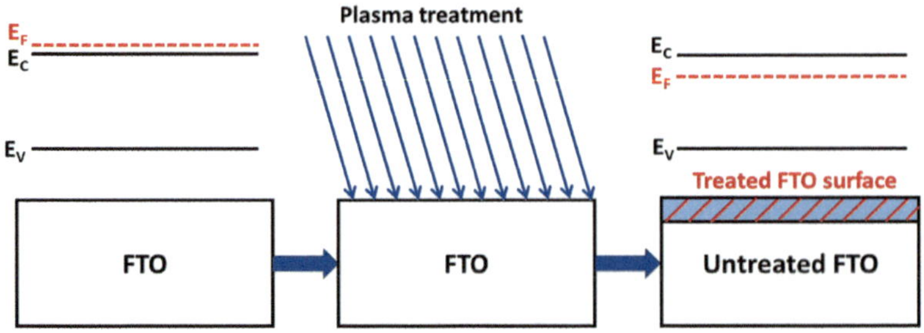

Fig. 4 Effect of oxygen plasma treatment on FTO surface

was an electron, and its concentration decreased while the O percentage increased at the FTO surface. As a result, the work function was expanded and the Fermi level was lowered.

Plasma implantation in surface passivation is a technique used to enhance the surface properties of materials, particularly semiconductors. This process involves introducing ions into the surface layer using a plasma source. Plasma, an ionised gas with free electrons and positive ions, is generated by applying an electric field to a gas. The ions in the plasma are then accelerated towards the material's surface and become embedded in it. The primary goal of plasma implantation is to reduce surface recombination of charge carriers, which can degrade the performance of electronic and optoelectronic devices such as solar cells, LEDs, and transistors. By passivating dangling bonds and other surface defects, plasma implantation decreases the number of states that can trap charge carriers, thereby enhancing the electrical performance of the devices.

Moreover, plasma implantation improves the chemical stability of the surface, making it less reactive to environmental factors. This results in more stable and durable semiconductor devices. The technique is valued for significantly improving device efficiency and longevity, making it a crucial step in fabricating high-performance electronic and optoelectronic components.

2.2 *Thermodynamics and Kinetics*

The thermodynamic properties of the metal and oxide, such as Gibbs free energy, are influenced by the oxide layers' stability and formation rate. The kinetics of oxidation involve factors like temperature, partial pressure of oxygen, and surface condition of the metal. Classifications of oxide formation processes include native oxides, thermally grown oxides, anodic oxides, chemical passivation, electrochemical passivation, application of organic layers, and inorganic coatings. Native oxides spontaneously form when metals are exposed to oxygen-rich environments, such as aluminium forming aluminium oxide (Al_2O_3) or titanium forming titanium dioxide (TiO_2). Thermally grown oxides are produced by heating metals in oxygen-containing atmospheres to accelerate oxidation, as seen with silicon wafers in semiconductor applications. Anodic oxides are made through anodising, where metals serve as the anode in an electrochemical cell, for example, anodising aluminium to create a thick, porous aluminium oxide layer. Chemical passivation involves forming protective oxide layers to prevent further oxidation, like passivating stainless steel with nitric or citric acid to generate a chromium oxide layer. Electrochemical passivation uses anodic or cathodic processes to develop protective films on metal surfaces, such as anodising aluminium, to form a durable oxide layer. Organic layers like paints and polymers act as environmental barriers, as seen when epoxy coatings are applied to steel structures for corrosion protection. Inorganic coatings include depositing thin films of materials such as silicon dioxide or titanium dioxide using chemical vapour deposition (CVD) or physical vapor deposition (PVD) techniques.

2.3 Surface Modification Techniques

A range of surface treatment techniques has been explored in the literature, focusing on enhancing corrosion resistance and wear protection. Vijay et al. and Singh et al. reviewed various methods, including laser surface coating, plasma-enhanced vapour deposition, and thermal spraying [27, 28]. These techniques have been found to improve the properties of titanium alloys and composites and protect metal components from wear and degradation. Bhushan et al. provide a comprehensive overview of the physics of tribological materials, including the classification of surface treatment techniques and their applications [29]. Park et al. highlight the importance of environmentally friendly surface treatment technologies in the aerospace industry, such as non-chromate anodising and sol–gel coating [30].

Mechanical treatments such as abrasive blasting, grinding, and polishing surfaces improve the hard coating applications for cutting tools, aerospace, and manufacturing. Abrasive blasting involves materials like sand or glass beads to clean and roughen surfaces, promoting mechanical interlocking for coatings. Grinding and polishing are employed to achieve the desired surface roughness and texture for optimal coating adhesion. Chemical treatments encompass acid etching and alkaline cleaning techniques. Acid etching utilises acidic solutions to create micro-roughness on surfaces, activating them for improved adhesion. Alkaline cleaning, on the other hand, removes organic contaminants and enhances surface energy to facilitate better coating adherence. Chromate and phosphate coatings enhance corrosion resistance and improve paint adhesion, and they are commonly applied to materials like steel and aluminium for added protection. Physical treatments involve plasma treatment, ion implantation, and laser surface treatment. Plasma treatment alters surface chemistry and boosts surface energy using ionised gas, while ion implantation introduces ions into the surface layer to modify properties such as hardness and wear resistance. Laser surface treatment employs laser beams to clean, texture, or alloy surfaces, enhancing their properties to promote better adherence to coatings. Galceran et al.

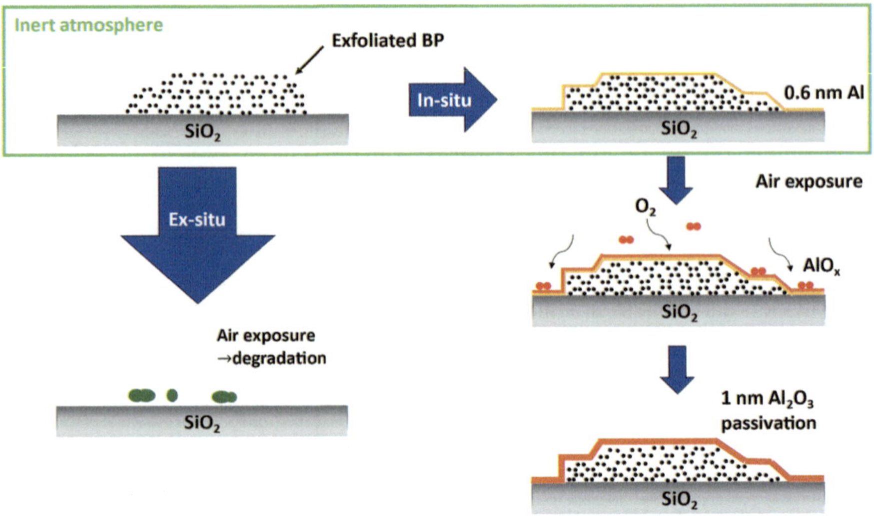

Fig. 5 Schematic presentation of steps to prepare the passivated black phosphorous samples

[31] described a large-scale passivation process that effectively preserves the structural characteristics of ultrathin black phosphorus flakes utilising an in-situ Al_2O_3 capping technique with a thickness as low as 1 nm shown in Fig. 5. On Si/SiO_2 substrates, 1 nm of Al_2O_3 effectively passivates exfoliated black phosphorus (below 5 layers). Such a transparent ultrathin passivation layer can function as a tunnel barrier, enabling the processing of black phosphorus devices without the need to remove the passivation layer.

2.4 *Surface Roughness*

Surface roughness refers to the irregularities on a material's surface, quantified by parameters like average roughness (Ra), root mean square roughness (Rq), and peak-to-valley height (Rz). It is a critical parameter in surface modification and treatment processes, significantly influencing the performance and longevity of materials. In various industrial and technological applications, surface roughness control can enhance material properties such as corrosion resistance, electrical conductivity, hydrodynamic performance, fouling behavior, material functionality, and overall durability. It plays a significant role in predicting the tactile properties of fabrics and influences physical and chemical phenomena at micro-/nanoscales. Measurement techniques for surface roughness can be broadly categorised into subjective and objective methods, including contact and non-contact approaches. Advanced measurement tools like atomic force microscopy (AFM) and optical interferometry are widely used for characterising surface roughness. Key parameters for describing surface roughness include the standard deviation of surface heights, the mean radius of curvature of asperity summits, and the surface roughness power spectrum. Proper reporting of measurement parameters, such as cutoff length, is crucial for comparing results across studies and facilitating interdisciplinary research. The following measurement techniques can evaluate these parameters.

Wang et al. [32] studied that the passivation of titanium alloys, particularly Ti6Al4V, modifies the surface roughness, wettability, and oxide layer chemistry. The study found that passivation affects the titanium alloy's surface roughness (Ra), with values ranging from 1.9 to 7.4 nm. A key finding is that the initial adhesion strength of fibroblasts, measured using a cytodetacher, is negatively correlated with surface roughness, meaning that smoother surfaces promote stronger cell adhesion. Additionally, the study observed that the morphology of fibroblasts differed based on the passivation treatment, further supporting the relationship between surface topography and cellular response. For titanium alloys used in dental implants, sandblasting and acid etching can create a desirable surface topography with large valleys, peaks, and micropores, promoting osseointegration [33]. For biomedical applications, controlled blasting followed by passivation is necessary in stainless steel to achieve optimal surface properties and corrosion resistance [34]. Research by Oh et al. highlighted that surface roughness variations from 100 to 1,000 nm significantly impact the corrosion resistance and re-passivation of super duplex stainless steel (SDSS) [35]. Additionally, Park et al. introduced ultrathin aluminium oxide (Al_2O_3) passivation layers ($\sim$10 nm) for PCB Cu pads and lines, emphasising the importance of passivation, adhesion, bond ability, and reliability in electronic applications [36].

For 5083 Al alloy, smaller surface roughness promotes the formation of passive films with lower carrier density and diffusion coefficient, enhancing corrosion resistance [37]. Furthermore, the study by Sunarso et al. discussed the surface roughness parameters of untreated Ti and $AgNO_3$-treated Ti, indicating the significance of surface roughness in different surface treatment methods [38]. Figure 6 shows the surface roughness texture of untreated Ti, Ag-coated Ti for 0.01 and 0.1 mol/L $AgNO_3$ produced by a roughness tester. For semiconductor surfaces, passivation treatments can reduce surface roughness and improve interface properties, with bromine-methanol solution containing sulphur showing promising results [39]. The relationship between surface roughness and cell adhesion is complex, with some studies indicating a negative correlation between nanoscale roughness and initial fibroblast adhesion strength. For silicon substrates, increased microroughness shifts optimal passivation towards stoichiometric silicon nitride films [40]. In passivated CdHgTe surfaces, treatment with a bromine-methanol solution containing sulfur significantly reduces surface roughness and improves interface properties, resulting in a lower density of fast interface states [41]. The study by Madhuri et al. [42] demonstrated that nitrogen-rich AlN films possess smoother surfaces with finer grain structures than those deposited under lower nitrogen flow rates by DC magnetron sputtering. Incorporating tungsten (W) into the AlN matrix significantly affects the surface roughness of the resulting $Al_{1-x}W_xN$ alloys. As the tungsten concentration increases, the surface roughness tends to change, reflecting alterations in the microstructure and growth dynamics of the film [43].

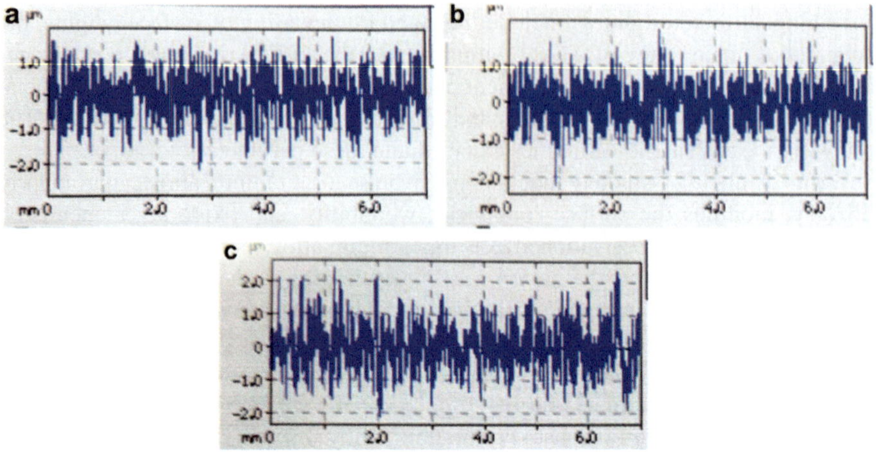

Fig. 6 Surface roughness texture produced by a roughness tester of **a** untreated Ti, **b** Ag-coated Ti for 0.01 mol/L $AgNO_3$, and **c** Ag-coated Ti for 0.1 mol/L $AgNO_3$ [38]

2.5 Role of Surface Modification and Treatment in Coating Adhesion

Surface modification and treatment are essential in enhancing coatings' adhesion to various substrates, ensuring the durability and performance of coated materials in a wide range of applications. The adhesion of a coating to its substrate is influenced by multiple factors, including surface roughness, chemical composition, and surface energy. Enhancing adhesion involves using surface treatments to promote strong adhesive bonds between coatings and substrates. Preventing corrosion is achieved by applying passivation layers that serve as protective barriers against environmental corrosion triggers. Improving wear resistance entails employing surface treatments to boost the hardness and longevity of coatings, thereby extending their operational life. Controlling surface energy involves treatments that effectively adjust surface energy levels to optimise wetting and bonding properties. The primary goal of surface modification is to create a surface that can form strong bonds with the coating material, which can be achieved through various methods. Mechanical treatments like sandblasting, grinding, or polishing are used to increase the surface roughness of a substrate. The increased roughness enhances mechanical interlocking, one of the key adhesion mechanisms, providing more surface area for the coating to grip onto the substrate. Zhang et al. [44] focus on these polymers composed of interlocked molecular architectures, such as rotaxanes and catenanes. These structures are not chemically bonded but mechanically linked with unique properties like flexibility, toughness, and dynamic responsiveness to external stimuli. The dense mechanical bonds in mechanically interlocked polymers enhance their mechanical properties, making them highly resilient and adaptable. This characteristic is precious in applications requiring materials that can withstand mechanical stress while maintaining structural integrity. Mandolfino et al. [45] discussed the critical role of sandblasting surface pretreatment in enhancing the mechanical properties and surface roughness of adhesive-bonded joints. Mechanical treatments enhance the roughness, remove contaminants, increase surface area, and improve wettability, all contributing to stronger adhesive bonds. Rudawska et al. [46] focused on how various surface treatments influence the strength and durability of adhesive joints. Surface roughness, created through mechanical processes such as grit blasting, sanding, or machining, significantly impacts the bonding strength by altering the surface energy and providing mechanical interlocking for adhesives. The surface preparation of the adherend is crucial in optimising the adhesion process, ensuring that the joint can withstand mechanical stresses and environmental factors over time.

Laser texturing uses laser beams to create micro- and nanoscale patterns on the surface, which can significantly improve adhesion by providing a more favourable topography for coating deposition. Chemical etching involves using acids or bases to remove the top layer of the substrate, thereby increasing surface roughness and modifying the surface chemistry. This can result in better wettability and stronger chemical bonds between the coating and the substrate. Applying adhesion promoters or coupling agents, such as silanes or phosphates, can chemically modify the surface to enhance bonding. These agents form chemical bridges between the substrate and the coating, improving adhesion at the molecular level. Plasma treatment is a versatile technique to clean, activate, or functionalise a substrate surface. Exposing the substrate to a plasma (ionised gas) removes contaminants and increases surface energy, leading to better wettability and stronger adhesion. Plasma treatment can also introduce functional groups onto the surface, which can react with the coating material to form strong chemical bonds. Friedrich et al. [47] provided a comprehensive resource on interface interactions and chemical bonding between metals and polymers and the significance of the chemical reactions that occur at these interfaces and also explored the effects of different functional groups on polymer surfaces, the influence of coupling agents, and the impact of various bonding mechanisms, including covalent and electrostatic bonds, on the overall strength and durability of metal-polymer systems.

Van der Waals forces are weak intermolecular forces contributing to adhesion, particularly in thin films and coatings. These forces arise from temporary dipoles created by electron fluctuations within atoms and molecules. Although individually weak, the cumulative effect of van der Waals forces over a large contact area can result in significant adhesion. This mechanism is crucial for materials where chemical solid bonds are impossible or substrate and coating materials are not reactive. In non-polar material applications, van der Waals forces can provide sufficient adhesion to keep the coating intact. Wager et al. [48] provided insights into how quantum mechanical tunnelling affects the stability and properties of chemical bonds across different materials. This study examined how tunnelling influences covalent, ionic, polar covalent, and transition metal bonding, concluding that tunnelling is essential for these bond types.

Conversely, bonds like van der Waals and free electron metal bonding do not depend on tunnelling. Jarvis et al. [49] studied the role of covalent bonding in enhancing adhesion between metal-oxide and oxide-oxide interfaces. The atomic-level interactions at interfaces such as SiO_2/Ni and ZrO_2/SiO_2 were examined using first-principles density functional calculations. The findings suggest that the presence of covalent bonding at these interfaces significantly strengthens them compared to interfaces formed with more ionic oxides like Al_2O_3. Li et al. [50] explored the complex interactions at the interface of epoxy adhesives and metallic oxides on galvanised steel, focusing on how these interactions influence corrosion resistance.

The research combines experimental and molecular dynamics (MD) simulations to understand the bonding mechanisms involved comprehensively. Key findings highlight the importance of water molecules at the interface, which can significantly impact the adhesion and overall adhesive bond performance. The study also delves into the lap-shear strength testing of epoxy-bonded joints after soaking in water, simulating real-world conditions. Results from MD simulations show how the epoxy adhesive interacts with different oxide surfaces, such as ZnO, Al_2O_3, and MgO, revealing the detailed molecular structure and interaction energies. Chemical bonding involves the formation of strong covalent, ionic, or metallic bonds between the atoms or molecules of the coating and the substrate. This type of adhesion is particularly effective when there are reactive functional groups on both the substrate and the coating material. For example, in the passivation of metals, the oxide layer that forms on the metal surface can chemically bond with additional protective coatings. Similarly, in polymer coatings, functional groups can react with those on the substrate, creating a durable chemical bond. Chemical bonding provides a robust adhesion mechanism, ensuring the coating remains firmly attached under various environmental conditions.

Surface treatments can remove impurities, modify surface energy, and create a more receptive surface to the coating material. Surface cleanliness is a fundamental requirement for good adhesion. Contaminants such as oils, oxides, or dust can be barriers between the substrate and the coating, leading to weak adhesion. Cleaning methods, including solvent, ultrasonic, and plasma cleaning, remove these contaminants and prepare the surface for coating. The presence of an oxide layer on a metal substrate can influence adhesion.

In some cases, a thin, stable oxide layer can enhance adhesion by providing a reactive surface for bonding. However, thick or unstable oxide layers can impede adhesion. Surface treatments, such as anodising or passivation, can control the thickness and composition of oxide layers to optimise adhesion. Surface energy is critical in determining the wetting behavior of coating material on a substrate. High surface energy substrates exhibit better wettability, allowing coatings to spread more evenly and form adhesive solid bonds. Treatments such as plasma or corona discharge can increase a substrate's surface energy, improving coating adhesion.

Here, we explore surface modification and treatment's various roles in enhancing coating adhesion, displayed in Table 1.

Table 1 Various roles of surface modification and treatments in enhancing coating adhesion

Technique	Surface treatments	Role
Cleaning and contaminant removal	Chemical cleaning, abrasive blasting, ultrasonic cleaning	Promotes better mechanical interlocking and chemical bonding, prevents defects in the coating
Surface roughening	Abrasive blasting, etching, mechanical abrasion	Increases surface area for better mechanical adhesion, distributes stresses more evenly across the interface
Chemical bonding	Silane coupling, phosphating, anodizing	Improves adhesion strength, provides a protective layer against corrosion
Oxide layer formation	Thermal oxidation, anodising	Provides stable and uniform surface for adhesion, Acts as a barrier against contaminants
Adhesion promoters	Application of adhesion promoters or primers	Creates strong chemical bonds at the interface, improves wettability for uniform coating application
Surface energy modification	Plasma treatment, UV/Ozone treatment	Enhances wetting and spreading of coating material and provides stronger adhesive forces
Stress relief and residual stress management	Shot peening, laser peening	Reduces risk of coating delamination and cracking, enhances overall durability and performance

2.5.1 Characterisation and Evaluation of Surface Roughness and Modifications

Surface energy measurement techniques involve utilising contact angle goniometry to evaluate surface energy and wettability characteristics. This assessment aids in understanding how materials interact with liquids on a surface, which is crucial for optimising wetting and bonding properties. Chemical analysis methods like X-ray photoelectron spectroscopy (XPS), Auger electron spectroscopy (AES), and Fourier-transform infrared spectroscopy (FTIR) are employed to scrutinise surface composition. These techniques unveil the chemical elements and functional groups present, offering insights into surface properties crucial for various applications. Topographical analysis uses tools like atomic force microscopy (AFM) and scanning electron microscopy (SEM) to delve into surface morphology. These techniques provide detailed information on surface roughness, texture, and topographical features, assisting in understanding how these factors influence adhesion and other surface properties. Adhesion testing methodologies involving peel tests, pull-off tests, and scratch tests are pivotal for gauging the efficacy of surface treatments in enhancing adhesion. These tests determine the strength of the bond between coatings and substrates, helping assess the durability and performance of the treated surfaces.

Contact profilometry involves using a stylus that physically traces the surface of a material. The stylus moves across the surface, and its vertical movements are recorded as deviations from a baseline, providing a surface topography profile. This method is widely used due to its simplicity and accuracy in measuring surface roughness on a micro-scale. However, it can be limited by the stylus size, potential surface damage, and slower measurement speeds. Recent advancements have led to improvements in profilometer design, with skidless styluses offering better measurement resolution and reduced dispersion of values compared to skidded styluses.

To enhance measurement accuracy, a new calibration method has been proposed to minimise the error caused by the sphericity of the scanning stylus, ensuring precise surface profile extraction while minimising stylus errors in contact profilometry. Atomic Force Microscopy (AFM) is a powerful surface characterisation technique that provides high-resolution 3D topographic maps with nanometer-scale precision, which scans the surface with a sharp tip attached to a cantilever. As the tip moves over the surface, forces between the tip and the surface cause deflections in the cantilever, which are recorded to create a three-dimensional surface profile. AFM can measure surface features down to the nanometer scale, making it ideal for detailed surface characterisation in nanotechnology and materials science. It provides highly precise topographical maps but can be time-consuming and requires careful sample preparation.

Optical methods, such as laser scanning and white light interferometry, are non-contact techniques that use light to measure surface roughness. Laser scanning projects a laser beam onto the surface, and the reflected light is analysed to determine surface variations. White light interferometry involves splitting a beam of white light, reflecting one part off the surface, and combining it with the other part to create an interference pattern that reveals surface details. As discussed in the research papers, these methods offer speed, large area coverage, and gentle treatment of delicate materials. However, challenges such as optical resolution and surface reflectivity limitations exist. Researchers have proposed innovative algorithms like FFT-LMA-PSI for enhanced accuracy and efficiency in surface recovery, and techniques like laser speckle interferometry for real-time, non-destructive surface morphology monitoring. It is fast, can cover large areas, and avoids physical contact with the surface, making it suitable for delicate or soft materials. However, they can be limited by optical resolution and surface reflectivity.

2.6 Applications of Surface Modification and Surface Treatment

Surface modification and surface treatment are crucial in various industries to enhance materials' and components' performance, longevity, and reliability. These treatments help to protect surfaces from corrosion, wear, and environmental degradation, ensuring the integrity and functionality of critical parts. Surface adhesion is crucial in passivation and treatment processes across various industries. Plasma treatment is widely recognised as an effective method for enhancing surface energy and improving the adhesion properties of materials, particularly polymers [51]. In chip packaging, argon plasma treatment has been shown to remove metal and fluorine residues, thereby improving the adhesion strength of dielectric passivation layers [52]. Surface preparation techniques for structural adhesive joints include degreasing, mechanical abrasion, chemical modification, and physical treatments like plasma and corona [53]. These methods alter surface free energy and introduce polar groups, creating receptive sites for adhesion. The importance of surface treatments extends to biological and microbiological adhesion in healthcare and food manufacturing industries, where techniques such as laser treatment, plasma treatment, and micro/nano printing are employed to modulate bio-adhesion properties [54]. Various methods, such as silicon oxide films, aluminium oxide films, and hydrogen-containing dielectric materials, have been developed to enhance adhesion in surface passivation for silicon solar cells.

Additionally, the surface passivation treatment liquid for hardware stamping parts includes components like inorganic corrosion inhibitors, rare earth salts, organic acids, and activators, contributing to high adhesion capability and good scratch resistance in the passivation film.

Furthermore, in the galvanised steel plate treatment context, the surface passivation process enhances corrosion and rust resistance and ensures strong adhesion of the protective layer, highlighting the importance of adhesion in surface treatments. Surface passivation can significantly improve adhesion and reduce surface contaminants, potentially replacing traditional tinplate in food packaging [55]. This research focuses on applying a Cr^{3+}-based passivation layer and polyethylene terephthalate (PET) film on steel surfaces.

2.6.1 2.6.1. Aerospace Industry

In the aerospace industry, surface passivation and treatment are vital for ensuring the durability and reliability of aircraft components. Aerospace materials are often exposed to harsh environments, including extreme temperatures, high pressure, and corrosive atmospheres. Aluminium and titanium alloys, commonly used in aircraft structures, are passivated to form protective oxide layers that prevent corrosion. Surface treatments such as anodising, hard chromium plating, and thermal spraying enhance the wear resistance of landing gear components, engine parts, and other critical assemblies. Surface treatments improve the adhesion of coatings, paints, and adhesives used in constructing airframes and other components, ensuring structural integrity and aesthetic quality. Shot peening and laser peening are applied to improve the fatigue resistance of high-stress components such as turbine blades and aircraft frames.

Park et al. and Bozdana et al. highlight surface treatment's importance in the aerospace industry. Park et al. emphasise the need for environmentally friendly alternatives to traditional surface treatment technologies, such as non-chromate anodising and sol–gel, which have been shown to deliver comparable or even superior performance [30]. Bozdana, on the other hand, provides a comprehensive review of mechanical surface enhancement (MSE) techniques, commonly used in the aerospace industry to improve components' surface characteristics and service properties [56]. These techniques include shot peening, laser shock peening, and low plasticity burnishing. Both studies underscore surface treatment's critical role in ensuring aerospace components' safety and performance. Coatings and corrosion inhibitors for aluminum-skinned aircraft provide survivability, corrosion protection, and cosmetic appearance. Citric acid can passivate stainless steel for aerospace applications. Gaydos et al. [57] observed the ability of citric acid to passivate stainless steel and suggested that it could eventually displace the nitric acid passivation solutions now employed in the aerospace sector. The aircraft constructors and operators are seeking new protective schemes and treatments to satisfy future environmental requirements.

2.6.2 Automotive Industry

In the automotive industry, surface passivation and treatment significantly enhance vehicles' performance, safety, and longevity. Steel and aluminium components, such as car bodies, chassis, and engine parts, are treated with zinc coatings, phosphate coatings, and e-coating to prevent rust and corrosion. Surface treatments like nitriding, carburising, and hard chrome plating improve the wear resistance of engine components, transmission parts, and brake systems. Electroplating and powder coating are applied to enhance the appearance and durability of exterior and interior parts, including bumpers, trim, and wheels. Thermal barrier coatings are used on exhaust systems and engine components to improve heat resistance and efficiency.

The automotive industry has seen significant advancements in surface treatments and passivation techniques, driven by the need for improved performance and durability. These advancements have been particularly evident in powertrain and engine applications, where traditional treatment-substrate systems are being replaced by more advanced treatments [58]. Heat treatment and surface modification technologies have also played a crucial role in enhancing the properties of automotive components [59]. These technologies can increase efficiency, reduce friction, and save energy and natural resources [60]. The development of innovative technologies for processing working surfaces of automotive parts has further contributed to these advancements.

Vetter et al. showed all vehicle parts needed for surface modification in Fig. 7. Plasma-treated polypropylene showed improved adhesion to a hot melt adhesive in an industrial headlamp application [61].

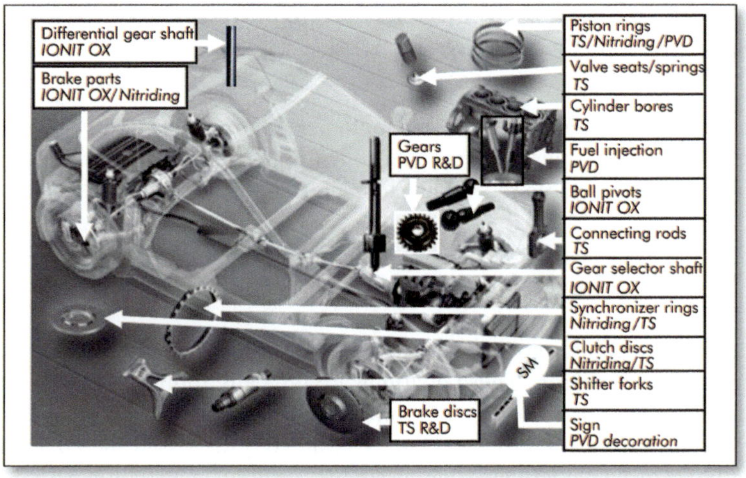

Fig. 7 A range of auto parts treated with various surface treatments [61]

2.6.3 Electronics Industry

Surface passivation and treatment are essential for protecting sensitive electronic components and ensuring their reliable operation in the electronics industry. Silicon wafers used in semiconductor devices are passivated with silicon dioxide or silicon nitride layers to protect against environmental contamination and improve electrical performance. Electronic assemblies are coated with thin layers of protective materials to prevent moisture, dust, and chemical ingress, enhancing reliability and longevity. Printed circuit boards (PCBs) are treated with ENIG to provide a smooth, solderable surface and prevent oxidation of copper traces. Metal connectors and contacts are plated with gold, silver, or nickel to avoid corrosion and to ensure reliable electrical connections. The sacrificial anode system method is commonly used in marine environments to protect ship hulls, offshore structures, and pipelines. The proposed current system method is suitable for large structures, such as underground pipelines, storage tanks, and reinforced concrete structures.

Surface passivation and treatment are crucial in the electronics industry, particularly in developing solar cells and semiconductor devices. Aberle et al. [23] highlight the significance of surface passivation in crystalline silicon solar cells, which has led to improved conversion efficiencies. This is further supported by Viktorovitch et al., who emphasised the importance of surface treatments and dielectric deposition in preventing the adverse effects of surface properties on electronic device performance [62]. Djokić et al. and Ohmi et al. underscore the use of electrochemical methods and specific passivation technologies, such as chromium oxide, in achieving desired surface properties for various applications in the electronics industry [63, 64].

2.7 Recent Advancements in Surface Modification and Surface Treatment

Surface modification and treatment technologies continuously evolve, with recent advancements to enhance performance, environmental sustainability, and cost-effectiveness. Some notable recent developments are listed below.

2.7.1 Nanotechnology in Surface Treatments

Advanced nanocoatings provide superior protection against corrosion, wear, and environmental degradation. These coatings are typically thinner, lighter, and more effective than traditional coatings. Applications span across the aerospace, automotive, and electronics industries. Hamdy et al. [65] discussed the advantages of different types of nanocoatings, such as sol–gel coatings, nanocomposite coatings, and nanoparticle-reinforced coatings, including improved adhesion to substrates, enhanced durability, and the ability to provide multi-functional protection (e.g., corrosion resistance combined with wear resistance). Nanocoatings also allow for thinner layers, which reduces material usage and weight.

Incorporating nano-capsules containing healing agents into coatings allows for automatic repair of microcracks and damage, significantly extending the lifespan of treated surfaces. Mirabedini et al. [66] explored the development and effectiveness of self-healing polymeric coatings that incorporate microcapsules filled with active materials. These coatings are designed to repair damage autonomously, extending the life and performance of the coated materials. The active materials used in the microcapsules include healing agents like epoxy resins, dicyclopentadiene, and other monomers or catalysts. These materials are selected based on their ability to polymerise or react upon release, forming a protective barrier that restores the integrity of the coating. Jolley et al. [67] discussed the microencapsulation process, where self-healing agents are enclosed within tiny capsules dispersed throughout a coating matrix. This technology is designed to release the healing agents when the coating is damaged, enabling autonomous repair and preventing corrosion progression. The study examined different types of self-healing agents, including polymeric materials, inhibitors, and other chemicals that can react to form protective barriers. The choice of agent depends on the specific corrosion environment and the type of substrate being protected. Upon damage to the coating, the microcapsules rupture, releasing the encapsulated healing agents. These agents then interact with the environment or other coating components to form a new protective layer, effectively sealing cracks and preventing corrosive elements from reaching the underlying material.

2.7.2 Advanced Surface Passivation Techniques

Oviroh et al. [68] explored emerging ALD methods and reviewed their diverse applications across multiple industries. In the semiconductor industry, ALD is crucial for fabricating advanced microelectronics, enabling the production of ultra-thin and conformal dielectric layers, barrier films, and high-k materials. In the energy sector, ALD is applied to improve the performance of batteries, solar cells, and fuel cells by enhancing electrode and electrolyte interfaces. Additionally, the ability of ALD to coat complex geometries makes it valuable in the biomedical field for creating biocompatible coatings on implants and sensors. Seiffe et al. [69] discussed optimising the PECVD process parameters, such as gas flow rates, plasma power, and deposition temperature, to achieve the desired stoichiometry and thickness of the SiO_xN_y and SiN_x layers. The silicon-rich SiO_xN_y layer provides excellent chemical passivation by reducing the density of interface states, while the SiN_x layer offers additional passivation through field-effect passivation due to its fixed positive charge. The combination of these layers aims to optimise the overall passivation quality with low surface recombination velocities achieved, leading to enhanced carrier lifetimes.

2.7.3 Environment-Friendly Treatments

Traditional hexavalent chromium-based passivation processes are highly toxic. Yang et al. [70] studied various chromium-free passivation techniques, focusing on environmentally friendly approaches such as organic coatings, inorganic nanocoatings, and rare earth compounds. These methods aimed to provide adequate corrosion resistance while eliminating the use of hazardous substances. The effectiveness of organic coatings, such as those based on carboxylic acids, thiols, and phosphates, was discussed in forming a protective barrier on copper surfaces. Additionally, inorganic nanocoatings, including silica, titania, and ceria, are explored for their ability to enhance corrosion resistance by forming a stable, protective oxide layer. The use of rare earth compounds, such as cerium and lanthanum-based compounds, has shown promise in providing effective passivation. These compounds promote the formation of a protective film on the copper surface, inhibiting corrosion without the environmental risks associated with chromium. Water-based coatings have reduced reliance on volatile organic compounds (VOCs), making surface treatments more environmentally friendly and compliant with stringent regulations. These coatings offer various advantages, such as lower VOC emissions, biocompatibility, and the ability to utilise diverse raw materials, making them ideal for applications in antibacterial coatings, wound dressings, and air filter treatments [71].

2.7.4 Advanced Materials and Coatings

Graphene, known for its exceptional strength and conductivity, is being explored as a surface coating material. Graphene coatings provide excellent corrosion resistance, thermal conductivity, and mechanical strength. Hasan et al. [72] deposited atomically thin graphene layers on single-crystal copper surfaces. With its uniform atomic structure, single-crystal copper is chosen to ensure a high-quality interface with the graphene layer, thereby maximising the effectiveness of the protective coating to prevent corrosive agents like oxygen and water from reaching the copper surface. The impermeability of graphene to most molecules makes it highly effective in protecting the underlying metal from chemical reactions that lead to corrosion. Superhydrophobic and superoleophobic coatings repel water and oil, preventing contamination and corrosion. They have applications in various industries, including aerospace, automotive, and electronics, to clean and dry surfaces. Zhou et al. [73] employed a combination of chemical etching and surface modification techniques to create the superhydrophobic coating with anticorrosion, antibacterial and antiadhesive properties.

2.7.5 Surface Functionalisation

Advanced surface treatments now include the functionalisation of surfaces to impart specific properties such as antibacterial, antifouling, or catalytic activity. Banerjee et al. [74] discussed different stages of fouling, starting with the initial adsorption of proteins onto surfaces, followed by bacterial colonisation, and finally, the attachment of larger organisms such as algae and barnacles. This is particularly important in medical devices, marine applications, and industrial processes. Laser surface texturing techniques create micro- and nanoscale textures on surfaces, enhancing properties like adhesion, friction, and hydrophobicity. Alsaigh et al. [75] reviewed the mechanism of surface modification through ultrashort-pulsed laser texturing. The high peak power of the pulses leads to rapid energy deposition and localised material removal, creating a range of surface features such as grooves, ridges, and dimples. The ability to control the laser parameters allows for the customisation of surface topography to achieve desired properties. This is useful in applications ranging from biomedical implants to industrial tooling.

2.7.6 Enhanced Mechanical Properties

Cold spray technology involves spraying fine metal particles at high velocity to create dense, high-strength coatings. Jibran et al. [76] explored the integration of cold spraying and friction stir processing to enhance the surface properties of aluminium–alumina metal matrix composites. Cold spray technology is used to repair and improve the mechanical properties of aerospace and automotive components. High-velocity oxy-fuel (HVOF) spraying produces high-density and adhesion coatings, improving the wear and corrosion resistance of treated surfaces, and is widely used in the aerospace and energy sectors. Lingappa et al. [77] compared the performance of HVOF and high-velocity air fuel (HVAF) sprayed WC-Co-Cr coatings on 21-4N steel, particularly regarding slurry and corrosion resistance. HVOF coatings exhibited slightly better slurry resistance due to their denser structure, which provides better protection against abrasive particles in slurry environments. HVAF coatings demonstrated superior corrosion resistance due to their lower porosity and reduced oxidation during spraying. This made them more effective in preventing corrosive elements from penetrating the coating.

2.7.7 Innovations in Surface Passivation for Electronics

Developing new dielectric materials for passivating semiconductor devices improves their performance and reliability. Materials such as high-k dielectrics are being used to enhance the capabilities of microelectronic devices. Wallace et al. [78] reviewed various high-κ materials, including hafnium oxide (HfO_2), zirconium oxide (ZrO_2), and aluminium oxide (Al_2O_3), which are considered promising candidates for replacing SiO_2 in MOSFETs. These materials have higher dielectric constants than SiO_2, enabling them to provide the necessary capacitance with a physically thicker layer, thereby reducing leakage currents. Surface modification techniques are being adapted for flexible and stretchable electronics, which require robust protection against environmental factors while maintaining flexibility. High-k gate dielectrics have been explored as key components in FSE, offering advantages over low-k alternatives in thin-film transistor applications.

2.8 Challenges and Future Directions

Enhancing adhesion involves employing surface treatments to foster robust adhesive bonds between coatings and substrates. This not only bolsters adhesion but also aids in addressing environmental and safety concerns. For instance, strategies involve minimising hazardous chemicals like hexavalent chromium in passivation processes and advancing towards eco-friendly and sustainable surface treatment methods. Passivation layers serve as protective barriers against environmental factors. Techniques also evolve to embrace advanced surface treatment technologies, such as nanotechnology innovations that lead to self-healing and smart coatings. Additionally, methods utilising advanced plasma and laser techniques allow for precise modifications of surfaces, enhancing the overall durability against corrosion and wear. Improving wear resistance encompasses applying surface treatments to enhance the hardness and durability of coatings, thereby prolonging their lifespans. These efforts are increasingly integrated with coating processes to optimise performance and efficiency. For example, integrated systems are being developed to combine surface treatment and coating applications, streamlining processes and ensuring better control and automation for enhanced consistency and productivity.

3 3. Conclusion

As industries continue to demand more from materials, the future of surface modification in coating technology is poised for significant advancements. Innovations in nanotechnology, self-healing materials, and environmentally friendly surface treatments are expected to significantly develop next-generation coatings. Moreover, integrating innovative coatings that respond to environmental stimuli will further enhance the functionality and durability of coatings, especially in high-performance applications. Adhesion is a fundamental property and a cornerstone of material performance in coatings. Strong adhesion ensures the longevity and effectiveness of coatings, whether used for protection, aesthetic purposes, or functional enhancements. The ability to control and optimise adhesion through surface modification is crucial for meeting the ever-increasing demands of modern applications. As technology progresses, understanding and improving adhesion will remain central to developing advanced materials and coatings, ensuring they perform reliably under various conditions.

References

1. Chen X, Shan W, Wu D, Patel SB, Cai N, Li, C, Ye S, Li, Z, Hwang S, Zakharov DN, Boscoboinik JA, Wang G, Zhou G (2023) Atomistic mechanisms of water vapor–induced surface passivation. Sci Adv 9:eadh5565. https://doi.org/10.1126/sciadv.adh5565
2. Xia J, Liang C, Gu H, Mei S, Li S, Zhang N, Chen S, Cai Y, Xing G (2021) Surface passivation toward efficient and stable perovskite solar cells. Energy Environ Mater 6:e12296. https://doi.org/10.1002/eem2.12296
3. Ma L, Pascalidou E, Wiame F, Zanna S, Maurice V, Marcus P (2020) Passivation mechanisms and pre-oxidation effects on model surfaces of FeCrNi austenitic stainless steel. Corros Sci 167:108483. https://doi.org/10.1016/j.corsci.2020.108483
4. Bae S, Collins RN, Waite TD, Hanna K (2018) Advances in surface passivation of nanoscale Zerovalent iron: a critical review. Environ Sci Technol 52(21):12010–12025. https://doi.org/10.1021/acs.est.8b01734
5. Yuan S, Wang X, Zhang H, Yuan S (2020) Atomistic insights into oxidation of chemical passivated silicon (100) surface: reactive molecular dynamic simulations. Chin J Chem 38. https://doi.org/10.1002/cjoc.202000476
6. Choi CH, Liu D, Evans JW, Gordon MS (2002) Passive and active oxidation of Si(100) by atomic oxygen: a theoretical study of possible reaction mechanisms. J Am Chem Soc 124(29):8730–8740. https://doi.org/10.1021/ja012454h
7. Chen X, Liu Z, Wu D, Cai N, Sun X, Zakharov DN, Hwang S, Su D, Wang G, Zhou G (2022) Passive oxide film growth observed on the atomic scale. Adv Mater Interfaces 9(11):2102487. https://doi.org/10.1002/admi.202102487

8. Van de Loo BWH, Macco B, Melskens J, Beyer W, Kessels WMM (2019) Silicon surface passivation by transparent conductive zinc oxide. J Appl Phys 125:105305. https://doi.org/10.1063/1.5054166
9. Chembath M, Balaraju J, Sujata M (2015) Surface characteristics, corrosion and bioactivity of chemically treated biomedical grade NiTi alloy. Mater Sci Eng C 56:417–425. https://doi.org/10.1016/j.msec.2015.06.051
10. Baron A, Simka W, Nawrat G, Szewieczek D (2008) Electropolishing and chemical passivation of austenitic steel. J Achiev Mater Manuf Eng 31:197–202
11. Bijimi D, Gabe DR (1983) Passivation studies using group VIA anions III anodic treatment of zinc. Br Corros J 18(3):138–141. https://doi.org/10.1179/000705983798273741
12. Mansfeld F, Lin SH, Kim S, Shih H (1989) Corrosion protection of Al alloys and Al-based metal matrix composites by chemical passivation. Corrosion 45(8):615–631. https://doi.org/10.5006/1.3579314
13. Callen BW, Lowenberg B, Lugowski S, Sodhi RN, Davies JE (1995) Nitric acid passivation of Ti6Al4V reduces thickness of surface oxide layer and increases trace element release. J Biomed Mater Res 29(3):279–290. https://doi.org/10.1002/jbm.820290302
14. O'Laoire C, Timmins B, Kremer LV, Holmes JD, Morris MA (2006) Analysis of the acid passivation of stainless steel. Anal Lett 39:2255–2271. https://doi.org/10.1080/00032710600755363
15. Yue YY, Liu C, Jiang M (2022) Evolution of passive film on 304 stainless steel during nitric acid passivation. Steel Res Int 93(8):2200026. https://doi.org/10.1002/srin.202200026
16. Grant NE, McIntosh KR (2009) Passivation of a (100) silicon surface by silicon dioxide grown in nitric acid. IEEE Electron Device Lett 30:922–924. https://doi.org/10.1109/LED.2009.2025898
17. Padhy N, Kamal S, Chandra R, Mudali UK, Raj B (2010) Corrosion performance of TiO_2 coated type 304L stainless steel in nitric acid medium. Surf Coat Technol 204:2782–2788. https://doi.org/10.1016/j.surfcoat.2010.02.047
18. Verdeguer P, Gil J, Punset M, Manero JM, Nart J, Vilarrasa J, Rúperez E (2022) Citric acid in the passivation of titanium dental implants: corrosion resistance and bactericide behavior. Materials 15(2):545. https://doi.org/10.3390/ma15020545
19. Costa AT, Oliveira MC, Antunes RA (2020) Interplay between the composition of the passive film and the corrosion resistance of citric acid-passivated AISI 316L stainless steel. Surf Interface Anal 53:374–384. https://doi.org/10.1002/sia.6927
20. Enos D, Wichhart D (2022) Refinement of citric acid passivation methods and comparison to traditional nitric acid based chemistries. In: AMPP CORROSION:D051S054R001
21. Conrad JR, Radtke J, Dodd RA, Worzala FJ, Tran NC (1987) Plasma source ion-implantation technique for surface modification of materials. J Appl Phys 62:4591–4596. https://doi.org/10.1063/1.339055
22. Yang Y, Tay B, Sun XW, Sze JY, Han ZJ, Wang J, Zhang X, Li Y, Zhang S (2007) Quenching of surface-exciton emission from ZnO nanocombs by plasma immersion ion implantation. Appl Phys Lett 91:071921. https://doi.org/10.1063/1.2772668
23. Aberle AG (2000) Surface passivation of crystalline silicon solar cells: a review. Prog Photovolt: Res Appl 8:473–487. https://doi.org/10.1002/1099-159X(200009/10)8:5%3c473::AID-PIP337%3e3.0.CO;2-D
24. Singh R, Vandana PJ, Singh PK (2016) Plasma assisted atomic layer deposited hafnium oxide films for silicon surface passivation. RSC Adv 6:97720–97727. https://doi.org/10.1039/C6RA19442G
25. Leng YX, Chen JY, Yang P, Sun H, Huang N (2003) Structure and properties of passivating titanium oxide films fabricated by DC plasma oxidation. Surf Coat Technol 166(2–3):176–182. https://doi.org/10.1016/S0257-8972(02)00780-6
26. Tang P, Liu C, Zhang J, Wu L, Li W, Feng L, Zeng G, Wang W (2018) Surface modification effects of fluorine-doped tin dioxide by oxygen plasma ion implantation. Appl Surf Sci 436:134–140. https://doi.org/10.1016/j.apsusc.2017.11.117

27. Vijay V, Arjun K, Kumar H, Kumar A, Chandrasekar P, Balusamy V (2007) Review of various surface treatment techniques on titanium alloys and their protective effects against corrosion. J Surf Sci Technol 23:49–58. https://doi.org/10.18311/JSST/2007/1955
28. Singh S, Kumar S, Khanna V (2023) A review on surface modification techniques. Mater Today: Proc 2023. https://doi.org/10.1016/j.matpr.2023.01.010
29. Bhushan B, Gupta BK (1991) Handbook of tribology: materials, coatings, and surface treatments
30. Park SY, Choi WJ, Choi HS, Kwon H, Kim SH (2010) Recent trends in surface treatment technologies for airframe adhesive bonding processing: a review (1995–2008). J Adhes 86(2):192–221. https://doi.org/10.1080/00218460903418345
31. Galceran R, Gaufres E, Loiseau A, Piquemal-Banci M, Godel F, Vecchiola A, Bezencenet O, Martin MB, Servet B, Petroff F, Dlubak B, Seneor P (2017) Stabilising ultra-thin black phosphorus with in-situ-grown 1 nm-Al_2O_3 barrier. Appl Phys Lett 111:243101. https://doi.org/10.1063/1.5008484
32. Wang CC, Hsu YC, Su FC, Lu SC, Lee TM (2009) Effects of passivation treatments on titanium alloy with nanometric scale roughness and induced changes in fibroblast initial adhesion evaluated by a cytodetacher. J Biomed Mater Res A 88A(2):370–383. https://doi.org/10.1002/jbm.a.31604
33. Strnad G, Chirilă N, Jakab-Farkas L (2015) Effect of surface preparation and passivation treatment on surface topography of Ti6Al4V for dental implants. AMM 809–810:513–518. https://doi.org/10.4028/www.scientific.net/AMM.809-810.513
34. Barboza AL, Kang KW, Bonetto RD, Llorente CL, Bilmes PD, Gervasi CA (2014) Blasting and passivation treatments for ASTM F139 stainless steel for biomedical applications: effects on surface roughness, hardening, and localized corrosion. J Mater Eng Perform 24(1):175–184. https://doi.org/10.1007/s11665-014-1300-5
35. Oh S, Kim D, Kim KC, Kim D, Chung W, Shin B (2023) The effect of surface roughness on re-passivation and pitting corrosion of super duplex stainless steel UNS S 32760. Int J Electrochem Sci 18(12):100351. https://doi.org/10.1016/j.ijoes.2023.100351
36. Park S, Kim G, Oh C, Yoo M, Lee K, Lee K (2021) Au-Less, Ni-Less & roughness-less PCB Cu surface treatment using All-in-One Al_2O_3 passivation. In: 2021 IEEE 71st electronic components and technology conference (ECTC):625–634
37. Chen J (2021) Effect of surface roughness on the characteristics of passive film formed on 5083 Al alloy in pH 8.4 Borate Buffer Solution. Int J Electrochem Sci 16(11):211122. https://doi.org/10.20964/2021.11.02
38. Sunarso S, Putra RJ, Theodorea CF, Pangesty AI (2022) Effect of hydrothermal treatment of titanium in high concentration of $AgNO_3$ solution on surface morphology and roughness. F1000Res 11:221. https://doi.org/10.12688/f1000research.79542.1
39. Jemat A, Ghazali MJ, Razali M, Otsuka Y (2015) Surface modifications and their effects on titanium dental implants. BioMed Res Int 2015(1):791725. https://doi.org/10.1155/2015/791725
40. De Wolf S, Agostinelli G, Beaucarne G, Vitanov P (2005) Influence of stoichiometry of direct plasma-enhanced chemical vapor deposited SiN_x films and silicon substrate surface roughness on surface passivation. J Appl Phys 97(6):063303. https://doi.org/10.1063/1.1861138
41. Rousière O, Lemoine D, Quémerais A, Assi CK, Granger R, Triboulet R (1998) Surface properties of passivated $Hg_{0.8}Zn_{0.2}Te$. Semiconductor Sci Technol 13(6):622–629. https://doi.org/10.1088/0268-1242/13/6/014
42. Madhuri A, Jena S, Gupta M, Swain BP (2023) Nitrogen flow rate dependent atomic coordination, phonon vibration and surface analysis of DC magnetron sputtered nitrogen rich-AlN thin films. Physica B 666:415141. https://doi.org/10.1016/j.physb.2023.415141
43. Madhuri A, Jena S, Gupta M (2023) Swain BP (2023) Tweaking the mechanical stability of $Al_{1-x}W_xN$ ternary nitride alloys for hard coating applications, and study on their structural, photoluminescence and surface chemical analysis. J Alloys Compd 968:171854. https://doi.org/10.1016/j.jallcom.2023.171854

44. Zhang Z, Zhao J, Yan X (2024) Mechanically Interlocked Polymers with Dense Mechanical Bonds. Acc Chem Res 57(6):992–1006. https://doi.org/10.1021/acs.accounts.4c00006
45. Mandolfino C, Lertora E, Gambaro C (2013) Effect of surface Pretreatment on the performance of adhesive-bonded joints. Key Eng Mater 554–557:996–1006. https://doi.org/10.4028/www.scientific.net/KEM.554-557.996
46. Rudawska A (2023) Mechanical surface treatment of Adherends for adhesive bonding. Progress in Adhesion and Adhesives 7:113–169. https://doi.org/10.1002/9781394198375.ch3
47. Friedrich JF (2017) Metal-polymer systems: interface design and chemical bonding. John Wiley & Sons
48. Wager JF (2024) Chemical bonding with or without tunneling. ACS Omega 9(4):5046–5050. https://doi.org/10.1021/acsomega.3c09577
49. Jarvis EA, Carter EA (2003) Exploiting covalency to enhance metal-oxide and oxide-oxide adhesion at heterogeneous interfaces. J Am Ceram Soc 86(3):373–386. https://doi.org/10.1111/j.1151-2916.2003.tb03309.x
50. Li S, Zhao Y, Wan H, Lin J, Min J (2023) Molecular understanding of the interfacial interaction and corrosion resistance between epoxy adhesive and metallic oxides on galvanised steel. Materials 16(8):3061. https://doi.org/10.3390/ma16083061
51. Wolf RA, Sparavigna AC (2010) Role of plasma surface treatments on wetting and adhesion. Engineering 2:397–402. https://doi.org/10.4236/eng.2010.26052
52. Suyama T, Suga K, Igarashi T, Fujimori N (2017) Influence of surface properties on adhesion strength of dielectric passivation in chip size packaging. In: 2017 international conference on electronics packaging (ICEP), 513–516. https://doi.org/10.23919/ICEP.2017.7939435
53. Purabgola A, Rastogi S, Sharma G, Kandasubramanian B (2020) Surface preparation for structural adhesive joints. In: Structural adhesive joints: design, analysis and testing, 1–34. https://doi.org/10.1002/9781119737322.ch1
54. Waugh DG, Toccaceli C, Gillett AR, Ng CH, Hodgson SD, Lawrence J (2016) Surface treatments to modulate bioadhesion: a critical review. Rev Adhes Adhes 4:1
55. Zumelzu E, de Melo HG, di Sarli AR, Tomachuk CR (2018) Effect of passivation treatment on adhesion and protective properties of steel coated with polymeric film. Mater Sci Forum 930:422–427. https://doi.org/10.4028/www.scientific.net/MSF.930.422
56. Bozdana AT (2005) On the mechanical surface enhancement techniques in aerospace industry—a review of technology. Aircr Eng Aerosp Technol 77(4):279–292. https://doi.org/10.1108/00022660510606349
57. Gaydos SP (2003) Passivation of aerospace stainless parts with citric acid solutions. Plat Surf Finish 90(3):20–25
58. Funatani K (2006) Emerging technology in heat treatment and surface engineering of automotive components
59. Suslov AG (2014) Functionally-oriented technologies of working surfaces treatment for automotive parts. Izvestiya MGTU MAMI 8(1–2):107–109. https://doi.org/10.17816/2074-0530-67786
60. Ben Salem D, Pappas DD, Buske M (2023) Atmospheric plasma activation and thin film deposition processes for adhesion improvement and corrosion resistance in the automotive industry. Plasma Process Polym 20(5):2200211. https://doi.org/10.1002/ppap.202200211
61. Vetter J, Crummenauer J, Barbezat G, Avissar J (2007) Surface treatments of automotive parts—research and applications. Vak Forsch Prax 19(1):18–23. https://doi.org/10.1002/vipr.200700311
62. Viktorovitch P (1990) Passivation des semiconducteurs III-V. Revue Phys Appl 25(9):895–914. https://doi.org/10.1051/rphysap:01990002509089500
63. Djokić SS (2012) Biomedical applications. Mod Asp Electrochem. https://doi.org/10.1007/978-1-4614-3125-1
64. Ohmi T, Ohki A, Nakamura M, Kawada K, Watanabe T, Nakagawa Y, Miyoshi S, Takahashi S, Chen MS (1993) The technology of chromium oxide passivation on stainless steel surface. J Electrochem Soc 140:1691–1699. https://doi.org/10.1149/1.2221625

65. Hamdy AS (2010) Corrosion protection performance via nanocoatings technologies. Recent PatS Mater Sci 3(3):258–267
66. Mirabedini SM, Alizadegan F (2020) Chapter 9–self-healing polymeric coatings containing microcapsules filled with active materials. Self-Heal Polym-Based Syst 2020:235–258. https://doi.org/10.1016/B978-0-12-818450-9.00009-X
67. Jolley ST, Li W, Buhrow JW, Calle LM (2011) Microencapsulation of self-healing agents for corrosion control coatings. In: 3rd international conference on self-healing materials bath, UK
68. Oviroh PO, Akbarzadeh R, Pan D, Coetzee RA, Jen TC (2019) New development of atomic layer deposition: processes, methods and applications. Sci Technol Adv Mater 20(1):465–496. https://doi.org/10.1080/14686996.2019.1599694
69. Seiffe J, Gautero L, Hofmann M, Rentsch J, Preu R, Weber S, Eichel R (2011) Surface passivation of crystalline silicon by plasma-enhanced chemical vapor deposition double layers of silicon-rich silicon oxynitride and silicon nitride. J Appl Phys 109:034105. https://doi.org/10.1063/1.3544421
70. Yang T, Chen W, Li X, Song J, Dong L, Fu YQ (2021) Environment-friendly and chromium-free passivation of copper and its alloys. Mater Today Commun 29:102826. https://doi.org/10.1016/j.mtcomm.2021.102826
71. Bertheau E, Simon V, Raynaud CD (2024) Emissions of volatile organic compounds (VOCs) as safety indicators in the development of wood-based binderless boards. Appl Sci 14(3):1266. https://doi.org/10.3390/app14031266
72. Hasan MM, Devadig R, Sigdel P, Lipatov A, Avci R, Jasthi BK, Gadhamshetty V (2024) Corrosion resistance of atomically thin graphene coatings on single crystal copper. Coatings 14(6):656. https://doi.org/10.3390/coatings14060656
73. Zhou W, Yang F, Yuan L, Diao Y, Jiang O, Pu Y, Zhang Y, Zhao Y, Wang D (2022) Construction of superhydrophobic coating on iron surface with enhanced anti-corrosion. Anti-Adhes Andantibact Prop Mater 15(23):8634. https://doi.org/10.3390/ma15238634
74. Banerjee I, Pangule RC, Kane RS (2011) Antifouling coatings: recent developments in the design of surfaces that prevent fouling by proteins, bacteria, and marine organisms. Adv Mater 23:690–718. https://doi.org/10.1002/adma.201001215
75. Alsaigh RA (2024) Enhancement of surface properties using ultrashort-pulsed-laser texturing: a review. Crystals 14(4):353. https://doi.org/10.3390/cryst14040353
76. Jibran W, Wanjara P, Baradari JG, Jarligo MO, McDonald A (2023) Localised surface modification of high-strength aluminium–alumina metal matrix composite coatings using cold spraying and friction stir processing. Adv Appl Ceram 122(3–4):181–196. https://doi.org/10.1080/17436753.2023.2227446
77. Lingappa R, Mamatha TG, Prabhuswamy MS, Singh V, Bansal A (2024) High-velocity oxy-fuel and high-velocity air fuel sprayed WC-Co-Cr coatings on solution-treated 21–4N steel for slurry and corrosion resistance. J Electrochem Sci Eng. https://doi.org/10.5599/jese.2246
78. Wallace RM, Wilk GD (2003) High-κ dielectric materials for microelectronics. Crit Rev Solid State Mater Sci 28(4):231–285. https://doi.org/10.1080/714037708

Chapter 5
Mechanism of the Wettability of Coating Materials

Aishwarya Madhuri, Siddharth Parthasarathy, and Bibhu Prasad Swain

Abstract The exploration of the wettability mechanisms in coating materials aims to elucidate the critical factors and underlying principles that govern this important property. By examining how wettability influences the performance and application of coatings, we can better understand the strategies to optimise and innovate in this dynamic field. Understanding the mechanism of wettability is crucial for the design, application, and performance of coating materials. It directly affects adhesion, durability, corrosion resistance, aesthetic qualities, ease of application, and environmental impact. This chapter looks into the wettability mechanism, surface wetting characteristics, adhesion effect, surface energy, surface roughness, chemical compatibility, surface modification, wettability in different applications, and the relevance of wettability. By controlling wettability, manufacturers can enhance the functionality and longevity of coatings, ensuring they meet the specific demands of various applications. Extremely wetting materials, like superhydrophilic and superhydrophobic surfaces, have garnered a lot of interest due to their potential applications in a variety of fields, including drag-reduction systems, anti-corrosive coatings, anti-fog windows, self-cleaning textiles, and effective water transportation.

Keywords Wettability · Surface energy · Surface roughness · Water contact angle · Hydrophilicity · Hydrophobicity

A. Madhuri (✉) · B. P. Swain
Department of Physics, National Institute of Technology Manipur, Langol, Imphal West, Manipur, India
e-mail: aishwaryamadhuri99@gmail.com

S. Parthasarathy
Department of Physics, Centurion University of Technology and Management, Bhubaneswar, Odisha, India

B. P. Swain (ed.), *Advances in Mechanical Coating*, Materials Horizons: From Nature to Nanomaterials, https://doi.org/10.1007/978-981-96-7484-8_5

1 Introduction

The wettability of coating materials is a fundamental property that influences their performance, application, and longevity across a wide range of industries, including automotive, aerospace, construction, electronics, and biomedical fields [1]. Wettability refers to the ability of a liquid to maintain contact with a solid surface, dictated by the balance of adhesive forces between the liquid and the solid and cohesive forces within the liquid itself [2]. This property determines how well a coating material can spread, adhere, and form a uniform layer over a substrate. Understanding the mechanism of wettability involves delving into the interactions at the molecular level between the coating material and the substrate. Factors such as surface energy, surface roughness, chemical composition, and environmental conditions play significant roles in influencing these interactions [3]. For instance, a high-energy surface will attract the liquid coating more effectively, resulting in better wetting and adhesion. Conversely, low-energy surfaces may repel the liquid, leading to poor wetting and potential defects in the coating.

In the realm of coating applications, optimal wettability is essential for achieving desired properties such as corrosion resistance, durability, aesthetic appeal, and functional performance. Properly wetted coatings can form robust barriers against environmental aggressors, enhance products' visual and tactile qualities, and impart specialised functionalities such as hydrophobicity or biocompatibility. Moreover, controlling wettability is vital for efficient application processes, reducing waste and ensuring consistent coverage with minimal effort. The mechanisms driving wettability also have significant implications for developing advanced coatings. Innovations in surface modification techniques, such as plasma treatments and nano-structuring, enable precise control over wettability, paving the way for next-generation coatings with tailored properties. Understanding and manipulating these mechanisms allows for the design of coatings that meet stringent requirements in demanding environments, from self-cleaning surfaces to protective layers in harsh industrial settings. Figure 1 shows the histogram chart of publications in the field of wettability. An exponential increase in research papers on engineered surfaces for wettability qualities can be attributed to the enormous interest in this field.

2 Mechanism of Wettability

Wettability is the capacity of a liquid to stay in contact with a solid surface when they are brought together and is the product of intermolecular interactions. The balance between cohesive forces inside the liquid and adhesive forces between the liquid and the solid surface determines the degree of wettability. Understanding wettability involves examining the mechanisms and factors that influence these forces.

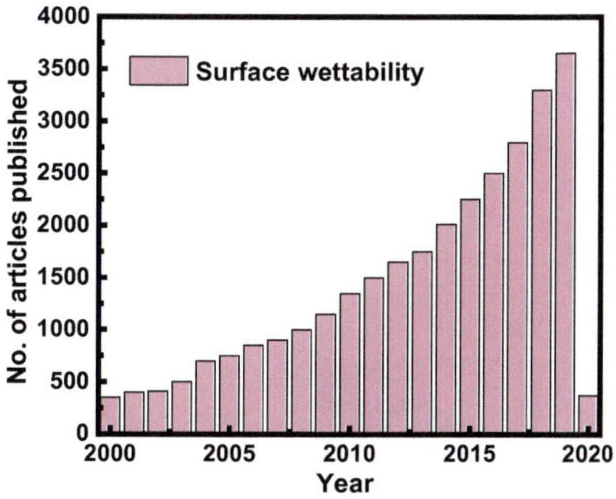

Fig. 1 Histogram chart of publications in the field of wettability

3 Properties of Surface Wetting

A macroscopic depiction of the interaction between the liquid and the solid substrate material is the solid surface's wettability. Measuring the contact angle (CA) of a sessile liquid drop on a solid surface in air is the most popular method for estimating surface wettability (Fig. 2a) [4]. Another frequently employed method is measuring the CA of a captive bubble (Fig. 2b). The advancing CA, denoted as θ_A (Fig. 2c), and the receding CA, denoted as θ_R (Fig. 2d), are the maximum and minimum CA of a liquid droplet with a given solid–liquid boundary, respectively. CA hysteresis (CAH) is typically defined as the difference between θ_A and θ_R (i.e., $\theta_A - \theta_R$). Because of pinning, CAH only occurs on surfaces with morphological or chemical heterogeneities. An alternative measure of the mobility of a liquid drop on a surface, θ_S, is defined as the tilting angle of a flat solid surface at which the liquid drop begins to slide (Fig. 2e). In this context, the frontier CA and the tail CA can also be used to determine the advancing and retreating CA (θ_A and θ_R). A surface with a high CA and a low CAH or θS is generally regarded as superhydrophobic or superoleophobic (super-repellent) (Fig. 2f).

Nevertheless, there are systems in place where these two conditions are not satisfied simultaneously. To prevent sliding or slipping off, a liquid drop with high CA and CAH can bead up on a vertical surface or even a reverse horizontal surface (Fig. 2g). Conversely, a system can have low CAH and low CA, as demonstrated by a highly slick surface to liquids with extremely low CA6 (Fig. 2h). The primary factors influencing the wettability modifications are surface free energy, roughness, chemical composition, porosity, surface tension, surface chemistry, and surface charge.

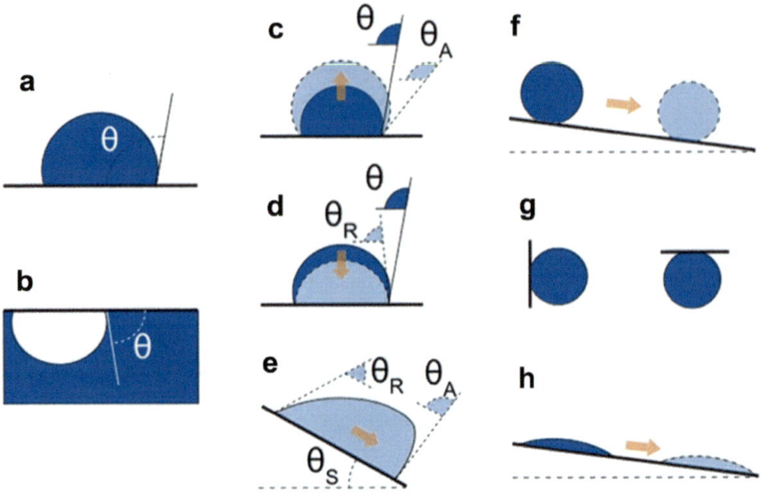

Fig. 2 **a** The contact angle (CA), denoted as θ, defined in terms of a sessile liquid drop on a solid surface. **b** CA defined using a captive bubble as an example. **c** The maximum CA of a sessile drop before its boundary enlarges with an increase in drop volume known as the advancing CA, or $\theta_A(>\theta)$. **d** Receding CA, $\theta_R(<\theta)$: when the drop volume decreases, the minimum CA of a sessile drop before its boundary contracts. **e** The sliding angle, or θ_S. **f** A super-repellent surface with ultralow θ_S and ultrahigh CA. **g** An extremely hydrophobic yet adhesive surface with very high θ_S and CA. A liquid drop may, in extreme circumstances, adhere to this type of vertical or even inverted horizontal surface. **h** Surface that is lyophilic but non-sticky due to extremely low CA and θ_S; liquid does not adhere to the surface or bead up [4]

3.1 Adhesion

Adhesion refers to the tendency of dissimilar particles or surfaces to cling to one another. In the context of coating materials, it describes the attraction between the liquid phase (e.g., water, oil) and the solid surface. The surface's wettability largely depends on the type and strength of this attraction. The attraction between molecules of the liquid and the solid surface is the adhesive force, whereas the cohesive forces are the attraction between molecules within the liquid. Strong adhesive forces promote wetting and strong cohesive forces can resist wetting if the adhesive forces are weaker in comparison. Good wetting properties can enhance the durability and lifespan of the coating by preventing peeling, cracking, and delamination. Adhesive forces are the interactions that occur at the interface between two different materials. These forces can vary widely in strength and origin. The main types include: (a) Van der Waals Forces: Weak, short-range forces arising from induced electrical interactions between molecules or atoms. (b) Hydrogen Bonding: Stronger than van der Waals forces, these arise from the attraction between a hydrogen atom, covalently bonded to a highly electronegative atom, and another electronegative atom. (c) Electrostatic Forces: Result from the attraction between oppositely charged ions or molecules. (d) Chemical Bonds: Strong interactions involving the sharing or transfer of electrons between atoms, such as covalent or ionic bonds. The relationship between adhesion and wettability is complex and influenced by several factors. Surfaces with high surface energy (e.g., metals, oxides) tend to have strong adhesive forces with liquids, resulting in low contact angles (high wettability). Surfaces with low surface energy (e.g., polymers, fluorinated compounds) exhibit weak adhesion, leading to high contact angles (low wettability). Polar surfaces attract polar liquids during liquid–solid interactions, while non-polar surfaces attract non-polar liquids.

Applying low-energy materials (e.g., silanes and fluoropolymers) reduces adhesion, resulting in superhydrophobic surfaces. Dong et al. [5] presented a method for fabricating superhydrophobic surfaces with micro- and nanoscale hierarchical structures on a Ti substrate. The research focused on achieving a surface with remarkable water repellency and durability, which can have significant applications in self-cleaning, anti-corrosion, and biomedical implants. The titanium substrate was first chemically etched to provide micro-scale roughness, and then it was modified with a low-surface-energy substance to increase hydrophobicity. The water contact angle measurements showed that the fabricated surface exhibited superhydrophobic properties, with contact angles exceeding 150° and low sliding angles. Wu et al. [6] introduced a novel anodisation process using a non-traditional electrolyte and specific anodisation conditions to form dense forests of nanowires on aluminium substrates. The nanoscale roughness and low surface energy materials contribute to the super-repellent properties. The hierarchical structure of the nanowire forests enhances the air pocket formation, reducing the contact area between the liquid and the surface. The surface maintains a Cassie-Baxter wetting state, where liquid droplets rest on top of the nanowires, minimising the adhesion. Hydrophilic treatments such as plasma or surface functionalisation increase surface polarity and adhesion, enhancing wettability. Certain biological surfaces, for example, rose petals and gecko toes, have a static contact angle greater than 150°, which allows the liquid droplets to be firmly stuck there, compared to superhydrophobic surfaces (such as lotus leaves and water strider's legs) with minimum adhesion.

Superhydrophobic surfaces are the combination of low-surface-energy materials and hierarchical micro/nanostructures that create air pockets, leading to high contact angles and low adhesion to water used as self-cleaning surfaces, water-repellent clothing, anti-corrosion coatings, and microfluidic devices. Superhydrophobic surfaces are high surface energy materials and roughness that promote rapid water spreading, resulting in low contact angles, used for anti-fogging coatings, oil–water separation membranes and biomedical devices for enhanced cell adhesion. Superoleophobic surfaces exhibit high contact angles with oils due to micro/nanostructures and low-surface-energy materials, preventing oil adhesion for anti-fouling coatings, oil-resistant fabrics, and protective coatings for electronics. Superoleophilic surfaces have a high affinity for oils due to surface chemistry and structure, leading to low contact angles with oils used for oil spill recovery, oil–water separation, and filtration systems. Lai et al. [7] used a straightforward electrochemical process to vary the diameter and length of the nanotubes to create nanostructured superhydrophobic TiO_2 films with adjustable surface adhesion. Surface roughness and capillary force were used to engineer the superhydrophobic porous nanostructures' water adhesive force. An efficient sol–gel method that involved first the surface chemical reaction of 3-Glycidyloxypropyl trimethoxysilane (A-187) and SiO_2 particles, followed by spray-coating onto the substrate, was used to successfully fabricate a novel superhydrophobic coating by Li et al. [8]. Increased surface fluorination and hardening processes made it possible for the SiO_2 coating, which had an ideal mass ratio of 2:1, to show excellent superhydrophobic characteristics and strong substrate adhesion. Our findings suggest that the mass ratio of SiO_2 and A-187 particles may play a major role in regulating the wettability and adhesion force of the superhydrophobic coating to substrates, as well as surface shape. Hydrophilic treatment (e.g., plasma treatment) increases the surface polarity, enhancing adhesion and wettability. Applying low-energy coatings (e.g., fluoropolymers) reduces adhesion, leading to superhydrophobicity.

3.2 Surface Energy

Surfaces with high surface energy, such as metals and ceramics, strongly attract liquids. When a coating material wets a high-energy surface, it spreads out and forms a thin, uniform layer that maximises contact with the substrate. This leads to better adhesion because the adhesive forces between the coating and the substrate are strong. Conversely, low-energy surfaces repel liquids like certain polymers and hydrophobic coatings. Poor wetting results in high contact angles, meaning the liquid forms droplets rather than spreading out. This limits the contact area and weakens the adhesive forces, leading to poor adhesion. The high specific surface area of micro- and nanostructured materials may impact their superwetting and super-antiwetting characteristics. The water droplet hovers over the surface due to surface alteration caused by the binding of low surface energy chemicals. This indicates that the surface's capillary action has failed, rendering it incapable of absorbing liquid.

On the other hand, creating micro and nanostructures such as spherical particles, tubes, rods, wires, and fibres increases surface free energy and provides a strong capillary force for liquid invasion [9, 10].

Furthermore, the liquid's altered surface tension and surface free energy contribute to the increase in the capacity of liquid droplets to disperse across the surface. A higher spread capacity indicates a more remarkable ability of different fluids to adhere to one another. Figure 3 shows that high surface tension and low surface energy result in hydrophobicity, whereas high surface energy and low surface tension contribute to hydrophilicity. By creating nanostructures with high surface energy materials, superhydrophilicity can be enhanced [11]. Similarly, hydrophobicity can be achieved by covering the nanostructure with a material with a low surface energy [12]. Yu et al. [13] have investigated the role of nanostructures and material surface energy in adjusting wettability. Liu et al. [14] examined the wettability of plasma-treated polyacrylonitrile fibres and found that increased plasma treatment time leads to better wettability, increasing surface energy from 27.2 to 65.1 mN/m. The polar component

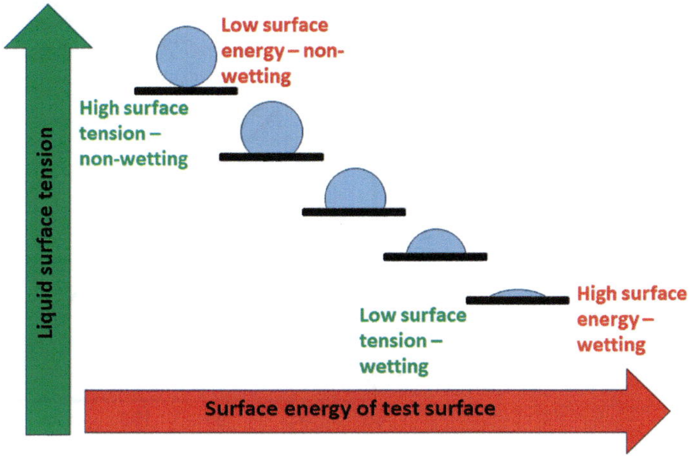

Fig. 3 Effect of surface energy and surface tension on wettability modification

of the test liquid affects the changes in wettability results. Ostrovskaya et al. studied the wettability of diamond films [15]. The water contact angle (WCA) of the surface treated with hydrogen plasma is 93°, while that of the oxidised diamond films is 32°. The surface energy of hydrogenated and oxidised diamond film increases from 47 to 65.8 mJ/m^2, which is the cause of this augmentation. Wang et al. [16] studied how changes in surface energy affect wettability in graphene films. Graphene oxide sheets were synthesised using Brode's method and reduced with hydrazine. WCA on graphite, graphene oxide, and graphene thin films were observed to be 98.3°, 67.4°, and 127°, respectively. Figure 4 shows the data available for graphene oxide, graphite and graphene thin films regarding surface energy, contact angle, and work of adhesion. Since surfaces with very high WCA are usually exclusively made of highly textured materials with low surface-free energy, fabricating such surfaces is not difficult. Other environmentally significant uses for interfacial materials with unique wettability include preventing corrosion [17], reducing drag [18], and catalysing surface reactions [19].

3.3 *Surface Roughness*

Surface roughness can enhance or reduce wettability depending on the intrinsic wettability of the material. In hydrophilic surfaces, roughness increases wettability by creating more surface area for the liquid to adhere to (Wenzel model). In contrast, in hydrophobic surfaces, roughness can trap air and lead to superhydrophobicity, where water droplets roll off easily (Cassie-Baxter model). Two conceivable states for a sessile liquid drop in air that comes into contact with a rough surface are the Wenzel and Cassie-Baxter states. When the liquid is in the Wenzel state, the

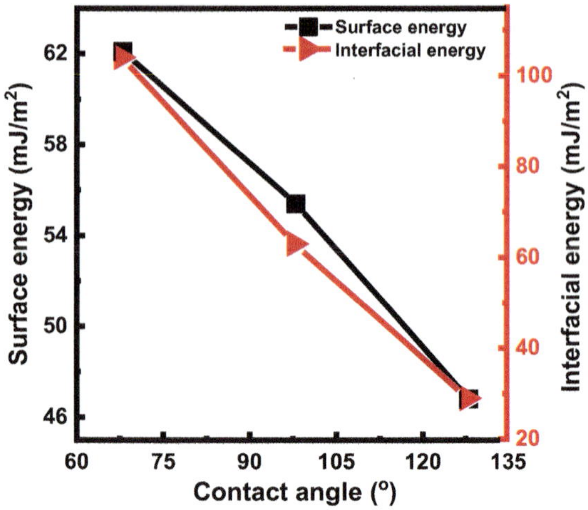

Fig. 4 Effect of surface energy on wetting ability

textured surface is entirely saturated, maximising the liquid–solid contact area. The intrinsic contact angle, θ_0, and the apparent contact angle for the Wenzel state, θ_W, are connected.

$$cos\,\theta_W = r\ cos\ \theta_0 \tag{1}$$

The surface roughness, denoted by r in this instance, is the ratio of the entire surface area to the projected surface area, and is always greater than unity. Equation (1) states that if the system is in a Wenzel state, the surface roughness will always increase a surface's intrinsic wettability [20], regardless of whether it is hydrophilic ($\cos\theta_0 > 0$) or hydrophobic ($\cos\theta_0 < 0$). For example, a Wenzel state can produce an ultrahigh CA if $\theta 0 > 90°$ and the surface roughness is high enough. However, pinning causes the liquid droplets in a Wenzel state system to be highly stationary, which results in significant contact angle hysteresis (CAH). The Cassie-Baxter hypothesis was developed for a chemically heterogeneous surface with a sessile drop supported by an air–solid composite surface (Fig. 5a, b). A rough surface can trap air in "pockets" between the liquid and solid substrate, allowing the Cassie-Baxter hypothesis to apply, where air is considered as the ultralow surface tension "chemical heterogeneity". In the Cassise-Baxter state, the apparent CA (θ_{CB}) can be related to the intrinsic CA (θ_0) as Eq. (2);

$$cos\ \theta_{CB} = f_{SL}\ cos\ \theta_0 - f_{LV} \tag{2}$$

The area fractions of the solid–liquid and liquid–air (vapor) interfaces, f_{SL} *and* f_{LV}, meet the equation $f_{SL} + f_{LV} = 1$. As part of the composite surface, air is the most hydrophobic and oleophobic substance, having an intrinsic CA of 180° for any liquid. Therefore, θ_{CB} is always larger than θ_0. Even if θ_0 is less than 90°, θ_{CB} may still be much higher than 90°. Hydrophobic or oleophobic rough surfaces can be produced in the Cassie-Baxter state but not in the Wenzel state by materials possessing hydrophilic or oleophilic smooth surfaces. Weak CAH, resulting from a shallow degree of pinning, is an essential characteristic of a system in the Cassie-Baxter state [21].

Consequently, the Cassie-Baxter state is crucial for creating super-repellent surfaces with extremely low surface adhesion and for attaining oleophobicity. Figure 5a, b depict the mechanism of both Wenzel and Cassie-Baxter states, respectively. Figure 5c, d exhibit structures that imply the existence of a Cassie-Baxter state even in the case of an intrinsic CA below 90°, a condition that can only be met by reentrant surface roughness. For a system to move from the Cassie-Baxter state (C − B1) to the Wenzel state (C − B2), which has a lower system free energy, it must cross an energy barrier as shown in Fig. 5e.

Surface roughness significantly affects wettability, and most researchers have explored how it affects wettability. Wu et al. [22] investigated the wettability properties of reactive and non-reactive surfaces as their roughness values changed. The AgCuTi/alumina and AgCu eutectic alloy/copper wetting systems were chosen for this study. The reactivity of liquid changes the spreading nature when it hits the surface. In this sense, a rougher surface improves the wetting of the non-reactive surface. Reactive liquids, in particular, impede capillary flow, which causes the spread to decrease because of surface irregularities. Therefore, the surface topography generates hydrophilicity and hydrophobicity depending on the surface chemistry [23]. Al liquid was utilised by Zhou et al. [24] to assess the surface wettability of Al_2O_3

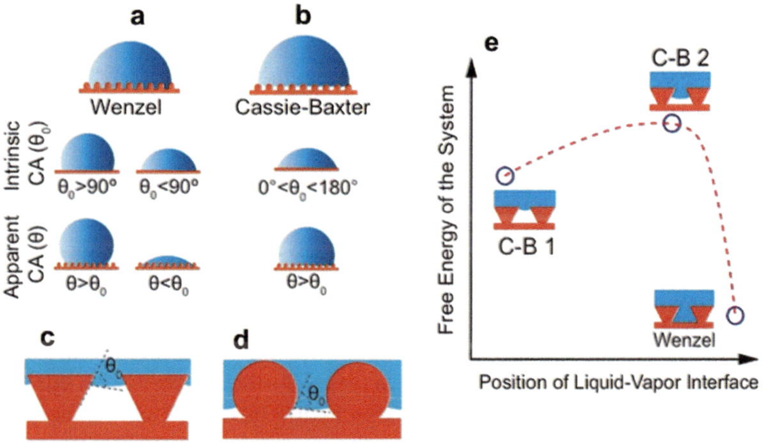

Fig. 5 **a** Wenzel state: The liquid droplet thoroughly wets the rough surface, amplifying the intrinsic wetting capabilities due to the surface roughness. **b** Cassie-Baxter state: Whether the intrinsic CA is greater or lower than 90°, surface roughness raises the CA. A composite surface consisting of the solid substrate and trapped air holds the liquid droplet in place. **c** A conventional reentrant configuration when the local CA is smaller than 90°. **d** Another common reentrant structure with a local CA of less than 90°, consisting of cylindrical fibers or spherical particles. **e** The system's free energy in relation to the liquid–air interface's position [4]

surfaces with varying levels of roughness. In order to connect the contact angle with traditional roughness measurements, a theoretical model has been put out that shows good agreement with experimental findings. According to Quere et al. [25], roughness increases hydrophilicity if the Young contact angle is less than 90°. Surfaces with micro- and nanostructures exhibit extreme wetting behaviours, such as superhydrophobicity (contact angles > 150°) and superhydrophobicity (contact angles $\approx 0°$). Depending on the material's intrinsic properties, roughness can amplify hydrophobicity and hydrophilicity. Utilising engineered Ti-surfaces, the impact of roughness on wettability change was investigated by Rosales-Leal et al. [26]. This study examined how the physical characteristics of Ti surfaces, specifically roughness, wettability, and morphology, affect osteoblast-like cells' adhesion, which is critical for bone integration in medical implants. Increased surface roughness provided more surface area and mechanical interlocking points for cells. Moderately hydrophilic surfaces (contact angles between 30° and 70°) promoted better cell adhesion than hydrophobic or superhydrophilic surfaces. To alter the roughness factor, surface treatments like blasting, acid etching, and polishing were used and increased surface wettability was observed with the roughness factor, as per the Wenzel theory. On the other hand, Kubiak et al. [27] confirmed that a drop and surge in wettability is observed when the surface roughness increases. More analysis is required to understand this random wettability behaviour with roughness change.

Surface tension is the force that acts along the surface of a liquid, minimising its surface area. The Marangoni effect is a phenomenon where variations in surface tension cause fluid motion. Surface tension gradients can be caused by differences in temperature (thermal Marangoni effect), concentration (solutal Marangoni effect), or the presence of surfactants. When there is a surface tension gradient, the liquid moves from lower surface tension to regions of higher surface tension. This movement helps to equalise the surface tension across the liquid surface. Marangoni flow can enhance or counteract other forces acting on the liquid, such as gravitational or capillary forces.

3.4 Chemical Compatibility

Another important element influencing wetting and solid/liquid adhesion is the chemical composition/compatibility with its range of surface energy. The difference in alkyl chain length or the ratio of high-energy hydrophilic domains can considerably modify the solid/liquid adhesion for a super-anti-wetting film with the same shape [28]. One can alter the water contact angle from hydrophilic to superhydrophobic by adjusting the surface micro-morphologies and chemical compositions [29]. As per Takeuchi et al. [30], the desorption of weakly bound water molecules from the surface of TiO_2 can occur due to the thermal energy of light irradiation. The surface tension of the H-bonded network and the water cluster decreases due to water molecules desorbing from the surface when the TiO2 thin film is heated, which is essential for surface wetting. The photocatalytic breakdown of organic pollutants, leading to hydrophilic conversion, happens concurrently with water desorption caused by the heating effect in the presence of UV radiation. Notably, there is no substantial drop in the free energy of cohesion of water with temperature. One possible explanation for the heating-induced change in contact angle by pointing out that the size of the water droplet would decrease as it evaporates. Heating-induced change in contact angle means that the water droplet size will decrease as it evaporates. The retreating contact angle will be less than the advancing contact angle due to the wide contact angle hysteresis.

Chemical compatibility is a critical factor influencing the wetting properties of coating materials, which determines how a liquid interacts with a surface, affecting adhesion, spreading, and eventual wetting behavior. This compatibility is governed by the surface energy of the coating material, the chemical nature of both the surface and the liquid, and their intermolecular interactions. Chemical compatibility affects how molecules from the liquid phase adsorb onto the surface, altering energy and wettability. The difference in surface free energy between the solid and liquid phases drives the wetting process. Compatible chemical groups can lower this difference, enhancing wetting. Localisation of the oxidised functional groups modified through the wet chemical etching method increases the surface energy. This approach is efficient due to the deep penetration of the solvent into the polymer matrix pores, hence permitting large-scale polymer treatment at much lower costs. Marchand-Brynaert et al. [31] introduced carboxyl groups onto the surface of PET films and membranes through controlled wet chemistry and characterised the chemical changes on the surface post-functionalisation. The functionalisation process improved the PET surfaces' wettability and chemical reactivity with high surface energy. Guo et al. [32] explored the development and properties of a sticky superhydrophobic surface, which combines the seemingly contradictory characteristics of strong water repellency and high adhesion to water droplets. Despite the high contact angle (> 150°), water droplets adhered strongly to the surface, resisting detachment even when tilted or subjected to mechanical forces. The ability to control water droplet adhesion and movement has implications in lab-on-a-chip technologies and other fluidic applications.

The liquid–solid interfacial interactions and wetting mechanisms are influenced by the nature of the interactions at the liquid–solid interface, transitions between different wetting states, and dynamic processes occurring during wetting. Dispersion forces are weak attractive forces arising from the interactions between induced molecular dipoles. They play a significant role in the adhesion of non-polar liquids to non-polar surfaces. Polar interaction forces involving permanent dipoles, are stronger than dispersion forces and influence the adhesion of polar liquids to polar surfaces. Hydrogen bonds are strong, directional interactions that significantly impact the wettability of surfaces that can donate or accept hydrogen bonds, such as water on hydroxylated or oxygen-containing surfaces. Charged interfaces are surfaces with ionisable groups that can interact with polar liquids through electrostatic forces. Depending on the liquid's polarity and ion content, surface charges can enhance or reduce wettability. Chemical reactions at the interface, forming covalent or ionic bonds, can lead to strong adhesion and significantly influence wettability. Surface functionalisation often leverages these interactions to modify wetting properties. Nayini et al. [33] reported on the functionalisation of CNTs using oxidation and chemical treatments to introduce various functional groups (e.g., carboxyl, hydroxyl, and amine groups) on their surfaces and observed that the carboxylated CNTs exhibited increased hydrophilicity compared to non-functionalised CNTs.

3.5 Surface Treatments

Surface treatments can significantly modify the wettability of a material. These treatments range from physical modifications to chemical alterations. Etching creates micro- and nanoscale roughness to alter wettability. Etching can be done using acids, alkalis, or plasma treatments. Deposition methods such as chemical vapor deposition (CVD) or physical vapor deposition (PVD) apply thin films of materials with desired wetting properties. Researchers have performed surface pre- and post-treatments to change the chemical composition of the surface by plasma, UV, and laser irradiation to induce reversible wetting. Numerous industrial applications, including water harvesting, self-cleaning panels, atmospheric water collection, and transportation, can benefit from pre- and post-treated surfaces. A macroscopic depiction of the interaction between the liquid and the solid substrate material is the solid surface's wettability. Measuring the contact angle (θ) of a sessile liquid drop on a solid surface in air is the most popular method for estimating surface wettability. Another frequently employed method is measuring the θ of a captive bubble.

The contact angle (θ) is a key parameter in wettability, defined as the angle formed at the junction where the liquid, solid, and vapor meet. $\theta < 90°$ indicates good wettability (hydrophilic surface) and $\theta > 90°$ indicates poor wettability (hydrophobic surface). Superhydrophobicity occurs when the contact angle exceeds 150° and the hysteresis is less than 10°, producing a surface that repels water [34]. The Young's equation relates the contact angle to the surface tensions of the solid–vapor (γ_{sv}), solid–liquid (γ_{sl}), and liquid–vapor (γ_{lv}) interfaces given in Eq. (3);

$$\gamma_{sv} = \gamma_{sl} + \gamma_{lv} \cos\theta \tag{3}$$

Therefore, the driving force D_f for the wetting can be defined as Eq. (4),

$$D_f = \gamma_{sv} - \gamma_{sl} \tag{4}$$

The work of adhesion, W_a provides a quantitative measure of the impact of wettability on contact adhesion, which is described as Eq. (5);

$$W_a = \gamma_{sv} + \gamma_{lv} - \gamma_{sl} \tag{5}$$

W_a represents the work needed to split one unit surface area of the two phases in a vacuum with dimensions of energy per unit surface area. It is possible to think of the effort of adhesion as a gauge for how strongly the two phases are bound together. One can write Eq. (5) in a vacuum and without any liquid components adhering to the solid's surface as Eq. (6);

$$D_f = -(\gamma_{lv} - W_a) \tag{6}$$

Using Eqs. (3) and (5), one can write Eq. (7) as follows;

$$W_a = \gamma_{lv}(1 + \cos\theta) \tag{7}$$

The work of adhesion (at equilibrium) can be experimentally determined when the surface tension lv of the liquid–vapour phase is known. This can be accomplished by measuring the contact angle at the edge of the liquid metal sessile drop resting on a solid flat surface, provided the solid surface is vacuum-free and uncontaminated. A high work of adhesion typically indicates excellent wetting, while a low work of adhesion typically indicates inadequate wetting.

In order to investigate the impact of surface wettability on the nucleation mechanism and boiling heat transfer, subcooled pool boiling experiments were conducted by Phan et al. [35]. The goal of using a nano-surface coating is to gently alter the surface's topography to change the static contact angle from 20° to 110°. The initial reference sample was made of stainless steel with a water contact angle of roughly 85°, and the other samples were prepared using MOCVD, PECVD, or NNBD to obtain different contact angles. More surface wettability was observed to increase the vapour bubble departure radius and decrease the bubble emission frequency for hydrophilic (wetted) surfaces. Furthermore, on hydrophobic (unwetted) surfaces, the first development of bubbles requires a lower superheat. Figure 6 shows the static contact angle of 2 μl sessile water droplets with and without nanoparticle depositions on stainless steel surfaces.

Kim et al. [36] explored the effects of ultraviolet (UV) and ozone treatment for one to four minutes on the band-bending properties of Ga-polar gallium nitride (GaN) when interfaced with aluminium oxide (Al_2O_3). The contact angles of the untreated and three-minute-treated wafers were 60° and 5.9°, respectively. The treatment time is extended from 3 to 4 min, resulting in a 12.9° contact angle. O_3 molecules react with the GaN wafer during plasma treatment, tending to enhance surface energy. A greater contact angle results from the O_3 reactivity with the wafer fluctuating with higher plasma exposure times. The calculated surface energies for the as-synthesised

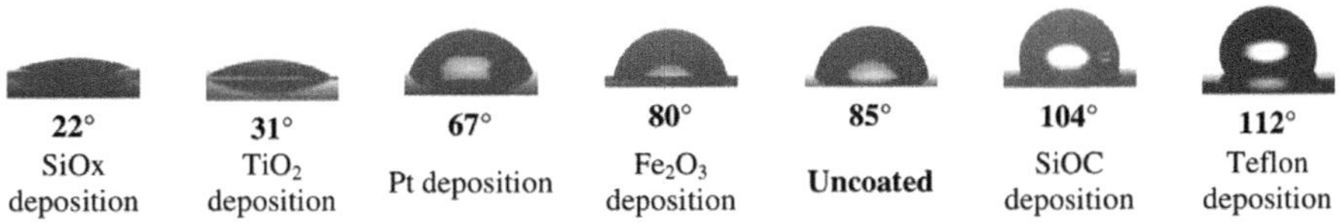

Fig. 6 Static contact angles of 2-μl sessile water droplets on stainless steel surfaces with and without nanoparticle deposition [35]

and three-minute plasma-treated wafers were 47.76 mJ/m^2 and 72.32 mJ/m^2, respectively. The reduction in band-bending at the GaN/Al_2O_3 interface was attributed to the passivation of surface states and the removal of contaminants that could otherwise act as charge traps. Gotoh et al. [37] investigated the impact of UV radiation and atmospheric pressure plasma (APP) on improving the wettability of polyethylene terephthalate (PET) films using a different methodology. The untreated PET film demonstrated hydrophobicity with a contact angle of 83.8o, while the APP-treated PET film displayed hydrophilicity with a contact angle 30.3°. After 90 s of exposure, the UV-treated PET film displays hydrophilic behaviour with a contact angle of 40.8°. When films are stored under atmospheric conditions, it has been found that APP-treated films retain their hydrophobicity more quickly than UV-treated films. The abovementioned study recommends sample post-treatment and wetting stability in atmospheric circumstances. The study by Madhuri et al. [38] demonstrated that nitrogen-rich AlN films possess smoother surfaces with finer grain structures than those deposited under lower nitrogen flow rates by DC magnetron sputtering. Bhujel et al. [39] used plasma treatment as a surface modification technique that can introduce functional groups, increase surface roughness, enhance wettability and improve the adhesion between SiNWs and PEDOT.

The superhydrophilic (SL)-superhydrophobic (SH) patterns of TiO_2 under UV light were investigated by Yanquing et al. [40] without the use of a photomask. An analysis of the importance of organosilicon adhesive in the lifetime of the SL-SH surface under real-time UV light (Solar irradiation) has been conducted. UV light has changed a surface that was initially SH (FAS-TiO_2) with a contact angle of 163° to 2°. TiO_2's photocatalytic activity breaks down the FAS exhibiting SL behaviour. Secondly, UV radiation has been applied to the adhesive-coated SH(FAS-TiO_2) surface. However, there is no change in the wetting behaviour on the surface. The encapsulation of TiO_2 by the adhesive has been identified as the cause of this behaviour. The surface remained wet even after 160 h of UV radiation. The method of wettability for the SL-SH patterned surface is shown in Fig. 7, and it is attributed to the low surface tension material from organosilicon that self-supplies after UV irradiation, giving the surface good endurance for outdoor applications. Zheng et al. [41] employed a one-step surface modification technique to create wettability patterns on aluminium substrates. In order to etch the surface, the substrates were first submerged in a $CuCl_2$ solution, which showed superhydrophilicity with $\sim 0°$ WCA. After that, siloxane was applied to the surface to produce superhydrophobicity with a WCA of $\sim 150°$. This result is comparable to previous researchers using low surface energy materials and a simple surface modification technique to create superhydrophobic patterns.

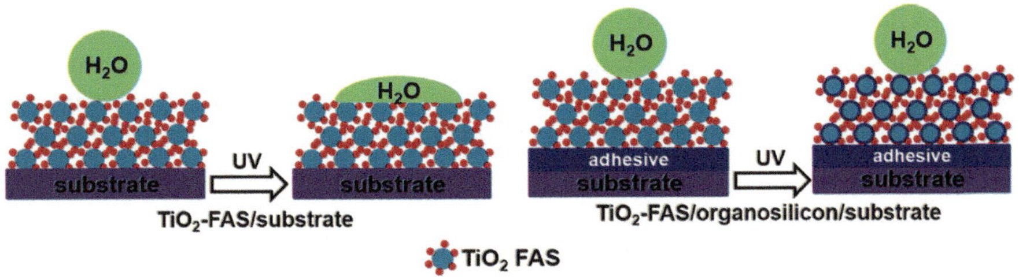

Fig. 7 Schematics of the mechanism of wetting caused by changes in the surface structure and chemical composition before and after UV exposure [40]

4 Wettability Alteration Using Surfactants

Wettability alteration of coating materials using surfactants is a technique used to modify the surface properties of materials to achieve desired levels of hydrophobicity or hydrophilicity. This alteration can significantly impact the performance and application of coating materials in various industries, such as anti-corrosion coatings, biomedical devices, and electronic components.

4.1 Surfactants and Their Function

Surfactants reduce surface tension between two substances, such as a liquid and a solid. They have a hydrophilic (water-attracting) head and a hydrophobic (water-repelling) tail, which allows them to arrange themselves at interfaces, modifying the surface energy and wettability. Different types of surfactants are based on the nature of their charge, which significantly influences their behavior and applications. (a) Anionic surfactants have negatively charged hydrophilic heads. They are highly effective in removing dirt and organic substances due to their strong cleaning properties. (b) Cationic surfactants have positively charged hydrophilic heads. They exhibit antimicrobial properties and bind to negatively charged surfaces such as skin and hair. (c) Nonionic surfactants do not carry any charge on their hydrophilic heads. They are less sensitive to water hardness and pH changes than ionic surfactants. (d) Amphoteric (Zwitterionic) surfactants can have positive and negative charges depending on the pH of the solution. They are mild and less irritating to the skin and mucous membranes.

Surfactants adsorb onto the surface of coating materials, forming a monolayer or multilayer that changes the surface energy. The findings from employing water-soluble surfactants to change the wettability of low-permeability chalk from oil-wet to water-wet were reported by Standnes et al. [42]. While anionic organic carboxylates in crude oil could not be irreversibly desorbed by anionic surfactants, cationic surfactants of type $R{-}N^+(CH_3)_3$ were able to produce an irreversible desorption of organic carboxylates from the chalk surface. Surfactants can facilitate chemical reactions on the surface that modify functional groups, thereby changing wettability. For example, surfactants can introduce or expose hydroxyl (-OH) groups to make the surface more hydrophilic. Boinovich et al. [43] examined the behaviour of hydrocarbon and fluoro-surfactants to create superhydrophobic coatings at the solid-water interface.

Specifically, how the wettability of these coatings changed over time when they were in long-term contact with aqueous media that differed in salinity and pH was compared. Four distinct surfactants were selected to be used as hydrophobic agents to lower the surface energy and facilitate comparability. While one pair of surfactants created chemical connections with the substrate, the other two were adsorbed physically on the coated surface. For one surfactant in each pair, the hydrocarbon tail served as the hydrophobic function, while the fluorocarbon tail served the hydrophilic purpose. Boinovich et al. also showed that the measured contact angle will rise if the hydrophobic agent is desorbed from the coating/droplet interface without causing a noticeable drop in wetting tension.

Conversely, droplet surface tension will be significantly reduced if the hydrophobic agent is transferred to the droplet/air interface. The surfactant binding process affected the endurance of the superhydrophobic coating, as these coatings containing physisorbed surfactants degraded rapidly in either acidic or alkaline environments. Surfactants can influence the formation of micro- and nanoscale roughness on the coating surface. Das et al. [44] investigated how non-ionic surfactants can alter the wettability of calcite surfaces, which is crucial for oil displacement in carbonate reservoirs, separation processes in froth flotation, and removal of hydrophobic contaminants from calcite-containing soils. This study focused on non-ionic surfactants, known for their stability and effectiveness in altering surface properties without introducing charges that could lead to unwanted interactions. Increased surfactant concentration generally led to higher hydrophobicity, as evidenced by increased contact angles. AFM and XPS analyses showed that surfactants formed a monolayer on the calcite surface, altering its energy and interaction with water. The wettability alteration was attributed to the adsorption of the surfactant molecules onto the calcite surface, which changed its surface energy. Non-ionic surfactants effectively reduced the surface tension, leading to a more hydrophobic surface. Figure 8 illustrates the schematic of the anticipated mechanism of wettability alteration.

5 Applications and Significance

In surface engineering, controlling the wettability of solid materials is a fundamental problem where two extremes are preferred. Complete wetting is the initial limit when a liquid that comes into contact with a solid naturally forms a film. For instance, in the case of a windshield, this film preserves the clarity of the glass. Moreover, when the automobile travels or stops, the film falls in the air owing to friction, carrying dust particles. Complete drying is the second limit when liquid droplets stay

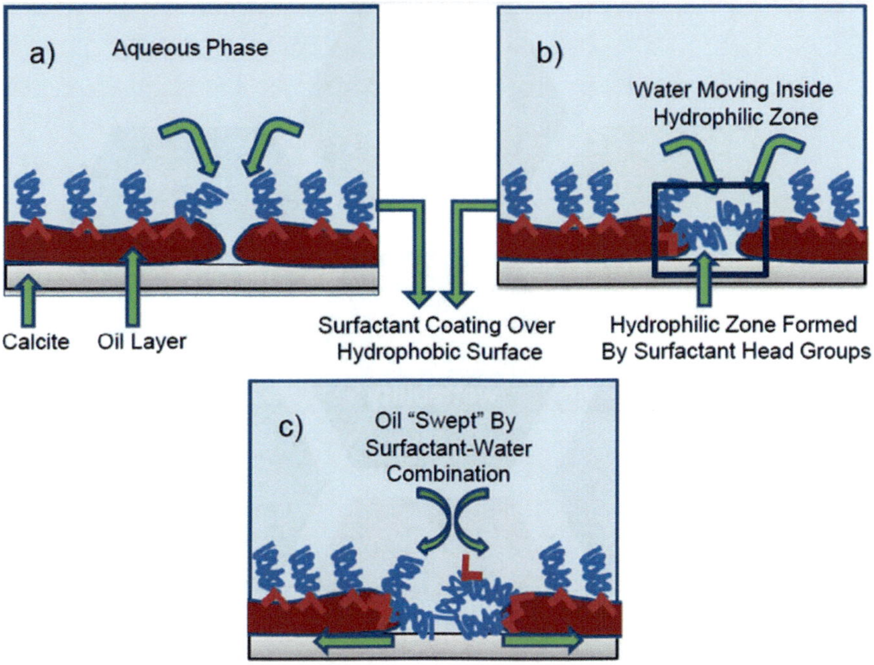

Fig. 8 Schematic of the suggested method of wettability alteration [44]

spherical and don't contact the substrate. As a result, they can be easily removed, avoiding liquid contamination of the solid surface. Figure 9 shows different applications of surface wettability. Proper wettability ensures strong adhesion of coatings to substrates, enhancing durability and resistance to environmental factors like UV radiation, moisture, and mechanical wear. As a droplet evaporates, the contact angle and contact area change, influenced by surface properties and ambient conditions; therefore, understanding evaporation dynamics is crucial for applications like inkjet printing and cooling technologies.

5.1 Self-Cleaning Surfaces

Superhydrophobicity (e.g., the lotus effect) is a technique used to generate surfaces that reject dirt and water, resulting in self-cleaning surfaces. The 'lotus effect' enables self-cleaning of superhydrophobic surfaces (SHSs) with a sliding angle of less than 10° and a WCA of more than 150° [45]. Superhydrophobic surfaces are typically associated with the self-cleaning property because the rolling motion of droplets makes it easy to remove dust particles from them. Self-cleaning can also be achieved on a regular surface using droplet impinging. Consideration is given to the sorts of

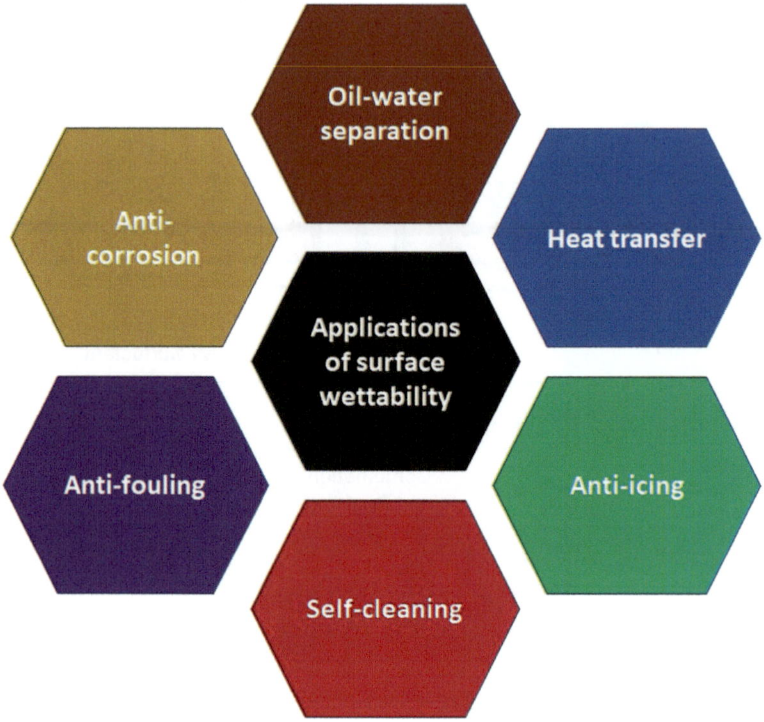

Fig. 9 Applications of surface wettability

dust particles and the impact of surface wettability. The process of attaining self-cleaning involves two steps: (a) an impinging droplet's water–air contact capturing particles, and (b) the droplet's release from the surface [46]. As the droplets recoil from the sloped surfaces, only their trailing edges are able to collect particles. Certain circumstances allow the hydrophilic surface to self-clean as well.

The car industry has employed self-cleaning coatings with antifogging qualities to provide glare-free, pristine glass, headlamps, and mirrors. On a superhydrophilic surface, water droplets tend to create a continuous coating that prevents light diffraction caused by condensed water droplets, allowing for a clean and unobstructed view. Using Ni^{2+} and Fe^{3+} dopants, TiO_2 thin films were created to provide visible light active self-cleaning building materials by Murugan et al. [47]. Wang et al. [48] revealed that the superhydrophilic TiO_2/SiO_2 composite film with 0.5 wt.% polyethylene glycol displayed exceptional antifogging properties due to increased surface roughness, with the WCA changing from 15° to 3° in just 0.16 s. Yilbas et al. [49] explored the combination of hydrophobicity with stretchability, which involves creating surfaces that can be deformed (stretched, compressed, and twisted) without losing their water-repellent properties. This study primarily utilised elastomeric polymers, such as polydimethylsiloxane (PDMS), and observed stretchable surfaces maintaining high contact angles even when deformed.

5.2 Coatings and Paints

Coatings and paints ensure good wetting for uniform coverage and adhesion. Polyester textiles are used in modern umbrellas and rainwear for a number of reasons. It is a reasonably priced cloth with adequate mechanical qualities, colourability, and hydrophobic qualities. Water droplets adhere comparatively effectively to the flat surface of monolithic polyester, according to a stated water contact angle of roughly 80° [50]. But when it rains, most droplets roll off the umbrella's cloth. Examining the droplet structure in more detail reveals a contact angle that seems to be greater than 120°, which is significantly larger than on a smooth polyester surface. This is only the outcome of the texture of the cloth. Apple has unveiled an oleophobic coating for the iPhone 3GS's glass screen that resists fingerprints. Although it doesn't altogether remove grease from the screen, it does make cleaning it simpler and more effective. Just wipe the screen down with a gentle, lint-free cloth to get rid of oily fingerprints. Wettability control in thermal barrier coatings (TBCs) can improve performance by ensuring uniform coverage and minimising defects. Gas turbines and aerospace engines use these coatings to improve energy efficiency. Reflective coatings with tailored wettability can enhance the reflectivity of surfaces and improve energy efficiency in buildings and solar energy applications. Pogorzelski et al. [51] explored the interaction between the metal substrate (Fe, Al, Cu, brass) and the paint layer, examining how different paint formulations and surface treatments alter wettability characteristics. The study highlights the importance of surface cleaning and pretreatment (e.g., acid etching, sandblasting) to remove contaminants and roughen the surface, enhancing paint adhesion, and also, surface energy modification techniques to modify the surface energy of metals, such as applying primers or surface-active agents, are discussed. Depending on the desired outcome, these treatments can make the surface more hydrophilic or hydrophobic.

5.3 Biomedical Devices

Incorporating functional groups (e.g., hydroxyl, carboxyl) or coatings (e.g., silanes, fluoropolymers) can tailor wettability to suit specific environments, such as biomedical or marine applications. Controlling wettability is crucial for applications in tissue engineering and implant compatibility. For a considerable time, scientists and engineers have drawn inspiration from biological organisms with distinct surface wettability properties. Specifically, manipulating chemical characteristics and structure to design surface wettability brings up new biomedical applications, from biomedical devices to high-throughput cell culture platforms. It is possible to create highly water-repellent superhydrophobic surfaces like lotus leaves by coating micro/nano hierarchical structures with low-surface-energy materials. These lotus leaf-inspired surfaces have been created using a variety of processes, such as layer-by-layer deposition, sol–gel method, lithography, electrochemical reaction, electrodynamics, wet chemical etching, and plasma treatment. Preventing unwanted biological debris from adhering to the surface is crucial in biomedical applications involving implanted medical devices. One important surface characteristic is its antibacterial nature, which helps avoid contamination during the implanted device's cell/tissue culture or inflammation. Due to its anti-fouling characteristic, which can stop bacteria from adhering to the surface, a bio-inspired superhydrophobic surface has drawn interest as a possible antibacterial surface [52]. Pipes and other random geometry can be effortlessly incorporated with liquid slippery surfaces. Leslie et al. [53] treated catheters and tubing of indwelling medical devices with slippery qualities to lower biofouling of the devices and morbidity and mortality from thrombosis, which affects platelets and fibrinogen in the blood. Manipulating liquid flow in small channels is done to achieve lab-on-a-chip devices. Figure 10 shows the biomedical device applications of surface wettability [54]. Tischer et al. [55] showed that a patterned treatment could facilitate high-density cell microarrays by applying antifouling polyethylene glycol (PEG) brushes to flat surfaces. Furthermore, Feng et al. [56] studied superhydrophobic/superhydrophilic micropatterning-derived 2D droplet microarray technology, which offers a perfect high-throughput, miniaturised platform for 2D and 3D cell cultures, as well as parallel addition procedures like medium addition and cell implantation.

5.4 Microfluidics

Microfluidic devices manipulate small volumes of fluids within channels, typically ranging from tens to hundreds of micrometers in size. These devices consist of intricate networks of channels, chambers, and valves. Common materials used in microfluidics include polymers (such as PDMS), glass, and silicon, each selected based on desired properties like transparency, chemical resistance, and mechanical stability. The design of microchannels often incorporates wettability patterns to guide fluids, prevent unwanted mixing, or ensure complete wetting of specific regions. Wettability and microfluidics are complementary fields that are growing together synergistically. When controlling microfluidic flows for fluid processing and manipulation at the microscale, surface wettability is crucial.

Conversely, microfluidics has become a flexible framework for controlling a material's wettability. In order to design materials with exquisite composition and morphology, microfluidics offers various liquid templates, laying the groundwork for precisely manipulating the wettability. Liquid templates are classified into single droplets, 1-D arrays, and 2-D or 3-D assemblies for the modular manufacturing of microfibers, microparticles, and monolithic porous materials, respectively, based on the degree of ordering. Saxena et al. [57] explored the critical role of surface wettability in the fabrication of microfluidic devices and their diverse applications. It delves into how surface properties influence fluid behavior at the microscale, impacting device performance and functionality. Ainla et al. [58] used pulse width

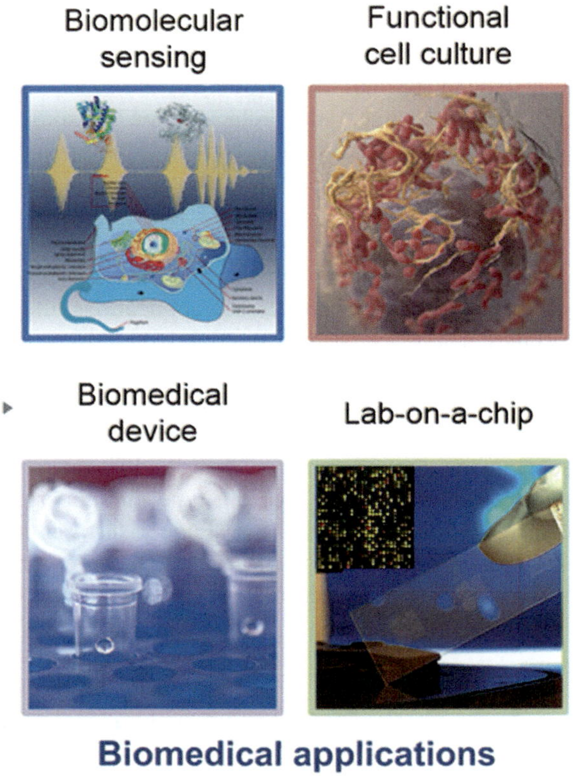

Fig. 10 Biomedical device, Lab-on-a-chip, and others [54]

flow modulation to design microdiluters by employing a microfluidic diluter to study the impact of Ca^{2+} concentration on phospholipid bilayer spread on a SiO_2 surface.

5.5 *Anti-Icing*

Innovative anti-icing techniques must be used immediately since unwanted surface icing can result in significant losses in human life and property. A growing body of research indicates that natural superhydrophobic coatings have strong anti-icing properties and excellent application potential. However, these properties are limited in their ability to be used in real-world applications. Kreder et al. [59] showed that dry and smooth self-assembled monolayers have lower performance than state-of-the-art and limited surface compatibility, whereas the bulk coatings are versatile and durable. In the case of dry and textured microstructured coatings, rapid shedding of droplets prevents ice nucleation with poor pressure and humidity tolerance, but nanostructured coatings possess improved humidity tolerance (jumping droplet effect) and pressure stability. The water-based lubricating layer's hydration coatings were poorly hydrophobic yet had minimal ice adhesion. Figure 11 shows the schematic diagram of the anti-icing theory of superhydrophobic coatings. Pan et al. [60] fabricated superhydrophobic coatings by spraying PMMA-SNs suspension on a steel surface to form a rough surface structure that enhances water repellency and evaluated these coatings for anti-icing and anti-corrosion applications. These superhydrophobic coatings delayed ice formation by reducing water droplet adhesion to the surface, allowing them to roll off before freezing. These coatings also facilitated easier ice removal once formed, requiring less mechanical force or energy input, which is particularly beneficial in applications like aircraft, power lines, and wind turbines. The research by Wei et al. [61] delved into the anti-icing capabilities of super-wetting surfaces, exploring both hydrophobic and slippery surfaces.

The study examined the mechanisms of icing resistance and ice-phobicity, evaluated the robustness of these surfaces under various conditions, and discussed potential applications. Super-wetting surfaces had water crystallisation points that are 3 °C lower than glass. Water freezing time on slippery surfaces took 20 times longer than on glass. Ice adhesion on slippery surfaces was three orders lower than on other surfaces. Hydrophobic surfaces repelled water due to their low surface energy and micro/nanostructured roughness, preventing water droplet adhesion and thus

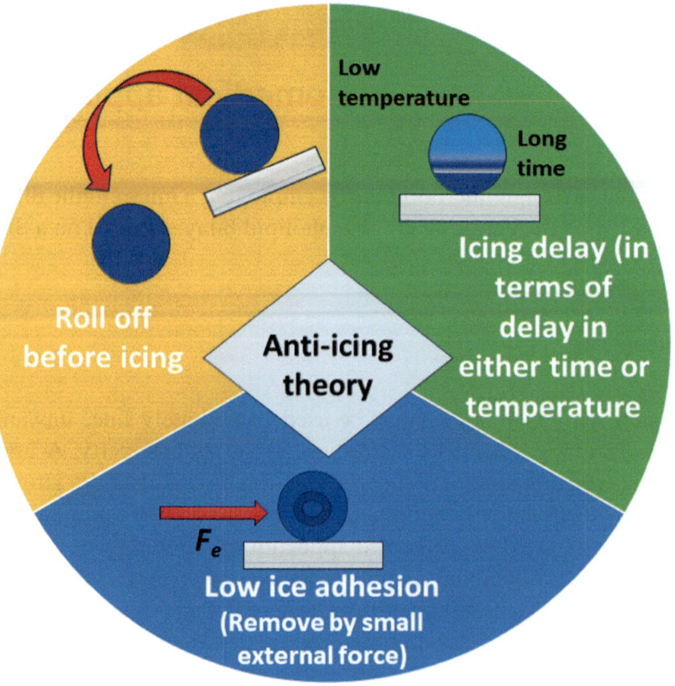

Fig. 11 Schematic of anti-icing theory of superhydrophobic coatings

delaying ice formation. However, the slippery surfaces minimise friction, allowing water and ice to slide off easily.

5.6 Heat Transfer

Surface energy and roughness affect the early nucleation and growth of droplets during condensation, affecting heat transfer efficiency in condensation processes. Marangoni effects improve cooling in thermal management systems through effective liquid dispersion and heat transfer. In systems involving phase changes (e.g., boiling, condensation), Marangoni convection can improve heat transfer efficiency. Hydrophilic coatings can increase the density of nucleation sites, promoting efficient boiling heat transfer. These surfaces promote film-wise condensation, where a continuous liquid film forms, leading to lower heat transfer efficiency due to the thermal resistance of the liquid film. Hydrophobic coatings can control bubble formation and detachment, influencing boiling performance. These surfaces favour dropwise condensation, where discrete droplets form and easily roll off the surface, providing higher heat transfer coefficients than film-wise condensation. Superhydrophobic coatings facilitate rapid droplet removal, reducing thermal resistance and improving condensation heat transfer. Asthana et al. [62] discussed the wettability of various coating materials on different substrates at elevated temperatures and the dynamics of **the** spreading behaviour of the molten coatings over the substrates. As the temperature increased, the viscosity of the molten coating decreased, and surface tension altered, affecting both wettability and spreading. High-temperature coatings are widely used in aerospace, automotive, and power generation industries to protect components from thermal degradation, oxidation, and corrosion.

6 Challenges and Innovations in the Wettability of Coating Materials

The mechanism of wettability in coating materials is a rich field of study with numerous research opportunities and challenges. Addressing the complexities of surface interactions, ensuring durability, and developing scalable manufacturing processes are key challenges that must be overcome. Innovations in smart coatings, nanostructured surfaces, bioinspired designs, and sustainable materials drive the field forward. Interdisciplinary approaches, combining insights from materials science, chemistry, biology, physics, and environmental science, are essential for advancing our understanding and application of wettability in coating materials. The future directions and research opportunities in this field are vast and promising. Heterogeneous engineered surfaces often have complex chemical compositions and structures, making predicting and controlling wettability challenging. Complex models and simulations are necessary because wettability is affected by interactions at several sizes, including atomic, molecule, micro, and macro. Developing coatings that respond to environmental stimuli (e.g., temperature, pH, light) is required to alter their wettability dynamically. Coatings often degrade under ecological conditions such as UV exposure, temperature fluctuations, and mechanical wear. They are incorporating self-healing mechanisms to repair damage and maintain wettability properties over time. Ensuring that wettability properties are kept over long periods is critical for practical applications. Developing cost-effective and scalable manufacturing processes for advanced coatings is a significant challenge. Rare or expensive materials can limit the widespread adoption of certain coating technologies. Developing more accurate and non-destructive techniques to measure wettability and surface properties is essential. Real-time monitoring of wettability changes during operational conditions remains a challenge.

7 Conclusions

The study of wettability presents intriguing prospects for creating cutting-edge surfaces and materials with specialised qualities. For these technologies to succeed in the future, it will be imperative that the issues of scalability, durability, environmental effect, and fundamental knowledge are addressed. New and improved wettability-controlled surfaces will be developed due to developments in materials science, manufacturing processes, and computational modelling. These developments will enable a broad range of applications across numerous industries. Wettability alteration using surfactants is a versatile and effective method for tailoring the surface properties of coating materials. By selecting the appropriate type of surfactant and application method, it is possible to achieve the desired level of hydrophobicity or hydrophilicity, thereby enhancing the performance and durability of the coatings in various applications.

References

1. Niedźwiedź M, Bara M, Korzekwa J, Kaptacz S, Sowa M, Olesiński A, Simka W (2020) Tribological properties and surface wettability of coatings produced on the Mg-AZ31B Alloy by plasma electrolytic oxidation. Coatings 14(7):780. https://doi.org/10.3390/coatings14070780
2. Kung CH, Sow PK, Zahiri B, Mérida W (2019) Assessment and interpretation of surface wettability based on sessile droplet contact angle measurement: challenges and opportunities. Adv Mater Interfaces 6:1900839. https://doi.org/10.1002/admi.201900839
3. Wang X, Zhang Q (2020) Role of surface roughness in the wettability, surface energy and flotation kinetics of calcite. Powder Technol 371:55–63. https://doi.org/10.1016/j.powtec.2020.05.081

4. Wang Z, Elimelech M, Lin S (2016) Environmental applications of interfacial materials with special wettability. Environ Sci Technol 50:2132–2150. https://doi.org/10.1021/acs.est.5b04351
5. Dong S, Wang Z, An L, Li Y, Wang B, Ji H, Wang H (2020) Facile fabrication of a superhydrophobic surface with robust Micro-/Nanoscale hierarchical structures on titanium substrate. Nanomaterials 10(8):1509. https://doi.org/10.3390/nano10081509
6. Wu W, Wang X, Wang D, Chen M, Zhou F, Liu W, Xue Q (2009) Alumina nanowire forests via unconventional anodisation and super-repellency plus low adhesion to diverse liquids. Chem Comm 9:1043–1045. https://doi.org/10.1039/B818633B
7. Lai Y, Gao X, Zhuang H, Huang J, Lin C, Jiang L (2009) Designing superhydrophobic porous nanostructures with tunable water adhesion. Adv Mater 21:3799–3803. https://doi.org/10.1002/adma.200900686
8. Li Y, Zhang Z, Zhu X, Men X, Ge B, Zhou X (2015) Fabrication of a superhydrophobic coating with high adhesive effect to substrates and tunable wettability. Appl Surf Sci 328:475–481. https://doi.org/10.1016/j.apsusc.2014.12.086
9. Balaur E, Macak JM, Taveira L, Schmuki P (2005) Tailoring the wettability of TiO_2 nanotube layers. Electrochem Commun 7:1066–1070. https://doi.org/10.1016/j.elecom.2005.07.014
10. Laurenti M, Cauda V, Gazia R, Fontana M, Rivera VF, Bianco S, Canavese G (2013) Wettability control on ZnO nanowires driven by seed layer properties. Eur J Inorg Chem 2013:2520–2527. https://doi.org/10.1002/ejic.201201420
11. Drelich J, Chibowski E, Meng DD, Terpilowski K (2011) Hydrophilic and superhydrophilic surfaces and materials. Soft Matter 7:9804. https://doi.org/10.1039/C1SM05849E
12. Nwankire CE, Favaro G, Duong QH, Dowling DP (2011) Enhancing the mechanical properties of superhydrophobic atmospheric pressure plasma deposited siloxane coatings. Plasma Process Polym 8:305–315. https://doi.org/10.1002/ppap.201000069
13. Yu E, Kim SC, Lee HJ (2015) Extreme wettability of nanostructured glass fabricated by non-lithographic, anisotropic etching. Sci Rep 5:9362. https://doi.org/10.1038/srep09362
14. Liu YC, Lu DN (2006) Surface energy and wettability of plasma-treated polyacrylonitrile fibers. Plasma Chem Plasma Process 26:119–126. https://doi.org/10.1007/s11090-006-9005-7
15. Ostrovskaya L, Perevertailo V, Ralchenko V (2002) Wettability and surface energy of oxidised and hydrogen plasma-treated diamond films. Diam Relat Mater 11(3–6):845–850. https://doi.org/10.1016/S0925-9635(01)00636-7
16. Wang S, Zhang Y, Abidi N, Cabrales L (2009) Wettability and surface free energy of graphene films. Langmuir 25(18):11078–11081. https://doi.org/10.1021/la901402f
17. Liu T, Chen S, Cheng S, Tian J, Chang X, Yin Y (2007) Corrosion behavior of superhydrophobic surface on copper in seawater. Electrochim Acta 52:8003–8007. https://doi.org/10.1016/j.electacta.2007.06.072
18. Ou J, Perot B, Rothstein JP (2004) Laminar drag reduction in microchannels using ultrahydrophobic surfaces. Phys Fluids 16:4635–4643. https://doi.org/10.1063/1.1812011
19. Sharma SD, Singh D, Saini KK, Kant C, Sharma V, Jain SC, Sharma CP (2006) Sol–gel-derived super-hydrophilic nickel doped TiO_2 film as active photo-catalyst. Appl Catal A 314(1):40–46. https://doi.org/10.1016/j.apcata.2006.07.029
20. Wenzel RN (1936) Resistance of solid surfaces to wetting by water. Ind Eng Chem 28(8):988–994. https://doi.org/10.1021/ie50320a024
21. Bhushan B, Jung YC (2011) Natural and biomimetic artificial surfaces for superhydrophobicity, self-cleaning, low adhesion, and drag reduction. Prog Mater Sci 56(1):1–108. https://doi.org/10.1016/j.pmatsci.2010.04.003
22. Wu M, Chang L, Zhang L, He X, Qu X (2016) Effects of roughness on the wettability of high temperature wetting system. Surf Coat Technol 287:145–152. https://doi.org/10.1016/j.surfcoat.2015.12.092
23. Wenzel RN (1949) Surface roughness and contact angle. J Phys Chem 53(9):1466–1467. https://doi.org/10.1021/j150474a015
24. Zhou XB, De Hosson JTM (1995) Influence of surface roughness on the wetting angle. J Mater Res 10:1984–1992. https://doi.org/10.1557/JMR.1995.1984

25. Quere D (2008) Wetting and roughness. Annu Rev Mater Res 38:71–99. https://doi.org/10.1146/annurev.matsci.38.060407.132434
26. Rosales-Leal JI, Rodríguez-Valverde MA, Mazzaglia G, Ramón-Torregrosa PJ, Díaz-Rodríguez L, García-Martínez O, Vallecillo-Capilla M, Ruiz C, Cabrerizo-Vílchez MA (2010) Effect of roughness, wettability and morphology of engineered titanium surfaces on osteoblast-like cell adhesion. Colloids Surf A: Physicochem Eng Asp 365(1–3):222–229. https://doi.org/10.1016/j.colsurfa.2009.12.017
27. Kubiak KJ, Wilson MCT, Mathia TG, Carval P (2011) Wettability versus roughness of engineering surfaces. Wear 271(3–4):523–528. https://doi.org/10.1016/j.wear.2010.03.029
28. Cheng Z, Lai H, Du Y, Fu K, Hou R, Li C, Zhang N, Sun K (2014) PH-induced reversible wetting transition between the underwater superoleophilicity and superoleophobicity. ACS Appl Mater Interfaces 6(1):636–641. https://doi.org/10.1021/am4047393
29. Liu TL, Kim CJ (2014) Turning a surface superrepellent even to completely wetting liquids. Science 346(6213):1096–1100. https://doi.org/10.1126/science.1254787
30. Takeuchi M, Sakamoto K, Martra G, Coluccia S, Anpo M (2005) Mechanism of photoinduced superhydrophilicity on the TiO_2 photocatalyst surface. J Phys Chem B 109(32):15422–15428. https://doi.org/10.1021/jp058075i
31. Marchand-Brynaert J, Deldime M, Dupont I, Dewez J-L, Schneider Y-J (1995) Surface functionalisation of Poly(ethylene terephthalate) film and membrane by controlled wet chemistry: chemical characterization of carboxylated surfaces. J Colloid Interface Sci 173(1):236–244. https://doi.org/10.1006/jcis.1995.1319
32. Guo Z-G, Liu W-M (2007) Sticky superhydrophobic surface. Appl Phys Lett 90:223111. https://doi.org/10.1063/1.2745251
33. Nayini MMR, Bastani S, Ranjbar Z (2014) Synthesis and characterisation of functionalised carbon nanotubes with different wetting behaviors and their influence on the wetting properties of carbon nanotubes/polymethylmethacrylate coatings. Prog Org Coat 77(6):1007–1014. https://doi.org/10.1016/j.porgcoat.2014.02.013
34. Liu M, Wang S, Jiang L (2017) Nature-inspired superwettability systems. Nat Rev Mater 2(7):17035-17036. https://doi.org/10.1038/natrevmats.2017.36
35. Phan HT, Caney N, Marty P, Colasson S, Gavillet J (2009) Surface wettability control by nanocoating: the effects on pool boiling heat transfer and nucleation mechanism. Int J Heat Mass Transf 52(23–24):5459–5471. https://doi.org/10.1016/j.ijheatmasstransfer.2009.06
36. Kim K, Ryu JH, Kim J, Cho SJ, Liu D, Park J, Lee IK, Moody B, Zhou W, Albrecht J, Ma Z (2017) Band-bending of Ga-polar GaN interfaced with Al_2O_3 through ultraviolet/ozone treatment. ACS Appl Mater Interfaces 9(20):17576–17585. https://doi.org/10.1021/acsami.7b01549
37. Gotoh K, Yasukawa A, Kobayashi Y (2011) Wettability characteristics of poly(ethylene terephthalate) films treated by atmospheric pressure plasma and ultraviolet excimer light. Polym J 43:545–551. https://doi.org/10.1038/pj.2011.20
38. Madhuri A, Jena S, Gupta M, Swain BP (2023) Nitrogen flow rate dependent atomic coordination, phonon vibration and surface analysis of DC magnetron sputtered nitrogen rich-AlN thin films. Physica B 666:415141. https://doi.org/10.1016/j.physb.2023.415141
39. Bhujel R, Rai S, Das S, Tiwari A, Chattopadhyay S, Swain BP, Deka U (2024) Unlocking the plasma advantage in elevating the performance of silicon nanowires/poly(3,4-ethylenedioxythiophene): Poly(styrenesulfonate) hybrid solar cell. Thin Solid Films 790:140202. https://doi.org/10.1016/j.tsf.2024.140202
40. Yanqing Z, Jifu S, Qizhang H (2017) A facile approach for TiO_2-based superhydrophobic–superhydrophilic patterns by UV or solar irradiation without a photomask. Chem Commun 53:2363–2366. https://doi.org/10.1039/C6CC09558E
41. Barthlott W, Neinhuis C (1997) Purity of the sacred lotus, or escape from contamination in biological surfaces. Planta 202:1–8. https://doi.org/10.1007/s004250050096
42. Standnes DC, Austad T (2000) Wettability alteration in chalk 2. Mechanism for wettability alteration from oil-wet to water-wet using surfactants. J Pet Sci Eng 28(3):123–143. https://doi.org/10.1016/S0920-4105(00)00084-X

43. Boinovich LB, Emelyanenko AM (2015) The behaviour of fluoro- and hydrocarbon surfactants used for fabrication of superhydrophobic coatings at solid/water interface. Colloids Surf A: Physicochem Eng Asp 481:167–175. https://doi.org/10.1016/j.colsurfa.2015.05.003
44. Das S, Nguyen Q, Patil PD, Yu W, Bonnecaze RT (2018) Wettability alteration of calcite by nonionic surfactants. Langmuir 34(36):10650–10658. https://doi.org/10.1021/acs.langmuir.8b02098
45. Zheng H, Huang S, Liu J, Chen F, Yang X, Liu X, Sun J, Xu W (2016) One-step modification method to fabricate wettability patterns on aluminium substrate. Micro Nano Lett 11(11):697–701. https://doi.org/10.1049/mnl.2016.0187
46. Quan YY, Zhang LZ, Qi RH, Cai RR (2016) Self-cleaning of surfaces: the role of surface wettability and dust types. Sci Rep 6:38239. https://doi.org/10.1038/srep38239
47. Murugan K, Subasri R, Rao TN, Gandhi AS, Murty BS (2013) Synthesis, characterisation and demonstration of self-cleaning TiO_2 coatings on glass and glazed ceramic tiles. Prog Org Coat 76(12):1756–1760. https://doi.org/10.1016/j.porgcoat.2013.05.012
48. Wang JJ, Wang DS, Wang J, Zhao WL, Wang CW (2011) High transmittance and superhydrophilicity of porous TiO_2/SiO_2 bi-layer films without UV irradiation. Surf Coat Technol 205(12):3596–3599. https://doi.org/10.1016/j.surfcoat.2010.12.033
49. Yilbas BS, Hassan G, Al-Qahtani H (2019) Stretchable hydrophobic surfaces and self-cleaning applications. Sci Rep 9:14697. https://doi.org/10.1038/s41598-019-50982-8
50. Drelich J, Kim JH, Payne T, Miller JD, Kobler RW (1999) Purification of polyethylene terephthalate from polyvinyl chloride by froth flotation for the plastics (soft-drink bottle) recycling industry. Sep Purif Technol 15(1):9–17. https://doi.org/10.1016/S1383-5866(98)00047-1
51. Pogorzelski S, Boniewicz-Szmyt K, Grzegorczyk M, Rochowski P (2022) Wettability of metal surfaces affected by paint layer covering. Materials 15(5):1830. https://doi.org/10.3390/ma15051830
52. Liu T, Dong L, Liu T, Yin Y (2010) Investigations on reducing microbiologically-influenced corrosion of aluminum by using superhydrophobic surfaces. Electrochim Acta 55(18):5281–5285. https://doi.org/10.1016/j.electacta.2010.04.082
53. Leslie DC, Waterhouse A, Berthet JB, Valentin TM, Watters AL, Jain A, Kim P, Hatton BD, Nedder A, Donovan K, Super EH, Howell C, Johnson CP, Vu TL, Bolgen DE, Rifai S, Hansen AR, Aizenberg M, Super M, Aizenberg J, Ingber DE (2014) A bioinspired omniphobic surface coating on medical devices prevents thrombosis and biofouling. Nat Biotechnol 32:1134–1140. https://doi.org/10.1038/nbt.3020
54. Shin S, Seo J, Han H, Kang S, Kim H, Lee T (2016) Bio-inspired extreme wetting surfaces for biomedical applications. Materials 9(2):116. https://doi.org/10.3390/ma9020116
55. Tischer T, Rodriguez-Emmenegger C, Trouillet V, Welle A, Schueler V, Mueller JO, Goldmann AS, Brynda E, Barner-Kowollik C (2014) Photo-patterning of non-fouling polymers and biomolecules on paper. Adv Mater 26:4087–4092. https://doi.org/10.1002/adma.201401006
56. Feng W, Ueda E, Levkin PA (2018) Droplet microarrays: from surface patterning to high-throughput applications. Adv Mater 30(20):1706111. https://doi.org/10.1002/adma.201706111
57. Saxena S, Joshi R (2020) Microfluidic devices: applications and role of surface wettability in its fabrication. In: 21st century surface science-a handbook. London, IntechOpen:147
58. Ainla A, Gözen I, Orwar O, Jesorka A (2009) A microfluidic diluter based on pulse width flow modulation. Anal Chem 81(13):5549–5556. https://doi.org/10.1021/ac9010028
59. Kreder MJ, Alvarenga J, Kim P, Aizenberg J (2016) Design of anti-icing surfaces: smooth, textured or slippery? Nat Rev Mater 1:1–15. https://doi.org/10.1038/natrevmats.2015.3
60. Pan S, Wang N, Xiong D, Deng Y, Shi Y (2016) Fabrication of superhydrophobic coating via spraying method and its applications in anti-icing and anti-corrosion. Appl Surf Sci 389:547–553. https://doi.org/10.1016/j.apsusc.2016.07.138
61. Wei C, Jin B, Zhang Q, Zhan X, Chen F (2018) Anti-icing performance of super-wetting surfaces from icing-resistance to ice-phobic aspects: Robust hydrophobic or slippery surfaces. J Alloys Compd 765:721–730. https://doi.org/10.1016/j.jallcom.2018.06.041
62. Asthana R, Sobczak N (2000) Wettability, spreading, and interfacial phenomena in high-temperature coatings. JOM-e 52(1):1–18

Chapter 6
Stress Behavior in Coating Materials

Ningthoujam Bidyapani Chanu and Bibhu Prasad Swain

Abstract In engineering building materials, residual stresses are both a challenge and an opportunity. When properly managed, they can have very positive impacts, enhancing the performance and reliability of products. However, when neglected, they can indeed have disastrous consequences. A fresh evaluation of conventional and contemporary measuring methods is desperately needed as concerns regarding the performance and dependability of products grow. Selecting the most valuable and efficient measuring technique for a particular application is essential to success. The content in Practical Residual Stress Measurement Methods gives the reader the knowledge they need to comprehend important residual stress concepts and make well-informed technical judgments regarding the best measurement approach. Written by invited professionals, each chapter is organised practically and focused, with subsections covering the measurement concept, residual stress evaluation, practical measuring techniques, sample applications, references, and more reading. This chapter provides a comprehensive overview of residual stress in mechanical coatings, discussing its nature, causes, measurement techniques, effects, mitigation strategies, and applications. It highlights the importance of understanding and controlling residual stress to optimise coating performance and reliability in various industries.

Keywords DLC · Wear Materials · Raman Spectroscopy · Barkhausen Noise Analysis

1 Introduction

Residual stress in mechanical coatings is technical and crucial in determining coated components' performance, durability, and reliability. These stresses, inherent to the coating process, significantly influence the coating and substrate's physical properties and failure mechanisms. Understanding residual stress's nature, causes, and effects is

N. B. Chanu · B. P. Swain (✉)
Department of Physics, National Institute of Technology Manipur, Langol, Imphal, Manipur, India
e-mail: bibhuprasad.swain@gmail.com; bpswain@nitmanipur.ac.in

B. P. Swain (ed.), *Advances in Mechanical Coating*, Materials Horizons: From Nature to Nanomaterials, https://doi.org/10.1007/978-981-96-7484-8_6

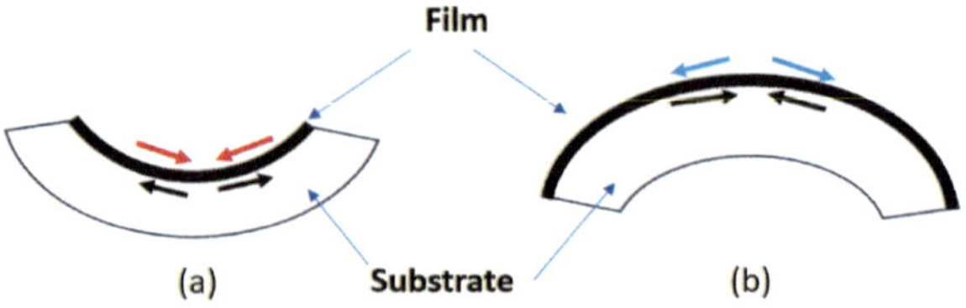

Fig. 1 **a** Tensile stress **b** Compressive stress in a film sputtered onto a substrate [1]

not just an academic exercise but an essential step in optimising coating performance in various applications.

1.1 Nature of Residual Stress

Residual stress refers to the internal stress that remains within a material after the original cause, such as mechanical loading or thermal treatment, has been removed. In mechanical coatings, these stresses can be either tensile or compressive, arising from various factors during the coating process, such as differences in thermal expansion, deposition techniques, and substrate-coating interactions (Fig. 1).

1.2 Types of Residual Stress

In mechanical coatings, residual stress can be tensile, pulling the material apart and potentially leading to cracking, delamination, and reduced fatigue life. Alternatively, it can be compressive, a force that pushes the material together and offers significant benefits such as enhancing fatigue resistance and inhibiting crack growth. The internal stress of the film is chiefly influenced by sputtering pressure, substrate temperature, and film thickness. However, factors like target alignment, bias voltage, and the mass ratio of cathode atoms to gas also play a role. The transition between compressive and tensile stress is primarily driven by sputtering pressure. At a specific critical pressure, as illustrated in Fig. 2, the stress transitions from compressive to tensile due to the influence of the deposition process on the film's stress.

Residual stress is also categorised by the scale at which it occurs:

- Type I stresses occur on a large scale, extending over distances from millimetres upwards, and are commonly observed in coated components. These are the primary focus of most stress measurement techniques in the coating industry.
- Type II stresses are more minor in scale, occurring at the micron level, often between grains in metals within the coating. Type I stresses can contribute to the development of Type II stresses.
- Type III stresses are highly localised at the atomic scale around dislocations and crystal interfaces within the coating.

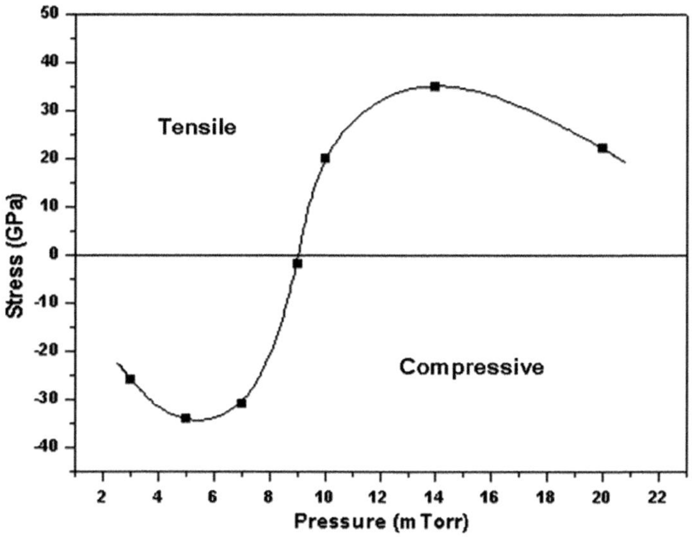

Fig. 2 Typical behaviour of the change in stress as a function of the gas pressure [1]

While most measurement methods primarily target Type I stresses, some can be adapted to detect Type II and, occasionally, Type III stresses. However, Type II stresses can sometimes complicate accurately measuring Type I stresses, particularly in specific diffraction methods (Fig. 3).

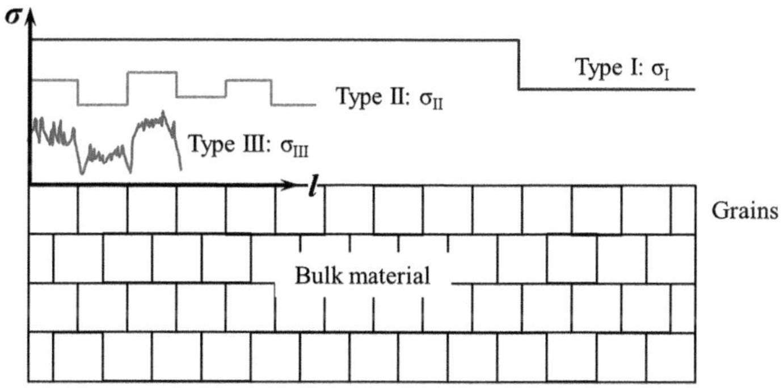

Fig. 3 Variation of length scale in Type I, II, and III residual stresses [2]

1.3 Causes of Residual Stress in Mechanical Coatings

Various factors, including the cooling process after deposition, variations in thermal expansion coefficients of the substrate and coating, mismatch in material properties, and the deposition process itself, can cause residual stress in mechanical coatings. These factors can cause tensile or compressive residual stresses to form inside the coating or at the interface of the coating and substrate, which can impact the mechanical properties and effectiveness of the coated system. Proper understanding and management of these residual stresses are crucial for ensuring the durability and reliability of coated components in practical applications. Residual stresses can arise from various sources during the coating process, including thermal, mechanical, and chemical factors.

1.4 Thermal Effects

The disparity in the coefficient of thermal expansion of the underlying material and the coating is called the thermal mismatch. During heating and cooling cycles, this variation can result in residual stress. Thermal gradients due to rapid cooling or uneven temperature distribution can lead to non-uniform stress distribution within the coating and substrate.

1.5 Mechanical Effects

Impact Treatments and peening treatments, such as shot peening, are mechanical processes that introduce compressive residual stresses to improve fatigue resistance. Substrate deformation refers to the mechanical properties of the substrate and deformation behaviour, which can influence the residual stress in the coating (Fig. 4).

1.6 Chemical Effects

The interaction or chemical reactions between the coating and substrate materials can lead to the development of residual stresses, which may profoundly influence the structural integrity and functionality of the coated materials. Consequently, these aspects must be carefully evaluated in materials design and selection to achieve the best performance of the coating-substrate system.

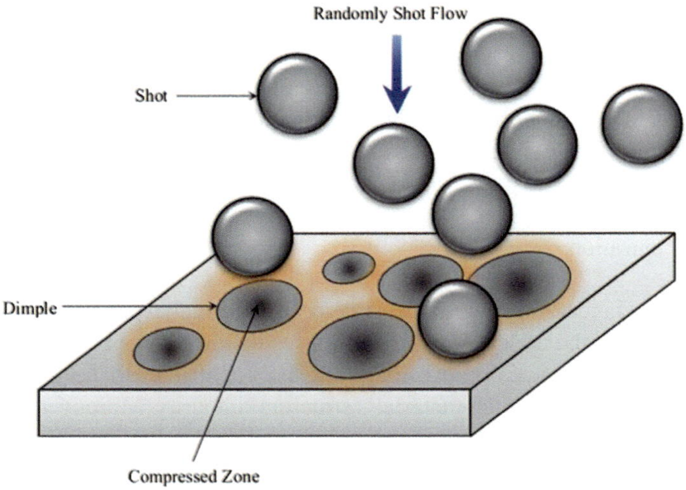

Fig. 4 Diagrammatic representation of the shot peening process [3]

1.6.1 Oxidation and Reduction

Chemical changes at the interface or within the coating can cause volume changes, leading to stress development. Oxidation and reduction (redox) processes can play significant roles in developing and managing residual stresses in materials. These processes can alter the material's microstructure, composition, and mechanical properties, influencing residual stresses. Here's how oxidation and reduction can affect residual stress.

Surface Layer Formation

Oxidation often leads to forming an oxide layer on the material's surface. For example, in silicon carbide (SiC), an oxide layer of silicon dioxide (SiO_2) can form. This layer can induce compressive residual stresses due to the volume expansion associated with the formation of oxides. Reduction can remove oxide layers or return metal oxides to their metallic state. This can change the surface and subsurface properties of the material. When an oxide layer is reduced, the removal can relieve compressive residual stresses induced by the oxide layer formation. However, this can also expose the underlying material to potential environmental degradation.

Protective Oxide Layers

In some cases, a thin and stable oxide layer can be a protective barrier, preventing further oxidation and corrosion, thereby maintaining mechanical strength and reducing surface stress concentrations in the material.

Stress Concentration

When an oxide layer on a material is brittle or has thermal expansion characteristics incompatible with the underlying material, it results in a stress concentration situation. This localised stress can cause micro-cracks or contribute to the growth of larger cracks, thereby weakening the material's structural integrity. Moreover, factors such as material impurities, surface roughness, and the level of applied stress can further intensify stress concentration, making it a critical factor in the design and production of components and structures.

Internal Oxidation

When oxidation penetrates below the surface of the material, it can cause changes in volume, leading to the development of internal residual stresses. This issue can become complicated if the oxidation occurs unevenly or produces new phases with different characteristics.

Phase Transformation

Oxidation can induce phase transformations in materials, altering their crystal structure and mechanical properties and leading to residual stresses.

Thermal Oxidation

Due to thermal gradients developed by high-temperature oxidation, the materials may undergo differentiable expansion and contraction, which causes residual thermal stresses.

Thermal Mismatch

Different material parts may oxidise at different rates, creating thermal mismatches and complex residual stress patterns.

Hydrogen Embrittlement

Materials can suffer from hydrogen embrittlement in specific reduction environments, such as those involving hydrogen. Hydrogen atoms can diffuse into the material, creating internal stresses and weakening the material.

Hydrogen-Induced Cracking

Hydrogen can lead to cracking and blistering, significantly altering the residual stress state within the material.

Thermochemical Treatments

Reduction processes are often part of thermochemical treatments used to modify the surface properties of materials, such as carburising or nitriding in metals. By enhancing fatigue resistance and mechanical performance, these treatments may provide advantageous compressive residual stresses on their surfaces.

Practical Implications

Controlled Oxidation and Reduction: By carefully controlling oxidation and reduction processes, engineers can tailor the residual stress profiles in materials to enhance their performance. For instance, controlled oxidation can induce beneficial compressive stresses on the surface.

Protective Coatings: Protective coatings that undergo controlled oxidation can protect the underlying material and maintain a desirable residual stress state.

Heat Treatment: Heat treatments involving redox processes can be optimised to balance forming and removing oxide layers, minimise adverse residual stresses, and enhance material properties.

2 Measurement and Characterisation of Residual Stress

Accurate residual stress measurement and characterisation are essential for assessing coating performance and predicting potential failure modes. Various techniques have been developed to evaluate residual stresses, including X-ray and neutron diffraction, ultrasonic velocity, and hole drilling. These methods are categorised based on their characterisation scale and whether they are destructive or nondestructive. Mechanical methods like sectioning and hole-drilling assess type I macro-stresses over large distances and are destructive, measuring deformation during or after stress release. Nondestructive physical processes, such as diffraction analysis, are better suited for evaluating type II and type III stresses at the grain or atomic level. However, measurement accuracy can vary depending on the material, particularly in titanium alloys, where diffraction overlap poses challenges.

2.1 *X-Ray Diffraction (XRD)*

The powerful, non-destructive technique of X-ray Diffraction (XRD) can measure the residual stresses in materials. XRD provides detailed information about the crystalline structure's macro and micro stress states by assessing strain in the crystal lattice. The overview of how XRD is used for residual stress measurement, the principles behind it, and the practical considerations involved.

2.1.1 Analysis of Crystalline Structure

X-ray Diffraction (XRD) measures a crystal's distances between atomic planes. Under stress, these distances change, causing shifts in the X-ray diffraction angles.

2.1.2 Bragg's Law

The fundamental principle behind XRD is Bragg's Law, which relates the wavelength of X-rays and the diffraction angle to the spacing between planes in the crystal lattice:

$$n\lambda = 2d\sin(\theta) \tag{1}$$

where n, λ, d, and θ denote the diffraction order, the X-ray wavelength (1.54 Å), interplanar spacing, and the Bragg angle, respectively. Figure 5 shows the XRD patterns of rGO and Fe_3O_4/ZnO/rGO nanocomposites from the study of Dutta et al. [4].

2.1.3 Stress-Induced Shifts

When residual stresses are present, they cause strain in the material, altering the interplanar spacings. The changes in d spacing indicate that residual stresses have been quantified.

By analysing a scattered intensity, I(Q) line, the position of exchange momentum Q_0 (or angle θ) provides the periodicity distance, $d_0 = 2\pi/Q_0$. If a small material region undergoes deformation, the distance d will deviate slightly from the nominal d_0. In a large sample, different regions may experience either tension ($\varepsilon > 0$) or compression ($\varepsilon < 0$). Under tension, the line shifts towards smaller Q values, whereas

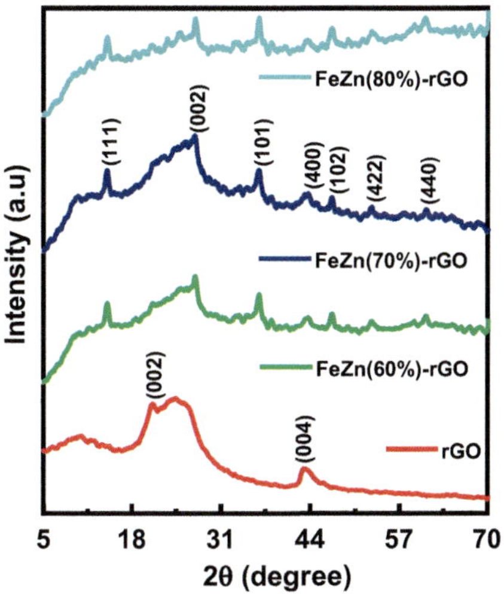

Fig. 5 XRD patterns of rGO and Fe_3O_4/ZnO/rGO nanocomposites [4]

under compression, it shifts towards larger Q values.

$$\varepsilon = \left(\frac{d - d_0}{d_0}\right)_{hkl} = \left(\frac{\Delta d}{d_0}\right)_{hkl} = -\left(\frac{\Delta Q}{Q_0}\right)_{hkl} = -\frac{(\Delta 2\theta)\cot\theta_0}{2} \quad (2)$$

The sensitivity of the method ($-d\theta/\varepsilon = \tan\theta$) enables the measurement of ε, over a range of approximately 10^{-2} to 10^{-5}.

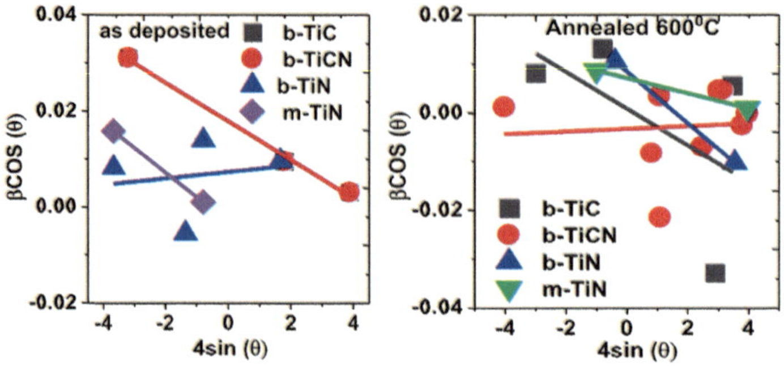

Fig. 6 Williamson-Hall plot of as-deposited and post-annealed (600 °C) m-TiN, b-TiN, b-TiCN, and b-TiC thin films [5]

According to G. K. Williamson and his student W. H. Hall, broadening due to crystallite size, β_L, and broadening due to lattice strain, β_e, show notable differences depending on the Bragg angle (θ) (Fig. 6).

$$\beta_L = \frac{K\lambda}{L\cos\theta} \tag{3}$$

$$\beta_e = C\varepsilon\tan\theta \tag{4}$$

$$\beta_{tot} = \beta_L + \beta_e = \frac{K\lambda}{L\cos\theta} + C\varepsilon\tan\theta \tag{5}$$

Simplifying the equation:

$$\beta_{tot}\cos\theta = C\varepsilon\sin\theta + \left(\frac{K\lambda}{L}\right) \tag{6}$$

The work of Jena et al. [5] demonstrates that annealing effectively reduces microstrain in thin films, as shown in the Williamson-Hall (WH) plot in Fig. 1. Before annealing, the TiN and TiC/Ti films exhibited compressive strain, which lessened afterwards, suggesting some release of strain. In contrast, the TiN/Ti and TiCN/Ti films experienced a shift from lower to higher tensile strain after annealing, likely due to the replacement of carbon atoms with nitrogen atoms. Overall, the reduction in microstrain across all samples is due to the migration of defects and phase changes occurring during annealing, leading to a more relaxed lattice structure. The stress in Cu/Cu_xO thin films, as studied by Madhuri et al. [6], is influenced by the $Cu(NO_3)_2$ concentration. As the concentration increases, the films exhibit varying types of stress: compressive strain is observed at lower concentrations (10–30 mL), while tensile strain occurs at higher concentrations (40 and 50 mL). The lattice strain increases with $Cu(NO_3)_2$ concentration, with the full width at half maxima (FWHM) of the diffraction peaks broadening, indicating higher strain. The stacking

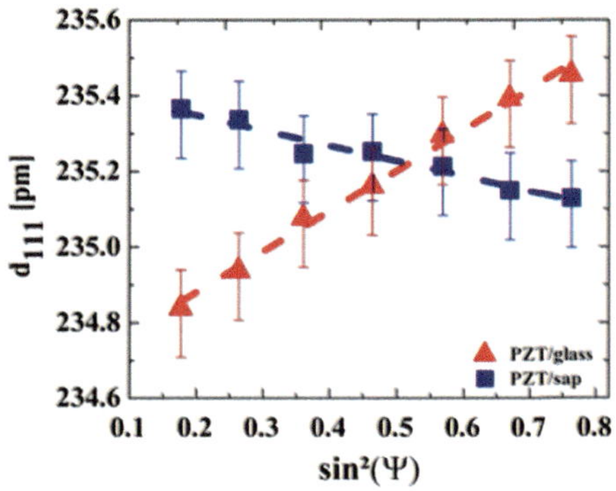

Fig. 7 Plot of d_{111} vs $\sin^2\Psi$ for PZT on glass and sapphire substrates, error bars represent the uncertainties in the 2θ values of the (111) diffraction peaks [8]

fault values increase with $Cu(NO_3)_2$ concentration, reflecting higher defect levels and stress within the films. So, their work shows that the strain component and size component can be obtained by the Williamson-Hall plot.

2.1.4 The Relationship Between D and $\sin^2\Psi$

The $\sin^2\Psi$ X-ray diffraction (XRD) method is essential for measuring residual stresses in PVD coatings. It works by measuring the spacing between crystal planes (d-spacing) at various angles (Ψ) and plotting these values against the $\sin^2\Psi$. The slope of this plot indicates the lattice strain, which correlates with the residual stress in the coating.

Although effective, this method has challenges, especially for thin films, where weak and broad diffraction peaks can lead to inaccuracies. The glancing incident XRD technique is used to overcome this by varying the incident angle. However, it can still introduce uncertainties due to differences in X-ray penetration and material anisotropy. The residual stress of thin films can be analysed by

$$\begin{aligned}\varepsilon = \frac{\Delta d}{d} = \frac{s_2}{2}\big[\sigma_{11}\cos^2\Phi\sin^2\Psi \\ + \sigma_{22}\sin^2\Phi\sin^2\Psi + \sigma_{33}\cos^2\Phi \\ + \sigma_{12}\sin^2\Phi\sin^2\Psi + \sigma_{13}\cos\Phi\sin^2\Psi \\ + \sigma_{23}\sin\Phi\sin^2\Psi + s_1[\sigma_{11} + \sigma_{22} + \sigma_{33}]\end{aligned} \quad (7)$$

To simplify strain equations

$$\varepsilon = \frac{\Delta d}{d} = \frac{1+\upsilon}{E}\sigma_1\sin^2\Psi - \frac{\upsilon}{E}(\sigma_1 + \sigma_2) \quad (8)$$

where υ, Φ, and Ψ are Poisson's ratio, azimuthal angle, and polar or inclination angle, respectively [7] (Fig. 7).

Fig. 8 Experimental setup of Holl drilling machine [9]

2.2 Hole Drilling Method

The Hole-Drilling Method (HDM) is a semi-destructive technique for assessing residual stresses by drilling a small hole and measuring the strains around it. This method, effective for analysing stress profiles up to approximately 1 mm deep, primarily utilises a strain gauge rosette to capture the strains relieved during the drilling process. HDM can be executed as a through-hole, typically used for thin sheets, or more commonly as an incremental blind hole, where strain readings are recorded progressively and converted into stress distributions (Fig. 8).

The ASTM E837 standard governs this method, providing detailed guidelines on best practices and calibration procedures. However, it may not fully address factors that impact accuracy, such as plastic deformation, material thickness, hole chamfers, surface irregularities, and hole alignment.

Various regularisation techniques have been developed to improve HDM's accuracy, including Tikhonov regularisation, featured in recent ASTM E837 editions. Additionally, HDM can be enhanced by analysing surface displacements near the drilled hole using techniques like electronic speckle pattern interferometry (ESPI) or digital image correlation (DIC). These advanced methods have proven effective, particularly in additive manufacturing, where they assess residual stresses by monitoring surface displacements post-drilling. For instance, Santus et al. introduced a high-resolution numerical approach that integrates axisymmetric harmonic finite elements with direct influence functions to refine HDM accuracy in conjunction with DIC [10].

2.3 Barkhausen Noise Analysis

Barkhausen noise analysis for residual stress measurement is a non-destructive technique that detects changes in stress-related magnetic properties. Magnetic

Barkhausen Noise, or Barkhausen Emission, involves measuring electromagnetic pulses within a frequency range of (10^2–10^6) Hz across a sensor coil and is particularly useful for ferromagnetic materials [11].

This method is based on magnetic effects in ferromagnetic materials, which have regions called magnetic domains. The concept of magnetic domains was first explored in 1907, and Prof. H. G. Barkhausen provided the initial experimental proof in 1919 [12]. During magnetisation, the orientation of these domains changes, causing fluctuations in the overall magnetisation of materials. These fluctuations generate voltage pulses in a measuring coil, which Barkhausen detected and amplified using a loudspeaker. This signal is known as Barkhausen Noise (BN).

Alterations in material characteristics, including those at a microscopic scale, are often reflected in the Magnetic Barkhausen Emission (MBE) profile. This profile can reveal changes in magnetic and mechanical properties related to dislocation density, alloy composition, secondary phases, grain size, and applied or residual stresses. For example, a high density of dislocations in quenched steel reduces MBE levels because magnetic domain walls move less freely. Conversely, tempering the steel decreases dislocation density, leading to higher MBE levels. Microstructural changes, such as variations in grain size and secondary phase precipitates, also impact the magnetisation process, as these factors influence domain wall movement and nucleation. MBN measurements are typically divided into high-frequency and low-frequency [13]. High-frequency MBN utilises external magnetic field excitation frequencies greater than 10 Hz and examines the MBN signal within the 2–1000 kHz range, setting a lower cut-off frequency to avoid interference from higher harmonics of the excitation. Low-frequency MBN, on the other hand, operates with excitation frequencies below 1 Hz and analyses the MBN signal within the 0.5–100 kHz range.

2.4 Neutron Diffraction

Neutron diffraction, like the well-known X-ray technique, determines internal lattice stress in a material by measuring the elastic lattice strain in its crystallites using Bragg's law of diffraction. This non-destructive method penetrates deeper into materials than XRD, making it suitable for thick coatings and bulk materials. Assuming no change in the path length, lattice strain (ε) can be calculated from the slight shift in time-of-flight Δt_{hkl}, for the maximum diffraction.

$$\Delta\lambda_{hkl} = 2\Delta d_{hkl}\sin\theta_0 \tag{9}$$

And since the time of flight is proportional to neutron velocity [14],

$$\frac{\Delta d_{hkl}}{d_{hkl}} = \frac{\Delta\lambda_{hkl}}{\lambda_{hkl}} = \frac{\Delta t_{hkl}}{t_{hkl}} \tag{10}$$

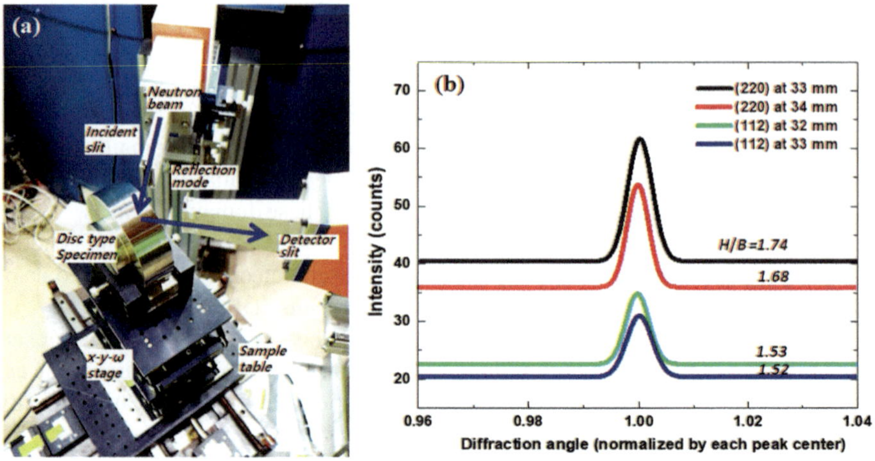

Fig. 9 **a** Neutron diffraction setup in reflection geometry **b** Gaussian fitted diffraction peaks from the (220) and (112) planes at each location, with peak height-to-background ratio (H/B) indicated [15]

Neutron diffraction is a powerful technique for measuring residual stress in industrial metals and alloys due to its deep penetration capability. However, it is limited by the weak brightness of neutron sources, restricting the maximum path length in materials like steel to around 60 mm. To enhance this, researchers have focused on using specific neutron wavelengths near the Bragg edges, which reduces attenuation and increases the depth of measurement. Advances in this area have allowed stress measurements in steel specimens up to 83 mm thick using focused neutron beams, with results validated against destructive methods. Additionally, improvements in instrumentation, such as using bent perfect crystals for silicon monochromators, have further enhanced the technique's effectiveness (Fig. 9).

2.5 *Raman Spectroscopy*

Raman spectroscopy is a practical, non-invasive technique for analysing gaseous, liquid, and solid materials. It is commonly used to assess stress in bulk materials, thin films, coatings, and single and polycrystalline substances. Based on the interaction between monochromatic light and chemical bonds, as discovered by C. V. Raman in 1928, this method generates spectra that reveal the energies of chemical bonds, aiding in phase differentiation within a material, such as distinguishing between diamond and graphite. Additionally, it provides insights into strain by detecting frequency shifts in the spectra caused by changes in bonding due to internal or external stresses. This technique is instrumental in solid-state physics for evaluating stress in semiconductors like silicon, where stress can affect device performance.

Raman spectroscopy offers high spatial resolution, from 100 to 600 nm, and typically requires minimal sample preparation.

Utilising the Raman effect to measure stress in materials exhibiting cubic crystals involves studying vibrational modes to gather stress-related microscopic data. The Raman effect, characterised by the inelastic scattering of light interacting with phonons or molecular vibrations, reveals how mechanical stress alters these vibrational frequencies. This alteration results in shifts in peak positions compared to those observed in stress-free materials. An analogy is tuning a guitar string, where adjusting the tension changes the vibrational frequency. Similarly, mechanical stress changes vibrational frequencies in materials, leading to shifts in Raman peak positions. Accurate stress resolution depends on the sensitivity of Raman peaks and the signal-to-noise ratio. Based on the study of Louden [16], silicon possesses three Raman tensors. Within the crystal coordinate system, where x = [100], y = [010], and z = [001], the Raman tensors are given by

$$\Delta_1 = \begin{bmatrix} 0 & 0 & 0 \\ 0 & 0 & d \\ 0 & d & 0 \end{bmatrix}, \Delta_2 = \begin{bmatrix} 0 & 0 & d \\ 0 & 0 & 0 \\ d & 0 & 0 \end{bmatrix}, \Delta_3 = \begin{bmatrix} 0 & d & 0 \\ d & 0 & 0 \\ 0 & 0 & 0 \end{bmatrix} \quad (11)$$

Here, d is the non-zero component. These components determine the intensity of specific Raman peaks. For silicon, d is the same for all three tensors and is considered 1 [17]. Ganesan et al. [18] proposed that the frequencies of the three optical modes with strain can be derived by solving

$$\begin{pmatrix} p\varepsilon_{11} + q(\varepsilon_{22} + \varepsilon_{33}) - \lambda & 2r\varepsilon_{12} & 2r\varepsilon_{13} \\ 2r\varepsilon_{12} & p\varepsilon_{22} + q(\varepsilon_{33} + \varepsilon_{11}) - \lambda & 2r\varepsilon_{23} \\ 2r\varepsilon_{13} & 2r\varepsilon_{23} & p\varepsilon_{33} + q(\varepsilon_{11} + \varepsilon_{22}) - \lambda \end{pmatrix} = 0 \quad (12)$$

where p, q, and r denote the coefficients for the photon deformation potential; ε_{ij} (i, j = 1, 2, 3) represent the strain components in anisotropic materials. Thus, the frequency variation of each mode, $\Delta\omega_j$ (j = 1, 2, 3), from the unstrained condition, ω_{j0}, to strained state ω_j, can be determined from the eigenvalues:

$$\Delta\omega_{ij} = \frac{\lambda_j}{2\omega_{j0}} \quad (13)$$

Regarding silicon, techniques for determining stress tensors from Raman spectra have been introduced for single crystals with established orientations and polycrystalline materials with varied orientations. In simple cases, such as when uniaxial stress is applied along the <110> direction of a silicon crystal, Wolf et al. [19] polarised the incident laser light along the <110> direction of the crystal. They recorded the Raman signal in the back-scattering geometry along the <001> direction.

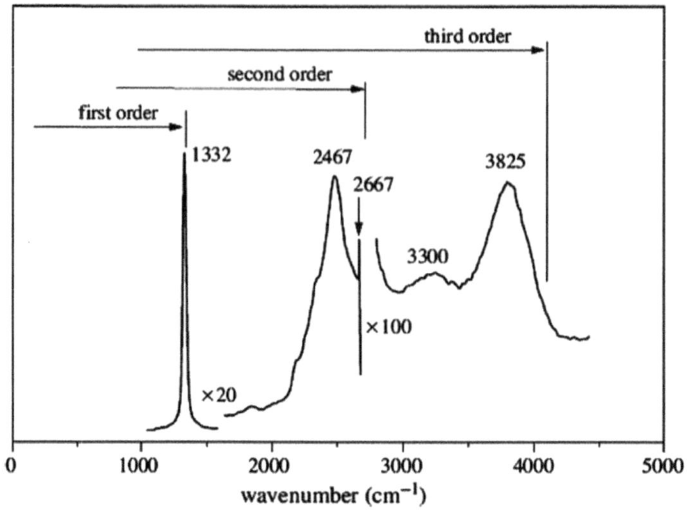

Fig. 10 A general Raman spectrum of gem-quality diamond measured at room temperature with a 228.9 nm excitation wavelength [21]

$$\Delta \upsilon\left(cm^{-1}\right) = -1.93 \times 10^{-9}\left(\frac{\text{cm}^{-1}}{\text{Pa}}\right)\sigma(Pa) \tag{14}$$

The diffraction limits of the optical system constrain the resolution in conventional micro-Raman spectroscopy. Typically, the laser beam's spot size on the material's surface measures between 1 and 2 μm in diameter. This limitation results from directing the incident laser beam onto the material's surface via a set of lenses and analysing it using an optical microscope.

Raman spectroscopy is widely utilised to examine the phonon modes of various carbon allotropes and non-crystalline carbon phases due to its ability to offer distinct signatures for each allotrope. For example, single-crystal diamond exhibits a peak at 1332 cm^{-1}, single-crystal graphite shows a G peak at 1575 cm^{-1}, and disordered graphite displays a D peak at 1355 cm^{-1}, as illustrated in Fig. 10. The G and D peaks in amorphous carbon become noticeably wider and manifest within the 1500–1550 cm^{-1} spectral range, providing identifiable characteristics through Raman spectra. Jena et al. [20] examined the Raman spectra of GaP nanostructures with varying levels of Si doping (0.5–2.5 wt %). Their analysis, illustrated in Fig. 11b, identified two primary Raman features: the transverse optical (TO) mode at 364.64 cm^{-1} and the longitudinal optical (LO) mode at 401.44 cm^{-1}. As the Si doping increased, a slight upward shift in the TO mode indicated a reduction in the bond angle of Ga-P and a stronger interaction between Si and GaP. Conversely, the LO mode showed a minor increase in wavenumber with higher doping, suggesting the presence of compressive residual stress, while the Si-modified LO mode decreased, reflecting tensile residual stress. These spectral shifts highlight how the amount of Si doping influences residual stress and structural changes in GaP nanostructures.

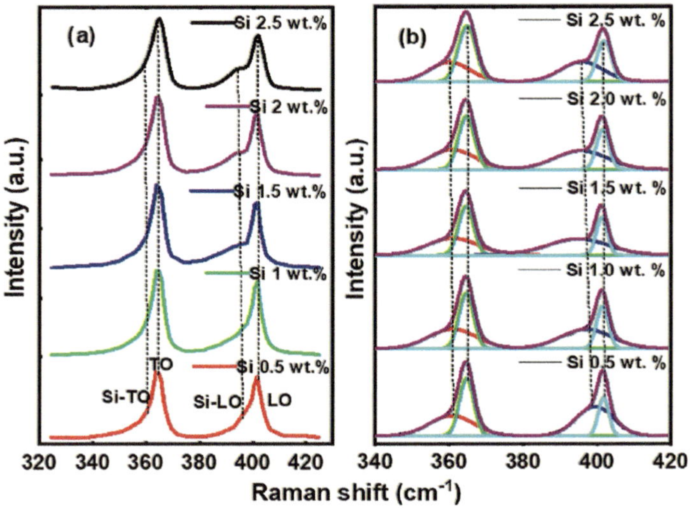

Fig. 11 **a** Raman spectra of GaP: Si nanostructures and **b** their convoluted Raman spectra for Si doping levels of 0.5, 1.0, 1.5, 2.0, and 2.5 wt % [20]

2.6 Curvature Method

Curvature measurement is a non-destructive technique for evaluating overall residual stresses, applicable specifically to laminated composites with asymmetric ply configurations. Unlike symmetric laminates, these composites naturally curve after curing and remain flat. The resulting curvature is influenced by factors such as the lay-up design, thickness, curing temperature, and the structural properties of materials. The total stress in thin films can be expressed as

$$\sigma_{tot} = \sigma_{ext} + \sigma_{th} + \sigma_{t} \tag{15}$$

where σ_{ext} denotes the external stress caused by external loading, σ_{th} denotes the stress generated by thermal change, and σ_{i} denotes the internal stress. The thermal component can be determined by the equation provided (16).

$$\sigma_{th} = E_f(\alpha_f - \alpha_S)(T_S - T_M) \tag{16}$$

Here, T_S represents the temperature of coating, T_M is the temperature at which the thermal stresses are assessed or during service operation of the coating system, E_f denotes Young's modulus of the film, α_f and α_s are the thermal expansion coefficients of the film and the substrate, respectively. The residual stress is typically determined using the Stoney formula:

$$\sigma = \frac{E_s}{(1 - \upsilon_s} \frac{h_s^2}{h_f} \left(\frac{1}{R} - \frac{1}{R_0} \right) \tag{17}$$

where E_s, v_s, h_s, h_f, R, and R_0 represent Young's modulus, Poisson's ratio, substrate thickness, film thickness, radius of curvature of the substrate after deposition, and radius of curvature before deposition, respectively [22]. The Stoney equation is utilised for stress measurement in thin films and is expressed as

$$\sigma = -\left(\frac{E \times h}{6 \times (1 - \upsilon)} \right) \times \left(\frac{\Delta\lambda}{\lambda} \right) \tag{18}$$

where σ represents the film stress, E is Young's modulus of the film material, h is the film thickness, υ is the Poisson's ratio of the film material, Δλ is the difference in the principal radii of curvature of the substrate before and after film deposition, and λ is the principal radius of curvature of the substrate.

3 Residual Stress on Mechanical Coating

Understanding and improving the performance of engineered surfaces significantly depends on the study of residual stress in mechanical coatings. Researchers have investigated a variety of coating materials and deposition parameters to elucidate their effects on residual stress profiles and mechanical properties. Investigations into residual stress in coatings such as DLC, metal carbides and nitrides, carbonitrides, MoS_2, and many others have provided valuable insights into optimising their mechanical properties and durability.

3.1 DLC (Diamond-Like Carbon) Coatings

Several scientists have focused on exploring diamond-like carbon (DLC) thin film coatings due to their remarkable mechanical, tribological, optical, and biological attributes. DLCs are characterised by an amorphous carbon structure, where atoms are organised in a random, continuous network that includes both sp^3 (diamond-like) and sp^2 (graphitic-like) bonds, with the latter representing the defective carbon. Residual stress in diamond-like carbon coatings ranges from −0.03 to −4 GPa. Let's discuss how residual stress affects DLC coatings and the various development processes used to mitigate these disadvantages.

Sharma et al. [23] deposited DLC via Pulsed laser ablation on Si and Ti-6Al-4 V substrates at varying temperatures and reported residual compressive stress of nearly 3 GPa on Si, as measured by Raman spectroscopy. This value is consistent with the previously reported results and is compared to the stress in a free-standing film peeled off from a titanium alloy substrate. They also noted that the residual compressive

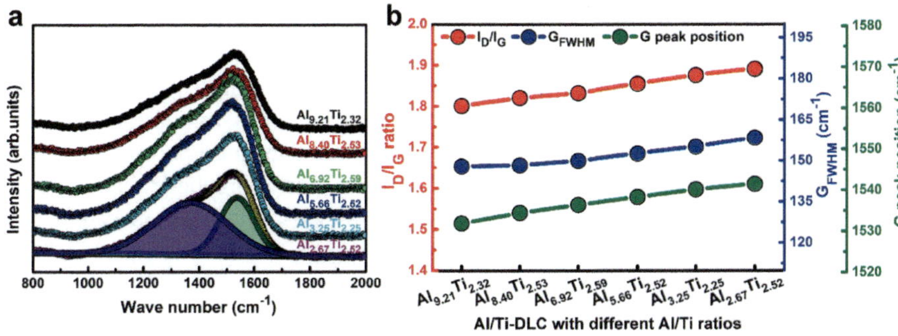

Fig. 12 **a** Raman spectra and **b** fitting results from I_D/I_G, GFWHM, and G peak position of Al/Ti-DLC films with varying Al/Ti ratios [29]

stress in DLC films, which typically measures a few GPa, contributes to their weak adhesion to different substrates. Xiao et al. [24] studied how a DLC coating on a steel base responded mechanically. They observed that increased surface roughness lowered the contact stiffness and the maximum load applied during indentation on peak asperities. On the other hand, higher compressive residual stress resulted in greater indentation loads at a fixed penetration depth. This residual stress and roughness combined led to cracking under shear stress during loading and separation under tensile stress during unloading.

Zhang et al. [25] investigated plasma-sprayed thermal barrier coatings (TBCs), examining nonlinear elastic behaviour and residual stresses using curvature-temperature plots during and after spraying. They revealed the interplay of microstructure, residual stresses, and mechanical properties. Microstructures and elastic moduli can be correlated with increased temperature or thickness. Their study showed that the coating modulus rose with compressive stress, possibly due to microcrack closure and inter-splat sliding. The axial force, reflecting residual stress and coating thickness, captured their combined impact on the coating modulus. The adhesion strength of nitrogen-doped diamond-like carbon (N-DLC) coating is higher than that of pure DLC coating due to reduced residual stress resulting from creating C-N bonds [26, 27]. Pei et al. [28] explored the advantages of creating nitrogen-doped diamond-like carbon (N-DLC) coatings on nickel-titanium alloys using plasma immersion ion implantation and deposition techniques. Their research demonstrated that nitrogen forms chemical bonds with carbon within these coatings, improving adhesion strength by reducing residual stress. With the Al/Ti ratio decreasing from 4.0 to 1.1, Al/Ti-DLC film shows steady changes in the residual stress from 0.46 ± 0.01 GPa to 0.93 ± 0.01 GPa. This rise is primarily due to the higher proportion of TiC crystallites, which introduces more disorder into the DLC structure, as shown in Raman analysis in Fig. 12. This disorder leads to easier detachment, more corrosion channels during the process of immersion, and a faster rate of diffusion of corrosive ions [29].

The accumulation of silver and gold within the DLC matrix during deposition posed a significant challenge to achieving a uniform distribution of Ag or Au nanoparticles. To address this, Paul et al. [30] synthesised uniformly dispersed Au and Ag-containing DLC films for surface plasmon and field emission applications and observed the reduction in internal stress. They also studied the modulation of mechanical properties upon Au incorporation into DLC films, attributing these changes to the sp^2/sp^3 ratio alteration upon the addition of Au into the DLC matrix. Conversely, Guo et al. [31] studied the effect of copper concentration on the residual stress in Cu-DLC films. They observed a notable decrease in residual stress in Cu-DLC films (less than 0.60 GPa) compared to pure carbon film (2.0 GPa). This reduction was attributed to the formation of antibonding interactions between Cu and C and the relaxation of distorted bond angles and lengths. The residual stress and microhardness of the DLC coatings strongly depend on the substrate ion current density, which was set after the deposition of the TiC interlayer. According to earlier research by Guo et al., adding elements like Al and Ti, particularly Al, which do not chemically bond with carbon, significantly reduces the misaligned bond angles and lengths within carbon structures. This reduction in distortion leads to a notable drop in residual compressive stress [32, 33]. Undoped DLC films demonstrate a residual stress of 0.16 GPa and a hardness of 7.1 GPa. On the other hand, DLC films with an oxygen flow rate of 0.6 standard cubic centimeters per minute show a maximum residual stress of 0.29 GPa and a hardness of 19.6 GPa [34]. Incorporating a titanium interlayer reduced residual stress by addressing the thermal expansion mismatch between the substrate and the DLC film. Similarly, including a Si-DLC interlayer also contributed to lowering residual stress. Utilising multiple interlayers increased the overall thickness, which helped in mitigating the rise in residual stress [35]. In the study of Won et al., it was presented that a graded film with a lower concentration of diamond-like carbon (DLC) content exhibits a noticeable decrease in residual stress relative to a pure DLC film of identical thickness. Specifically, while the direct numerical reduction in residual stress between the films is notable (16.7 times lower for the graded film), it's important to note that even when compared to a 150 nm thick pure DLC film, the graded film still shows a substantial 9.8 times decrease in residual stress. This reduction is considered significant for applications involving thin films. [36]. Ghadia et al. [37] studied N-DLC coatings applied to silicon substrates through thermal CVD with varying N_2 flow rates. Analysis with AFM revealed that the surface roughness of the coatings increased as the N_2 flow rate went up, likely due to surface erosion. Raman spectroscopy showed that the I_D/I_G ratio grew with higher N_2 flow, indicating reduced sp^3 bonding within the films. The coatings achieved maximum hardness and elastic modulus of 25.4 GPa and 180.1 GPa, respectively, at a 70 sccm N_2 flow rate, with both properties decreasing at higher flow rates. The coatings also exhibited a low friction coefficient of 0.04, making them suitable for mechanical and biomedical applications. According to Stoney's equation, the coatings experienced compressive stress, which dropped from 1.2 to 0.65 GPa as the N_2 flow rate increased (Fig. 13).

Reducing the modulation period (bilayer thickness) will increase the interface number for multilayer DLC films of the same thickness. More interfaces will thus be able to absorb more energy from the hard DLC layers, resulting in multilayer films

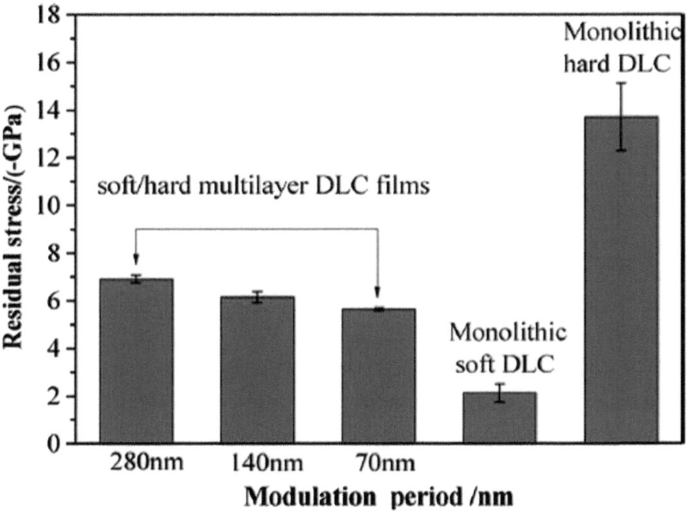

Fig. 13 Relative stress for soft/hard multilayer DLC films with varying modulation periods and monolithic DLC films [38]

with less residual stress. As a result, the compressive stress of the multilayer DLC film with a short modulation period is lower than that of the film with a large modulation period [38]. Based on ion bombardment, sp^3 bond formation, subplantation, densification, and thermodynamic potential, several models have been presented to examine the underlying mechanisms of residual stress in amorphous carbon films by McKenzie et al., Lifshitz et al., Robertson, and Kelires [39–43]. Deng et al. [44] showed that the ion current density and substrate bias voltage substantially impacted the residual stress value. The molecular dynamics simulations on DLC film deposition revealed that sp^3 hybridisation clusters initiate within sp^2 carbon formations on the film surface, transitioning via $sp^2 \rightarrow sp^3$ bond conversions during growth. This structural change, where sp^3 bonds exhibit approximately an 8.5% longer bond length than sp^2 under stress-free conditions, generates increasing compressive stress within the growing sp^3 clusters [45].

3.2 Metal Carbide and Nitride Coatings

Metal carbides and nitrides are gaining prominence across various fields, including electrochemical devices, gas sensing, photocatalysis, ceramics, environmental remediation, and medicine. Their appeal lies in their exceptional thermal stability, corrosion and wear resistance, and impressive electronic and catalytic properties. Notably, materials like ZrC, TiC, SiC, and HfC are highly valued for their durability in ceramics due to their covalent carbon networks [46].

The residual stress in these coatings can significantly impact their performance. Typically, metal carbide and nitride coatings experience compressive residual stress from deposition processes. However, tensile stresses can also develop due to thermal mismatch or defects. Understanding these stresses is key to optimising coating performance.

Recent studies have shed light on how residual stress affects coating behaviour. Zhang et al. studied the impact of powder feed rate on the mechanical characteristics and microstructure of Ni–Cr-B-Si–C coatings sprayed by supersonic plasma. They found that increased PFR led to greater porosity, yet residual stress first declined, eventually reached a minimum, and subsequently increased in response to PFR [47]. Similarly, Dummer et al. used a non-destructive method to analyse residual stresses in PVD-Ti (C, N)-graded coatings, revealing that compressive residual stresses improved load-bearing capacity and fracture strength [48]. Lugscheider et al. explored (Cr, Al)N coatings deposited via arc ion plating and observed that high compressive stresses resulted from elevated ionisation rates and substrate bias voltage. Their study correlated coating thickness with residual stress using stripe bending tests and X-ray diffraction (XRD) and assessed the effect of compressive stresses on wear behaviour through impact testing. [49].

In addition, Wang et al. [50] studied residual stresses in WC–Co coatings applied via detonation gun (D-Gun) spraying at various thicknesses, using the Stoney formula and accurate curvature measurements. They observed a transition from tensile to compressive stresses as the coating thickness increased, with stress nearing a stress-free state at approximately 365 μm. The research highlighted the role of peening stress in influencing the final stress state due to the dynamics of D-Gun spraying. It also demonstrated how compressive stresses can improve coating properties, while tensile stresses can have a detrimental effect. Cavaleiro et al. [51] explored W-Si–N coatings created through reactive sputtering in an $Ar + N_2$ environment using a tungsten target with different silicon levels, followed by annealing at 900 °C. The films were deposited on various iron-based substrates with distinct thermal expansion coefficients. Their research examined the Young's modulus of films, residual stress, hardness, structure, and chemical composition. The incorporation of silicon and nitrogen influenced whether the films were crystalline or amorphous. A significant connection between hardness and residual stress was observed, with certain films demonstrating tribological benefits as indicated by H/E ratios exceeding 0.1. Bouaouina et al. [52] investigated a multilayer Mo_2N/CrN coating deposited on a silicon substrate through reactive magnetron sputtering. They analysed the impact of varying period thickness on the coating. Mo_2N was found to have a face-centred cubic structure, while CrN had an orthorhombic structure. The study measured compressive residual stresses in Mo_2N and tensile residual stresses in CrN, correlating these with the structural properties of coatings. Nanoindentation was also used to evaluate the mechanical performance of the coating. Ang et al. highlighted the effectiveness of cold spray nanostructured tungsten carbide cobalt (WC–Co) coatings for wear applications. They analysed the residual stress profiles of these coatings, which were applied using advanced KM systems. The residual stress introduced during the coating process considerably impacts the fatigue properties of these critical components [53]. Meghwal et al. investigated electrolytic hard chrome (EHC) plating coatings, mainly made up of WC and Co, with slight compositional differences. They conducted quantitative Rietveld analysis and discovered that these coatings exhibited compressive residual stresses primarily caused by thermal contraction rather than peening. The wear resistance, measured using the pin-on-disk technique, was on par with traditional WC–Co coatings [54]. Dang et al. analysed the mechanical properties of titanium nitride (TiN) thin films deposited via HiPIMS on cold-sprayed titanium (Ti) layers. The TiN/Ti duplex coatings were applied to Ti substrates with three different cold spray Ti layer configurations. Surface roughness, hardness, and Young's modulus were determined using nanoindentation and atomic force microscopy (AFM). The ring-core drilling method was employed to measure residual stress. Scanning electron microscopy and focused ion beam milling were also applied to examine stress at varying depths [55].

Fan et al. investigated four WC–10Co–4Cr feedstock powders with varying WC particle sizes. These powders underwent fast sintering and agglomeration processes before being sprayed with High-Velocity Oxygen Fuel (HVOF) onto steel substrates. They observed that larger WC particles led to higher compressive stress, offering valuable insights for optimising feedstock powder design to enhance the durability of coatings [56]. Gopalan et al. [57] investigated residual stresses in multi-metal carbide coatings, revealing that deposition and thermal factors influence these stresses. According to Fig. 14, residual stresses are highly compressive at the lowest growth temperature of 573 K, with a value of -4830 ± 240 MPa. As the temperature increases, the magnitude of these stresses decreases, eventually transitioning to a tensile stress of 570 ± 90 MPa at 1023 K.

Moreover, the disparity in thermal expansion rates between the coating and the substrate leads to significant tensile stress as the coating undergoes cooling. For instance, in a sample deposited at 873 K, the coating's higher thermal expansion coefficient relative to the silicon substrate and silicon nitride barrier layer results in considerable tensile residual stress. The nanoindentation tests performed by Hamid et al. indicated that the indenter penetration depth increases with applied load, with ZrN/ZrCN coatings showing depths within the recommended range (10% of coating thickness) while Zr/ZrN coatings have deeper penetrations (11–20% of thickness) due to their reduced hardness. This implies that Zr/ZrN coatings are softer and more prone to deformation. Consequently, ZrN/ZrCN coatings, being harder and denser, are likely to experience lower residual stress compared to Zr/ZrN coatings. These studies collectively showcase the significance of residual stress control in fabricating and using metal nitride and carbide coatings.

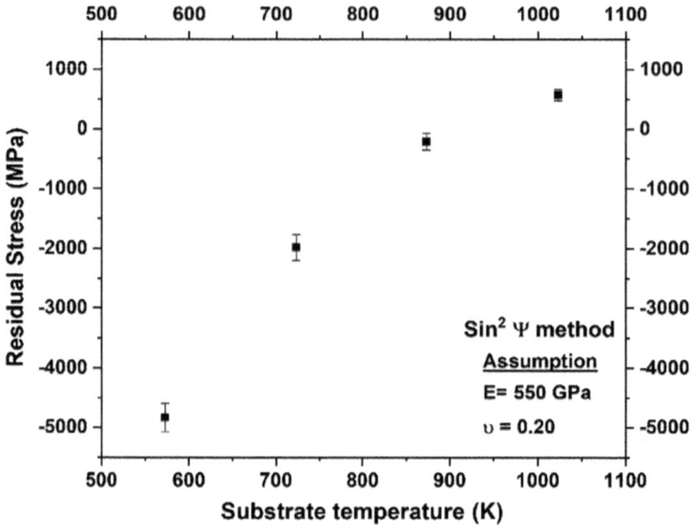

Fig. 14 Residual stress determined as a function of deposition temperature for (HfNbTaZr)C coatings [57]

3.3 *Carbonitride Coatings*

Carbonitride coatings, a specialised class of hard coatings incorporating carbon and nitrogen, have become a focal point of research due to their exceptional surface properties. These coatings are typically applied to enhance the surface properties of metals and alloys, improving hardness, wear, and corrosion resistance. As usual, residual stress in carbonitride coatings also arises from the deposition process and subsequent thermal treatments. During deposition, carbon and nitrogen atoms are incorporated into the coating material, often leading to lattice distortions resulting in residual stresses. Residual stresses in these coatings can be either compressive or tensile, depending on the deposition process and differences in thermal expansion between the coating and the substrate. Understanding and controlling these residual stresses is crucial for optimising the performance and longevity of carbonitride coatings in practical applications.

Recent advancements in coating technologies have unveiled a range of fascinating findings on residual stress in various thin films and coatings. Perry et al. primarily focused on measuring residual stress in Ti(CN) films made by reactive ion plating and their steel substrates, mainly stainless steel for decorative parts and high-speed steel (HSS) for tooling applications. Due to differential thermal contraction effects, the compressive residual stress for Ti(CN) films deposited onto HSS by Chemical Vapour Deposition (CVD) is 920 MPa. Ion-plated films on stainless steel had much greater residual stress, most likely due to shot peening effects and trapped argon [58]. During the 1990s, the Ti-C-N coating proved successful in machining settings with low process temperatures. However, industrial coating practices of the time highlighted many challenges with the materials and processes that came up during the deposition of Ti-C-N [59].

Karlsson et al. investigated how residual stress and composition affect arc-evaporated TiC_xN_{1-x} thin films. They found that decreasing the substrate bias and increasing the carbon content led to higher intrinsic compressive stress, up to −7 GPa, limited by internal cracking. Hardness improved with stress and carbon content, peaking at 44 GPa with −5.5 GPa stress. For TiN films, hardness varied between 33 GPa at −5.8 GPa stress and 26 GPa at −1.2 GPa stress. Young's modulus remained around 610 GPa, with minor reductions at lower stress levels [60]. Further advancements were made by varying the reactive gas flow ratio, significantly reducing biaxial residual stress while improving thermal stability. This demonstrated that the initial hardness and stress are closely linked to microstructural characteristics rather than macroscopic stress alone [61].

Murotani et al. employed the arc ion plating (AIP) technique to deposit polished TiCN thin films. Their research highlighted several significant findings. The residual stress in these films was predominantly compressive, ranging from −4.6 to −5.9 GPa. The TiCN thin films exhibited a fiber texture along the <111> direction. Moreover, the compressive residual stress in the films relaxed with increasing annealing temperature, decreasing stress levels. These results underscore the influence of deposition conditions and thermal treatments on the mechanical performance and structural characteristics of TiCN thin films [62]. In the study of Peters et al., tetrakis-dimethylamino titanium, an organo-metallic chemical vapor deposition (CVD) precursor, was used to create titanium carbo-nitride (Ti-C-N) coatings. The coatings exhibited hardness values ranging from 8 to 14 GPa, assessed through nanoindentation, while their residual stresses, determined via wafer curvature techniques, fell within the range of −440 to −100 MPa. The coefficient of friction ranged from 0.2 to 0.45, with coatings containing less titanium showing the lowest values [63]. Kanchanomai et al. investigated carbonitrided AISI 1015 steel under cyclic loading, revealing significant residual stress and fatigue behaviour findings. The surface exhibited a high compressive residual stress of approximately 900 MPa, transitioning to a stable low tensile stress of about 20 MPa deeper within the material. This residual stress distribution significantly enhanced the material's fatigue resistance. Compared to untreated AISI 1015 steel, the carbonitrided variant demonstrated superior fatigue performance, with a measured fatigue limit of 340 MPa, while untreated steel exhibited a lower limit of 300 MPa [64]. In Gonzalo et al.'s study, residual stresses in ZrCN and TiN coatings on WC tools were analysed, revealing highly compressive states ranging from −4.58 GPa to −7.99 GPa for ZrCN and from −5.8 GPa to −11.44 GPa for TiN after deposition, influenced by bias voltage. Thicker coatings generally exhibited higher compressive stresses, correlating with increased wear resistance in machining tests on AISI 1045 and AISI 4340 materials. However, greater compressive stresses were associated with shorter tool life and more excellent flank wear (VB), contrasting with findings from other studies [65]. Wang et al.'s investigation into CNx coatings revealed that raising the N2/Ar flow ratio from 0.05 to 10 resulted in a steady decline in residual stress, from 4.4 to 2.3 GPa. Initially, the hardness remained constant at 28.0 GPa, with the ratio increasing from 0.05 to 0.1,

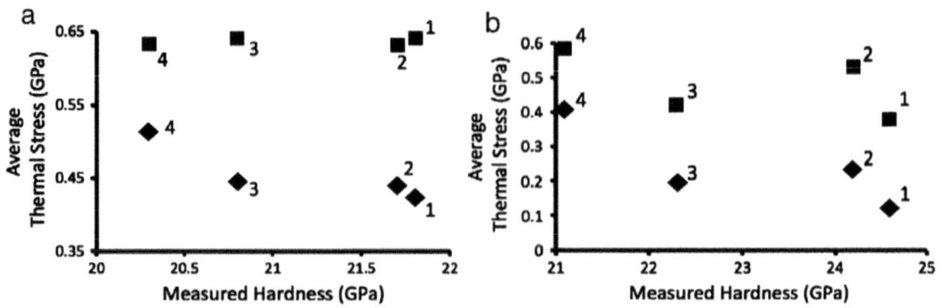

Fig. 15 Relationship between hardness and average residual thermal stress for **a** TiC_xN_{1-x} and **b** α-Al_2O_3 [67]

but then experienced a significant drop to 24.0 GPa, stabilising after that as the ratio continued to rise. The introduction of nitrogen into the gas mixture influenced several factors, including chemical sputtering, reduction in residual stress, modification of hardness, and changes in surface texture during deposition. Replacing carbon atoms with nitrogen likely reduced the average coordination number, leading to decreased residual stress [66]. Chien et al. studied the microstructures of four α-Al_2O_3 and TiC_xN_{1-x} multilayer coatings on WC–Co cemented carbide tool bit inserts. They developed realistic models to compute residual thermal stresses using finite elements. Their findings showed that the coatings' channel fissures reduced thermal stresses, with the degree of stress reduction increasing as film thickness grew. Furthermore, they found an inverse relationship between the computed mean thermal stress and the coating's hardness and fracture toughness as determined by nanoindentation. In other words, the hardest sample exhibited the lowest mean thermal stress [67] (Fig. 15).

Wang et al. fabricated CN_x coatings using a hybrid coating method of DC magnetron sputtering and radio frequency plasma-enhanced chemical vapor deposition, varying sputtering power and substrate voltage (200W 400 V, 100W 400 V, 200W 800 V, and 100W 800 V). Raising the substrate bias voltage and lowering the sputtering power decreased the film growth rate, hardness, and nitrogen-to-carbon atomic ratio of the CN_x while increasing the residual stress. The highest hardness of 15.0 GPa and the lowest residual stress of 3.7 GPa were achieved at 400 V 200W. In contrast, the lowest friction coefficient, 0.12, was observed at 800 V, 100W. Raman and X-ray photoelectron spectroscopy (XPS) analyses showed decreased sp3 carbon bonding and increased sp^2 carbon bonding with the changes in substrate bias voltage and target sputtering power [68]. The thin films consistently display tensile residual stress, decreasing as the reaction gas proportion increases. The highest tensile stress, reaching around 91 MPa, occurs with an $Ar/N_2/C_2H_2$ ratio of 15/4/2, while the lowest, approximately 32 MPa, is observed with a ratio of 15/6/4. These variations in residual stress are influenced by the structural characteristics inherent to the coatings [69]. To assess the residual stress in TiCN/TiNbCN coatings, Caicedo et al. evaluated the relationship between residual stress and varying thicknesses of silicon wafers (275–350 μm) with a Poisson's ratio of 0.223 and a Young's modulus of 162.91 GPa. Their findings indicate that as the number of bilayers in the coatings increases, the residual stress, initially high and compressive, gradually decreases and trends towards a tensile state. In other words, the stress transitions from being strongly compressive to less compressive and begins to shift towards tensile [70]. In this study, CrCN films with varying carbon content were deposited using RF-reactive magnetron sputtering. Higher carbon content resulted in the formation of carbon and CN_x and increased compressive residual stress. Combined with solid solution strengthening, this enhanced the hardness from around 17 GPa at 0 at.% carbon to approximately 19 GPa at 6.3 at.% carbon [71]. These evolving techniques highlight the significant advancements in regulating residual stress in carbonitride coatings. So, the exploration of residual stress in thin films and coatings has yielded numerous insights into their mechanical and structural properties.

3.4 *MoS_2 (Molybdenum Di Sulphide) Coatings*

MoS_2, known as the "king of lubrication," features a hexagonal close-packed crystal structure, making it an exceptional solid lubricant. MoS_2 is a layered two-dimensional transitional metal dichalcogenides (TMDs) family member. Its crystal structure comprises covalently bonded sheets stacked together through weak van der Waals forces, akin to the arrangement seen in graphite and hexagonal boron nitride. With its low friction coefficient, MoS_2-based coatings are extensively utilised [72]. MoS_2, especially in thin layers, demonstrates low friction in dry conditions. Naturally occurring as molybdenite, MoS_2 exhibits strong in-plane bonding and weak out-of-plane bonding, leading to highly anisotropic mechanical properties and easy shearing of the stacks. Single-layer or few-layer MoS_2 can be produced and studied, and it has various structural forms depending on the bonding within and between the sheets. It serves as an effective lubricant in various applications: as a powder blended with greases or oils, or applied via spray coating, pulsed laser ablation, or chemical deposition. Nano-sized MoS_2 particles (n-MoS_2) are used in solid lubrication, oil additives, and self-lubricating polymers, offering superior tribological characteristics compared to conventional particles. The performance and morphology of ultrafine MoS_2 are influenced by its preparation methods. While MoS_2 films can deteriorate in oxygen-rich environments, they help reduce frictional heat. Studies indicate that MoS_2 nanotubes and nanosheets enhance friction reduction and wear resistance more effectively than larger particles or commercial micro-MoS_2. Additionally, MoS_2 composites in high-density polyethylene improve tribological performance. The process of electrophoretic deposition of MoS_2 into base oils can be fine-tuned to enhance film properties, with kinetic models aiding in the understanding and optimisation of the deposition process (Fig. 16).

J. K. Lancaster wrote in his paper in 1996 that the hardness of the substrate metal significantly affects the performance of MoS_2 as a solid lubricant. Soft-metal substrates enhance the adhesion of MoS_2 films, while hard-metal substrates improve their cohesion and uniformity. By pre-coating hard substrates with a thin film of a softer metal, such as Pb, Au, or Ag, the performance of MoS_2 as a lubricant is

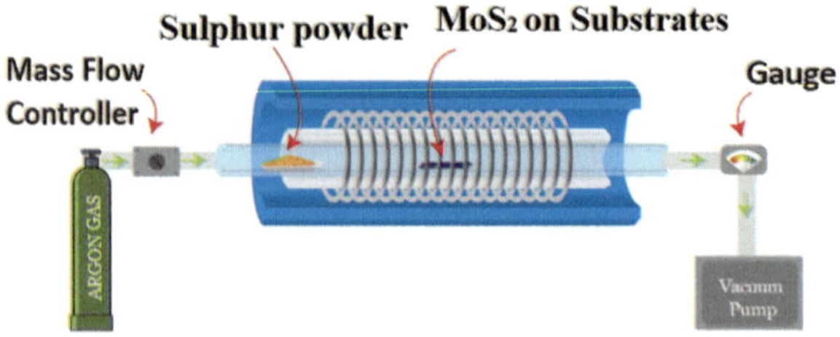

Fig. 16 Schematic diagram of MoS_2 growth using the chemical vapor deposition (CVD) system [73]

further enhanced. Moreover, the primary cause of wear in lubricated metal surfaces is the abrasion by MoS_2 itself [74]. According to Holmberg et al., differential thermal contractions brought on by the deposition temperature resulted in compressive strains of 160 MPa on silicon wafers and tensile stresses ranging from 800 to 1300 MPa on steel substrates [75]. Additionally, compressive internal stresses between 100 and 500 MPa were noted by Weise et al. Pa [76] on silicon wafers. These stresses were relatively consistent for argon pressures between 1.3 Pa and 4.0 Pa, and they reduced when argon pressure dropped from 0.2 Pa to 1.3.

Residual stress in MoS_2 coatings is akin to the hidden force driving a high-stakes performance. Every tweak in the coating process, from substrate selection to doping, meticulously refines the coating's resilience. Just as a conductor fine-tunes an orchestra, scientists manipulate these stresses to ensure MoS_2 coatings perform flawlessly under extreme conditions, enhancing their wear resistance and longevity. Guozheng et al. [77] utilised a nano electro-brush plating method to create a Ni/MoS_2-C coating on AISI 52,100 steel. Their research indicated that adding nano graphite particles improves the mechanical performance of the coatings by lowering the residual stress to 16 MPa and increasing resistance to moisture. XRD analysis revealed that the residual stresses for the Ni/MoS_2-C and Ni/MoS_2 coatings were 16 MPa and 45 MPa, respectively. Mechanical stress measurements conducted by Vierneusel et al. [78] revealed that the films exhibited compressive stress levels ranging from 55 to 386 MPa. Films with higher compressive residual stresses showed lower wear rates in both air and vacuum environments. These films also demonstrated the highest hardness values, up to approximately 19 MPa, and were highly wear-resistant in air and vacuum conditions. Additionally, they exhibited high values of the hardness-to-modulus ratio, approximately 0.3. Overall, they concluded that residual stresses and the H/E ratio are reliable indicators of wear resistance in MoS_2 coatings. Vierneusel et al. [79] focused on how compressive stresses, in particular, might enhance wear resistance and highlight the vital role that residual stresses play in the tribological performance of sputtered coatings. They noted that these stresses are tied to the coating's microstructure, and their effects are typically considered during tribological evaluations. To clarify the specific impact of residual stresses on coating performance, the study aimed to separate this effect from microstructural influences. They achieved this by mechanically pre-stretching MoS_2-coated substrates during coating deposition and subsequently relaxing them to minimise macroscopic residual stresses in the coating.

Saha et al. [80] study highlights that structural mismatch, growth-induced stress, and thermal factors primarily influence residual stress in coatings. Despite similar crystal structures and room temperature deposition, residual stress variations do not significantly impact the critical load, which correlates more strongly with the hardness of the coatings. Coatings with higher titanium content exhibit greater hardness and brittleness, while those with higher MoS_2 content are softer and more prone to spallation. Adding MoS_2 nanosheets to Ni-MoS_2 composite coatings in a Watts bath significantly refines the microstructure of Ni deposits and concurrently reduces compressive residual stresses. This refinement is beneficial for improving protective properties, such as retarding hydrogen permeation and enhancing resistance to hydrogen embrittlement in QP steel. The concentration of MoS_2 in the electrodeposition bath can be adjusted to control the MoS_2 content in the coatings, optimising their microstructure and managing residual stresses effectively [81]. Residual stresses in MoS_2 nanocrystalline films were analysed using the $\sin^2\psi$ method, revealing positive slopes indicating tensile stresses and negative slopes indicating compressive stresses. Changes in residual stress were observed with varying pulsing bias voltage, where lower voltages facilitated ion relaxation and the formation of a softer or harder multilayer structure, reducing stress levels. Higher bias voltages, however, intensified ion bombardment and ion pinning effects, resulting in increased residual stress due to structural irregularities and voids formed during growth. Effective stress reduction strategies included subplantation and ion pinning techniques, particularly effective around −600 V bias. In contrast, higher biases like −800 V led to higher residual stress and structural aberrations in the films [82] (Fig. 17).

Jue et al. [83] computed using Stoney's equation that the Mo-S–N film with 34.9 at.% nitrogen has a compressive residual stress (σ) of around −2.1 GPa, whereas the MoS_x film is −1.1 GPa. This stress's two primary components are residual internal and thermal stress. The addition of nitrogen induces lattice distortion, which contributes to an increase in residual internal stress within the films. Long et al. improved the friction qualities at high temperatures by modifying CrN coatings with Mo and S co-doping and varying MoS_2 concentrations via unbalanced magnetron sputtering. They measured the residual stress of K1–K4 coatings using X-ray elastic

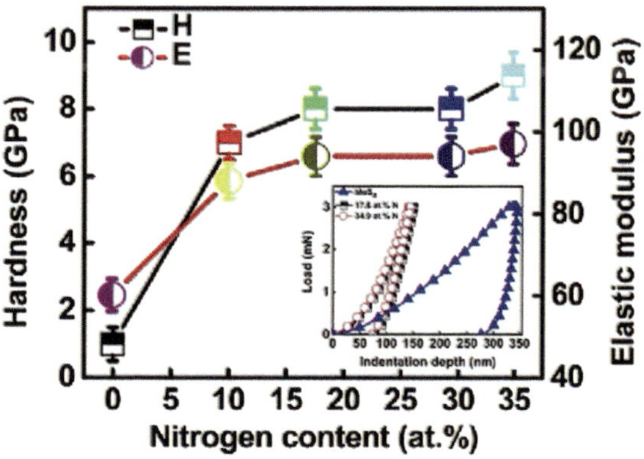

Fig. 17 Hardness (H) and elastic modulus (E) of Mo-S–N films with varying nitrogen content. Inside: Load vs. displacement curves for MoS_x and Mo-S–N films at 17.6 and 34.9 at.% nitrogen [83]

strain analysis and applied constitutive equations based on generalised Hooke's law. With a larger MoS_2 concentration, the coatings increased compressive internal stress from − 0.93 GPa to − 0.12 GPa and − 0.25 GPa. The stress is caused by high-energy ions that are implanted within a few nanometers of the coating surface during development, producing outward expansion from the substrate, according to d'Heurle's model, which describes how compressive stress creation in coatings is caused by ion bombardment [84]. Zhang et al. [85] investigated residual stresses in grinding processes, noting that dry grinding with a friction coefficient of 0.592 leads to higher grinding temperatures (289.9 °C) and lower surface residual stress (272.9 MPa) due to increased sliding friction. In contrast, flood cooling (friction coefficient 0.513) optimises cutting force and achieves lower temperatures (156.7 °C), resulting in higher surface residual stress. The research emphasises the combined influence of mechanical and thermal factors on residual stress levels. Experimental validation demonstrates that MoS_2 nanoparticles in minimum quantity lubrication enhance lubrication and cooling, lowering friction coefficients and temperatures. Future work aims to refine predictive models and explore initial residual stress impacts from manufacturing processes.

MoS_2 was effectively integrated into the DLC matrix when Chen et al. synthesised MoS_2-DLC composite films with varied quantities of MoS_2 pellets, with the Mo content in the films ranging from 0 to 6.15 at.%. When deposited, the MoS2-DLC composite films showed a dense, featureless structure. MoS_2 doping enhanced their adhesion strength and internal stress while decreasing their hardness and elastic modulus [86] (Fig. 18).

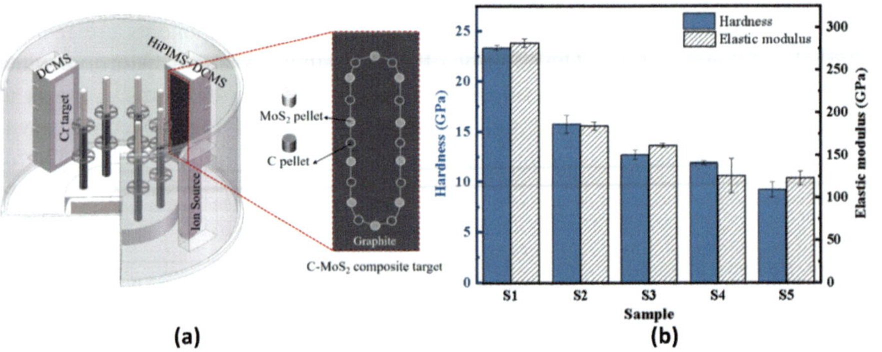

Fig. 18 **a** Schematic of the HiPIMS-DCMS deposition system with a C-MoS_2 composite target **b** Hardness and elastic modulus of samples (S1to S5) [86]

3.5 Other Types of Coatings

The continuously advancing field of coating technologies keeps revealing a deeper understanding of the complexities of residual stress and its significant effects on performance. This exploration expands beyond the previously mentioned coatings, addressing various materials and applications where residual stress is crucial. A comprehensive understanding of these stresses is essential for optimising the performance of individual coatings and enhancing the overall longevity and dependability of coatings across various materials and industries. Li et al. [87] investigated plasma-sprayed yttrium-stabilised zirconia (YSZ) splats and coatings on 316L stainless steel at different temperatures. They found that deposition temperature significantly affected the flattening behavior of YSZ splats and the interfacial bonding due to Zr cation diffusion. While phase composition was minimally affected, higher deposition temperatures improved surface smoothness, reduced microcracks, and densified microstructures. This led to notable enhancements in hardness, toughness, and cohesive strength. Adjusting deposition temperatures also enabled the desired compressive residual stress levels in the coatings. Ghara et al. [88] studied plasma-sprayed alumina coatings by modifying the temperature and velocity of the particles. They utilised the X-ray $\sin^2\psi$ technique to measure the residual stress profiles. Increasing particle temperature notably enhanced hardness and indentation modulus. Varied velocity initially decreased coating porosity, but at high speeds, porosity slightly increased. Mechanical properties were closely tied to porosity levels. Higher particle temperature is correlated with increased tensile residual stresses in the coatings. Additionally, an annealing effect during spraying led to partial recovery of the compressive substrate conditions caused by grit blasting. Residual stress is critical in determining coating and material performance, impacting mechanical properties, wear resistance, and adhesion (Fig. 19).

Understanding and controlling residual stress is vital to optimise coating reliability. Various sources, including mechanical, thermal, and assembly factors, can induce residual stresses. Residual stress can have beneficial effects, like enhancing fatigue resistance or adverse effects, like promoting crack growth. Mitigation strategies include heat treatments, peening, and process optimisation to manage unwanted residual stresses. Various industries benefit from managing residual stress, ensuring the durability and performance of critical components [90]. Das et al. [89] studied the impact of laser remelting on plasma-sprayed alumina and chromia coatings. They observed that remelting caused tensile residual stress in alumina coatings and compressive stress in chromia coatings. The tensile stress in alumina increased by up to 500%, while compressive stress in chromia decreased by up to 80%. Higher cooling rates led to increased tensile stress in the remelted layers. The initial nature of the residual stress was preserved in the remelted layers, and the residual stress decreased with a more significant overlap of remelted tracks. Hamedani et al. [91] studied the influence of residual stress on the tribological behavior of calcium-phosphate (CaP) coatings on titanium (Ti) substrates. Using an electrospinning and rapid heating and cooling (EMRHC) process, they measured residual stress with X-ray methods

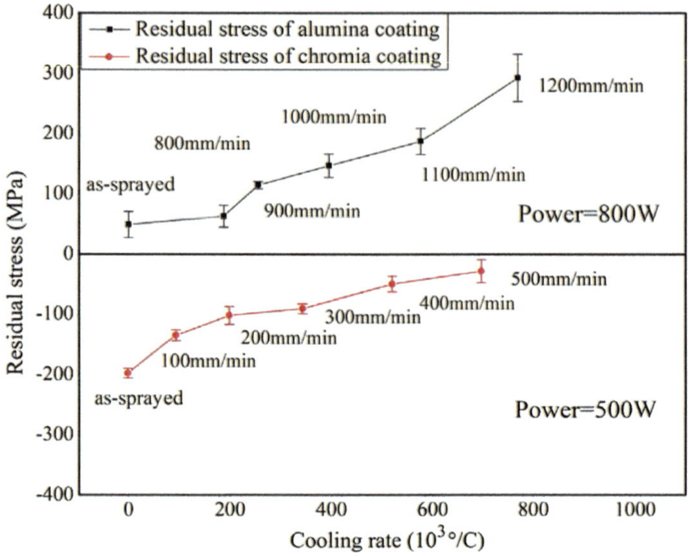

Fig. 19 Variation in surface residual stress of the remelted coatings with variation in cooling rate [89]

at both halfway and fully completed stages. Hydroxyapatite on Ti transformed into several compounds during the process. Tensile stress was $+591 \pm 89$ MPa at halfway, and compressive stress was -189 ± 42 MPa at completion. The EMRHC process enhanced Young's modulus, hardness, and peak tensile strength. Tribological tests in dry and simulated body fluid conditions revealed that the fully processed CaP coating exhibited the most negligible wear due to compressive residual stress, and the coefficient of friction significantly decreased in simulated body fluid.

4 Conclusion

This chapter has summarised how processes, temperature, substrate thickness, substrate characteristics, and coating influence residual stress in coatings. It provides a detailed outline of fundamental concepts related to residual stress, including its origins, measurement methods, and effects on mechanical coatings. Practical guidance on selecting and applying appropriate measurement techniques is also offered, enabling readers to make informed decisions based on their specific needs.

Several conclusions about residual stress in coatings can be drawn from the discussion. Hard PVD coatings typically exhibit high compressive residual stresses, ranging from 0.5 to 4 GPa, with some cases reaching up to 11 GPa. In contrast, hard CVD coatings show residual stresses ranging from 0.5 GPa in tension to 2 GPa in compression. When comparing coatings on different substrates, residual stress values are significantly higher for steel substrates than for silicon substrates. For example, TiN coatings on steel exhibit compressive stresses of 2.5–4.0 GPa, whereas on silicon, they are 0.9–1.3 GPa. DLC coatings show 0.03–0.24 GPa on steel compared to 0.1–0.7 GPa on silicon, and MoS_2 on silicon shows both compressive (0.16 GPa) and tensile (0.8–1.3 GPa) stresses. FEM modeling and scratch test simulations reveal that initial compressive residual stress reduces observed maximum tensile stresses in untouched regions of the coating but have minimal effects in deformed areas, suggesting stress relaxation under contact deformation. Fracture toughness data is limited and varies significantly, with TiN on steel ranging from 3.8 to 8.7 $MPam^{1/2}$ and DLC on silicon from 1 to 10.9 $MPam^{1/2}$. The study found 4–11.1 MPam1/2 fracture toughness values for TiN, 1–1.9 $MPam^{1/2}$ for DLC, and 1.8 $MPam^{1/2}$ for MoS_2-coated steel surfaces. Coatings with higher compressive residual stresses and elastic modulus tend to exhibit greater fracture toughness.

This chapter addresses the causes, implications, and mitigation strategies for residual stresses, highlighting the crucial role of stress management in enhancing coating performance and reliability across various industries. This comprehensive resource enriches the understanding of residual stresses and supports the practical application of measurement techniques to improve material performance and durability. Further research in this area will benefit many fields of coating technology.

References

1. Palmieri V, Caldarola G, Cisternino S, Pira C, Skliarova H, Stivanello F, Franco-Lespinasse DA (2017) The way of thick films toward a flat Q-curve in sputtered cavities. In: Schaa RWV, He Y, Li L, Zhao N (eds) Proceedings of the 18th international conference on RF superconductivity, Lanzhou China, 2017
2. Liu D, Flewitt PEJ (2014) Raman measurements of stress in films and coatings. In: Yarwood J, Douthwaite R, Duckett S (eds) Spectroscopic properties of Inorganic and organometallic compounds, 1st edn. Royal Society of Chemistry, pp 141–177
3. Maleki E, Unal O, Amanov A (2018) Novel experimental methods for the determination of the boundaries between conventional, severe and over shot peening processes. Surfaces and Interfaces 13:233–254
4. Dutta S, Madhuri A, Jena S, Swain BP (2024) Investigation of the chemical network, electronic environments and electrochemical performance of Fe_3O_4/ZnO/rGO nanocomposites. Bull Mater Sci 47:280
5. Jena S, Madhuri A, Gupta M, Swain BP (2024) Comparative assessment and experimental analysis of atomic coordination and mechanical properties of sputter deposited Ti-rich TiN, TiC, and TiCN thin films. J Alloy Compd 1002:175217
6. Madhuri A, Jena S, Dutta S, Swain BP (2025) Evolution of electroless Cu/CuxO nanostructures thin films: study of surface fractal, tuning band gap and electronic structure. Arab J Sci Eng 50:6789–6800
7. Swain BP (2007) Effect of residual stress on p doped nano-crystalline silicon deposited by HWCVD films. Adv Mater Res 26–28:653–656
8. Aruchamy N, Schenk T, Kovacova V, Glinsek S, Defay E, Granzow T (2021) Influence of tensile vs. compressive stress on fatigue of lead zirconate titanate thin films. J Eur Ceram Soc 41:6991–6999

9. Ghasemi AR, Tabatabaeian A (2022) A unified comparative study on the classical lamination theory, hole drilling, slitting, and curvature measurement methods for assessment of residual stress in laminated polymeric composites. Mech Mater 167:104267
10. Santus C, Neri P, Romoli L, Cococcioni M (2024) Residual stress determination with the Hole-drilling method on FDM 3D-printed precurved specimen through digital image correlation. Appl Sci 14:3992
11. Buttle DJ (2013) Magnetic methods. In: Schajer GS (ed) Practical residual stress measurement methods, 1st edn. John Wiley & Sons Ltd., Oxford, pp 225–258
12. Nahak B (2017) Material characterisation using Barkhausen noise analysis technique-a review. Indian J Sci Technol 10:1–10
13. Moorthy V, Shaw BA, Hopkins P (2006) Surface and subsurface stress evaluation in case-carburised steel using high and low frequency Barkhausen emission. J Magn Magn Mater 299:362–375
14. Bourke MAM (1990) Residual stress measurement in engineering components by neutron diffraction. Dissertation, University of London
15. Em V, Ivan K, Woo W, Mikula P (2024) Neutron diffraction measurements of residual stresses for ferritic steel specimens over 80 mm thick. Metals 14:638
16. Loudon R (2013) The Raman effect in crystals. Adv Phys 13(52):423–482
17. Becker M, Scheel H, Christiansen S, Strunk HP (2007) Grain orientation, texture, and internal stress optically evaluated by micro-Raman spectroscopy. J Appl Phys 101:063531
18. Ganesan S, Maradudin AA, Oitmaa J (1970) A lattice theory of morphic effects in crystals of the diamond structure. Ann Phys 56(2):556–594
19. Wolf ID, Maes HE, Jones SK (1996) Stress measurements in silicon devices through Raman spectroscopy: bridging the gap between theory and experiment. J Appl Phys 79:7148–7156
20. Jena S, Madhuri A, Dutta S, Swain BP (2024) Investigation of structural, phonon modes and electronic environment of GaP: Si nanostructure thin films. Phys B: Condens Matter 677:415664–415671
21. Liu D, Flewitt PEJ (2014) Raman Measurements of Stress in Films and Coatings. In: Douthwaite R, Duckett S, Yarwood J (eds) Spectroscopic properties of inorganic and organometallic compounds: technique, materials and applications, 1st edn. The Royal Society of Chemistry, p 141–177
22. Ardigo MR, Ahmed M, Besnard A (2014) Stoney formula: investigation of curvature measurements by optical profilometer. AMR 996:361–366
23. Sharma AK, Narayan RJ, Narayan J, Jagannadham K (2000) Structural and tribological characteristics of diamond-like carbon films deposited by pulsed laser ablation. Mater Sci Eng, B 77(2):139–143
24. Xiao Y, Wu L, Luo L, Zhou L (2020) Mechanical response of thin hard coatings under indentation considering rough surface and residual stress. Diam Relat Mater 105:107991
25. Zhang X, Watanabe M, Kuroda S (2013) Effects of processing conditions on the mechanical properties and deformation behaviors of plasma-sprayed thermal barrier coatings: evaluation of residual stresses and mechanical properties of thermal barrier coatings based on in situ curvature measurement under a wide range of spray parameters. Acta Mater 61(4):1037–1047
26. Koshinen J, Hirvonen JP, Levoska J, Torri P (1996) Tribological characterisation of carbon-nitrogen coatings deposited by using vacuum arc discharge. Diamond Relat Mater 5:669–673
27. Jan DJ, Ai CF, Lee CC (2004) Deposition of nitrogen-containing diamond-like carbon films on acrylic substrates by an ion beam process. Vacuum 74:531–538
28. Pei YL, Luan Y (2013) Surface modification of NiTi alloys using nitrogen doped diamond-like carbon coating fabricated by plasma immersion ion implantation and deposition. J Alloy Compd 581:873–876
29. Xu X, Zhou Y, Liu L, Guo P, Li X, Lee KR, Cui P, Wang AC (2019) Corrosion behavior of diamond-like carbon film induced by Al/ti co-doping. Appl Surf Sci 509:144877
30. Paul R (2017) Uniformly dispersed nanocrystalline silver reduces the residual stress within diamond-like carbon hard coatings. Nano Struct Nano Objects 10:69–79

31. Guo P, Li X, Sun L, Chen R, Ke P, Wang A (2017) Stress reduction mechanism of diamond-like carbon films incorporated with different Cu contents. Thin Solid Films 640:45–51
32. Guo T, Kong C, Li X, Guo P, Wang Z, Wang A (2017) Microstructure and mechanical properties of Ti/Al co-doped DLC films: dependence on sputtering current, source gas, and substrate bias. Appl Surf Sci 410:51–59
33. Li X, Guo P, Sun L, Zuo X, Zhang D, Ke P, Wang A (2017) Ti/Al co-doping induced residual stress reduction and bond structure evolution of amorphous carbon films: an experimental and ab initio study. Carbon 111:467–475
34. Safaie P, Eshaghi A, Bakshi SR (2016) Structure and mechanical properties of oxygen doped diamond-like carbon thin films. Diamond Relat Mater 70:91–97
35. Maruno H, Nishimoto A (2018) Adhesion and durability of multi-interlayered diamond-like carbon films deposited on aluminium alloy. Surf Coat Technol 354:134–144
36. Won YJ, Ki H (2014) Effect of film gradient profile on adhesion strength, residual stress and effective hardness of functionally graded diamond-like carbon films. Appl Surf Sci 311:775–779
37. Ghadai RK, Das S, Kalita K, Shivakoti I, Mondal SC, Swai BP (2020) Effect of nitrogen (N_2) flow rate over the tribological, structural and mechanical properties diamond-like carbon (DLC) thin film. Mater Chem Phys 260:124082
38. Xu ZY, Zheng YJ, Sun H, Leng YX, Huang N (2012) Numerical and experimental study of residual stress of multilayer diamond-like carbon films prepared by filtered cathodic vacuum arc deposition. IEEE Trans Plasma Sci 40:2261–2266
39. McKenzie DR (1996) Tetrahedral bonding in amorphous carbon. Rep Prog Phys 59:1611
40. McKenzie DR, Muller D, Pailthorpe BA (1991) Compressive-stress-induced formation of thin-film tetrahedral amorphous carbon. Phys Rev Lett 67(6):773–776
41. Lifshitz Y, Kasi SR, Rabalais JW (1989) Subplantation model for film growth from hyperthermal species: application to diamond. Phys Rev Lett 62:1290
42. Kelires PC (2000) Intrinsic stress and local rigidity in tetrahedral amorphous carbon. Phys Rev B 62:15686
43. Robertson J (1993) Deposition mechanisms for promoting sp^3 bonding in diamond-like carbon. Diamond Relat Mater 2:984–989
44. Deng J, Braun M (1996) Residual stress and microhardness of DLC multilayer coatings. Diamond Relat Mater 5:478–482
45. Kametani N, Nakamura M, Yashiro K, Takaki T (2022) Investigating residual stress evolution in the deposition process of diamond-like carbon film through molecular dynamics. Comput Mater Sci 209:111420–111426
46. Barraud E, Begin-Colin S, Caer GL, Barres O, Villieras F (2008) Mechanically activated solid-state synthesis of hafnium carbide and hafnium nitride nanoparticles. J Alloy Compd 456(1–2):224–233. https://doi.org/10.1016/j.jallcom.2007.02.017
47. Zhang XC, Zu BS, Wu YX, Xuan FZ, Tu ST (2008) Porosity, mechanical properties, residual stresses of supersonic plasma-sprayed Ni-based alloy coatings prepared at different powder feed rates. Appl Surf Sc 254(13):3879–3889
48. Dummer T, Lohe D (2002) PVD-wear resistant coatings of homogeneous and graded Ti (C, N): residual stresses and mechanical performance under hertzian load. SAE Trans 11:696–703
49. Lugscheider E, Bobzin K, Hornig Th, Maes M (2002) Investigation of the residual stresses and mechanical properties of (Cr, Al)N arc PVD coatings used for semi-solid metal (SSM) forming dies. Thin Solid Films 420–421:318–323
50. Wang TG, Zhao SS, Hua WG, Li JB, Gong J, Sun C (2010) Estimation of residual stress and its effects on the mechanical properties of detonation gun sprayed WC-Co coatings. Mater Sci Eng, A 527(3):454–461
51. Cavaleiro A, Marques AP, Fernandes JV, Carvalho NJM, Hosson JTD (2005) Evolution of the microstructure, residual stresses, and mechanical properties of W-Si-N coatings after thermal annealing. J Mater Res 20:1356–1368
52. Bouauina B, Besnard A, Abaidia SE, Haid F (2017) Residual stress, mechanical and microstructure properties of multilayer Mo_2N/CrN coating produced by R. F. Magnetron discharge. Appl Surf Sc 395:117–121

53. Ang ASM, Berndt CC, Luzin V, Crowe T, Gabel H (2018) Study of mechanical performance and residual stress in Kinetic MetallizationTM WC-Co coatings. In: Thermal spray 2018: proceedings from the international thermal spray conference, Orlando, May 2018
54. Meghwal A, Berndt CC, Luzin V, Schulz C, Crowe T, Gabel H, Ang ASM (2021) Mechanical performance and residual stress of WC-Co coatings manufactured by Kinetic Metallization^{TM}. Surf Coat Technol 421:127359
55. Dang NM, Lin WY, Wang ZY, Alidokht SA, Cromik RR, Chen TYF, Lin LMT (2022) Mechanical properties and residual stress measurement of TiN/Ti duplex coating using HiPIMS Tin on cold spray Ti. Coatings 12:759
56. Gopalan H, Marshal A, Hans M, Primetzhofer D, Cautaerts N, Breitbach B, Volker B, Kirchlechner C, Schneider JM, Dehm G (2022) On the interplay between microstructure, residual stress and fracture toughness of (Hf-Nb-Ta-Zr)C multi-metal carbide hard coatings. Mater Des 224:111323
57. Fan K, Jiang W, Luzin V, Gong T, Feng W, Ruiz-Hervias J, Yao P (2022) Influence of WC particle size on the mechanical properties and residual stress of HVOF thermally sprayed WC-10Co-4Cr coatings. Materials 15:5537
58. Perry AJ, Chollet L (1986) States of residual stress both in films and in their substrates. J Appl Phys A 4:2801–2808
59. Knotek O, Loffler F, Kramer G (1993) Deposition properties and performance behaviour of carbide and carbonitride PVD coatings. Surf Coat Technol 61:320–325
60. Karlsson L, Hultman L, Sundgren JE (2000) Influence of residual stresses on the mechanical properties of TiC_xN_{1-x} (x=0, 0.15, 0.45) thin films deposited by arc evaporation. Thin Solid Films 371:167–177
61. Almer J, Oden M, Hakansson G (2001) Microstructure, stress and mechanical properties of arc-evaporated Cr-C-N coatings. Thin Solid Films 385:190–197
62. Murotani T, Hirose H, Sasaki T, Okazaki K (2000) Study on stress measurement of PVD-coating layer. Thin Solid Films 377–378:617–620
63. Peters AM, Nastasi M (2002) Effect of carrier gas on the deposition of titanium carbo-nitride coatings by a novel organo-metallic plasma immersion ion processing technique. Vacuum 67(2):169–175
64. Kanchanomai C, Limtrakarn W (2008) Effect of residual stress on fatigue failure of carbonitrided low-carbon steel. J Materi Eng and Perform 17:879–887
65. Gonzalo O, Navas VG, Coto B, Bengoetxea I, Gopegi UR, Etxaniz M (2011) Procedia Eng 19:106–111
66. Wang P, Takeno T, Adachi K, Miki H, Takagi T (2012) Preparation and tribological characterisation of amorphous carbon nitride coatings in a RF PECVD-DC PVD hybrid coating process. Appl Surf Sc 258:6576–6582
67. Chien H, Diaz-Jimenez C, Rohrer GS, Ban Z, Prichard P, Liu Y (2013) The influence of residual thermal stresses on the mechanical properties of multilayer α-Al_2O_3/TiC_xN_{1-x} coatings on WC/Co cutting tools. Surf Coat Technol 215:119–126
68. Wang P, Takeno T, Fontaine J, Aono M, Adachi K, Miki H, Takagi T (2014) Effects of substrate bias voltage and target sputtering power on the structural and tribological properties of carbon nitride coatings. Mater Chem Phys 145:434–440
69. Li H, Wang Q, Zhuang M, Wu J (2015) Characterisation and residual stress analysis of TiN/TiCN films on AZ31 magnesium alloy by PVD. Vacuum 112:66–69
70. Caicedo JC, Guerrero A, Aperador W (2017) Physical properties evolution on ternary and quaternary carbonitride coatings. Vacuum 143:217–224
71. Liu C, Ju H, Han P, Quan C, Mo CP, Zhao Y, Geng Y, Xu J, Yu L (2020) The influence of carbon content on the microstructure, mechanical and frictional property of chromium carbon nitride composite films. Vacuum 178:109368–09377
72. Almohaimeed ZM, Karamat S, Akram R, Sarwar S, Javaid A, Oral A (2022) An effective route for the growth of multilayer MoS_2 by combining chemical vapor deposition and wet chemistry. Adv Condens Matter Phys 2022:3233252–3233258

73. Endo T, Iijima T, Kaneko Y, Miyakawa Y, Nishimura M (1995) Tribological characteristics of bonded MoS_2 films evaluated in rolling-sliding contact in a vacuum. Wear 190(2):219–225
74. Lancaster JK (1966) The influence of substrate hardness on the formation and endurance of molybdenum disulphide films. Wear 10(2):103–117
75. Holmberg K, Ronkainen H, Laukkanen A, Wallin K, Hogmark S, Jacobson S, Wiklund U, Souza RM, Stahle P (2009) Residual stresses in TiN, DLC and MoS_2 coated surfaces with regard to their tribological fracture behaviour. Wear 267(12):2142–2156
76. Weise G, Mattern N, Hermann H, Teresiak A, Bacher I, Bruckner W, Bauer HD, Vinzelberg H, Reiss G, Kreissig U, Mader M, Markschlager P (1997) Preparation, structure and properties of MoS_x films. Thin Solid Films 298:98–106
77. Guozheng M, Binshi X, Haidou W, Xiaohe W, Guolu L, Sen Z (2013) Research on the microstructure and space tribology properties of electric-brush plated Ni/MoS_2–C composite coating. Surf Coat Technol 221:142–149
78. Vierneusel B, Schneider T, Tremmel S, Wartzack S, Gradt T (2013) Humidity resistant MoS_2 coatings deposited by unbalanced magnetron sputtering. Surf Coat Technol 235:97–107
79. Vierneusel B, Benker L, Tremmel S, Goken M, Merle B (2017) Isolating the effect of residual stresses on coating wear by a mechanical stress relaxation technique. Thin Solid Films 638:159–166
80. Saha B, Dirckx M, Hardt DE, Tor SB, Liu E, Chun JH (2014) Effect of sputtering power on friction coefficient and surface energy of co-sputtered titanium and molybdenum disulfide coatings and its performance in micro hot-embossing. Microosyst Technol 20:1069–1078
81. Pengwei Z, Li W, Jin X (2017) The effect of MoS_2 content on the protective performance of Ni-MoS_2 composite coatings against hydrogen embrittlement in high strength steel. J Electrochem Soc 164:D23–D29
82. Xie R, Li Y, Huang H, Hu H, Li T, Guo B, Su L, Zhang X, Xu G, Wang L, Chi F, Liu H, Ma Y (2019) Fabrication, structural and vibrational properties, and physical and optical properties tailoring of nanocrystalline MoS_2 films. Ceram Int 45(15):18501–18508
83. Ju H, Wang R, Ding N, Yu L, Xu J, Ahmed F, Zuo B, Geng Y (2020) Improvement on the oxidation resistance and tribological properties of molybdenum disulfide film by doping nitrogen. Mater Des 186:108300–108308
84. Kong L, Huang K, Cao X, Lu Z, Zhang G, Hu H (2021) Effect of MoS_2 content on friction and wear properties of Mo and S co-doped CrN coatings at 25–600°C. Ceram Int 47(15):21450–21458
85. Zhang Z, Sul M, Li C, Zhou Z, Liu B, Chen Y, Sald Z, Debnath S, Sharma S (2022) Residual stress of grinding cemented carbide using MoS_2 nano-lubricant. Int J Adv Manuf Technol 119:5671–5681
86. Chen Y, Li H, Su F, Ma G, Li Q, Sun J, Lin S (2024) Friction and wear behavior of molybdenum-disulfide doped hydrogen-free diamond-like carbon films sliding against Al_2O_3 balls at elevated temperature. Wear 544–545:205296–205308
87. Li S, An Y, Zhou H, Chen J (2018) Plasma sprayed YSZ coatings deposited at different deposition temperatures, part 1: splats, microstructures, mechanical properties and residual stress. Surf Coat Technol 350:712–721
88. Ghara T, Bandyopadhyay PP (2022) Understanding the role of in-flight particle temperature and velocity on the residual stress depth profile and other mechanical properties of atmospheric plasma sprayed Al_2O_3 coating. J Eur Ceram Soc 42(10):4353–4368
89. Lomello F, Yazdi MAP, Sanchette F, Schuster F, Tabarant M, Billard A (2014) Temperature dependence of the residual stresses and mechanical properties in TiN/CrN nanolayered coatings processed by cathodic arc deposition. Surf Coat Technol 238:216–222
90. Das B, Nath AK, Bandyopadhyay PP (2018) Online monitoring of laser remelting of plasma sprayed coatings to study the effect of cooling rate on residual stress and mechanical properties. Ceram Int 44(7):7524–7534
91. Hamedani SK, Minoo Yadi, Hamid Esfahani (2024) Influence of residual stress on mechanical and tribology behaviors of calcium-phosphate coating on commercially pure titanium via integrated electrospinning and rapid heating and cooling process. J Mater Eng Perform 33:7689–7702

Chapter 7
Fracture and Toughness of Mechanical Coating

Armab S. Bhattacharyya

Abstract Fracture toughness is the most important mechanical property to be investigated for coatings used in mechanical applications. Indentation tests in static and sliding modes have been used to determine fracture-related properties. The nature of cracks observed under a microscope, qualitatively and sometimes even quantitatively, supplements the fractographic analysis of the coatings. Finite element modelling shows the stress distribution beneath the indentation contact point, which is especially challenging in scratch adhesion tests. The nature of carbon (graphitic) is sometimes found to provide a lubricative effect. The use of AI and ML in fractographic studies has also evolved, but it is in the primitive stages. The article discusses the issues mentioned earlier that can be applied to coating selection in harsh conditions requiring toughened coatings.

Keywords Fracture toughness · Coatings · Nanoindentation · Scratch

1 Introduction

The phenomenon of *fracture* is a brittle failure in materials, which usually manifests itself by means of crack propagation from pre-existing imperfections. It is one of the three mechanical failure modes in materials, the other two being *fatigue* (due to cyclic loading) and *creep* (due to high-temperature constant stress applied for a prolonged period). The term fracture is associated with fatigue and creep, as creep and fatigue fractures are elements of fracture that also set in during these failure modes. Fracture is, however, supposedly the quickest of the failure forms and is an outcome of an impulsive force. It occurs post-plastic deformation (yielding, σ_y) in metals and phases under uniaxial loading(compressive/Tensile) provided by a universal testing

A. S. Bhattacharyya (✉)
Department of Metallurgical and Materials Engineering, Central University of Jharkhand, Ranchi, India
e-mail: 2006asb@gmail.com; arnab.bhattacharya@cuj.ac.in

Centre of Excellence in Green and Efficient Energy Technology, Central University of Jharkhand, Ranchi, India

B. P. Swain (ed.), *Advances in Mechanical Coating*, Materials Horizons: From Nature to Nanomaterials, https://doi.org/10.1007/978-981-96-7484-8_7

machine (UTM). The resistance imposed towards fracture is called toughness and is typically the area under the stress–strain plot till fracture. Fractures can also occur without major yielding in ceramics or multicomponent composite materials hardened by different phase formations during their making. The phases, again, are the product of conditions initialised during the fabrication processes. As mentioned above, the crack generated in the fracture is followed by a plastic zone, which is more relevant to the crack length in case of ductile materials like metals. The features like crack tip opening displacement (CTOD) and J-integral come into effect at these stages. A much smaller plastic zone for brittle materials leads to categorising the fracture process in three modes (I, II, and III) for opening, sliding, and tearing, respectively, and the corresponding resistance being rendered as fracture toughness [1, 2].

In a layered system like coating/substrate, the uniaxial tests do not work. One needs to switch to another process called indentation, which plastically deforms a localised region and determines the mechanical parameters based on geometry, indenter shape, and indented regions. The fracture at the coating/substrate interface needs to be surmounted in these systems, and hence, the parameter interfacial toughness is the one that gets significant. For coatings of lower thickness (preferably called thin films), indentations at the nanoscale (nanoindentation) are performed [3–6]. The following sections reflect the fracture process and evaluation taking place in coatings of different configurations based on their applicability.

2 Applications and Configurations of Mechanical Coatings

Hard coatings protect against abrasions, wear, and environmental conditions. They are developing N/MEMS-based devices and sensors and testing them with nanoscale indentations before use in electronic packaging applications. Metal-CuS/MoS2 composite coatings with improved tribological properties were created by selecting the appropriate metal components and fine-tuning the heat treatment technique. CuS and MoS2 were co-sputter-deposited on copper or aluminium metals to form the composite coatings [7]. CuS and CuO nanoparticles were used to create Cu nanofluids, engine oils designed to minimise friction at the piston ring-liner interface. Cu nanoparticles added to engine lubricants can significantly reduce friction between the piston ring and liner, increasing fuel efficiency [8]. A new technique for non-destructive evaluation involving sub-Å-depth indentation measurements of two-dimensional materials and ultra-stiff and ultra-thin films for in-situ elasticity research employing topography imaging at the nanoscale was recently reported [9]. The effects of several sample and indenter configurations, like friction and tip bluntness, were investigated [10].

Multicomponent coatings, such as polymer-derived ceramics (PDC), high and medium entropy alloys, and ceramics, are essential, but determining and correlating their structure and properties is often difficult because they form multiple phases and are associated with intermetallics during their formation. Their application has switched, or shared, from being utilised as a protective coating in aerospace, automation, and mechanical industries to microelectronics, bio-implantation, sensors, and energy and environmental applications. PDCs such as Si–O–N, Si–O–C–N, and even polymers are being explored, and deposition techniques such as sputtering or molecular beam epitaxy (MBE) are being used [11, 12]. The hardness variation of heat-treated ceramics was attributed to the phase separation and crystallisation of TiB_2, TiC, CNx, SiC, and Si_3N_4. An observation was also recorded when the carbon phase split out in its softer sp2-hybridized state, giving a decreased film hardness. Nanocrystalline biomedical alloys contain non-toxic metals such as Ti, Ta, Mg, and Zn. The effect of Ta, Mg, and Zn concentrations on Ti alloys and the application of the hot-pressing method were investigated. It was discovered that adding magnesium and zinc positively influenced the consolidation process, resulting in bulk alloys with reduced porosity and increased mechanical properties due to microstructure refinement [13, 14].

Coatings such as Si_3N_4, SiC, TiN, cBN, Al_2O_3, ZrO_2, and nanocomposites such as Si–C–N, Ti–Al–N, Ti–B–N, and Ti–B–Si–C–N are utilised as protective coatings for industrial equipment and microelectromechanical systems (MEMS) [15, 16]. Materials used in the fabrication procedure include Si3N4, SiC, single/polycrystal silicon, and diamond-like tetrahedral amorphous carbon (ta-C). These materials are very resistant to climatic conditions, thermally stable, and rigid. These materials are combined with polymeric compounds to increase flexibility. Si–C–N ceramic, made of polymers, is used in MEMS as an electrostatic actuator, pressure transducer, thermal actuator, as well as micro grippers [17]. Electrostatic actuation uses materials such as SiN, SiC, and TiN, as well as nanocomposites of Si, C, Ti, B, and N, in addition to thin film PZT and AlN. Micromachining is essential in the production of MEMS. As a result, different microstructures, internal and residual stress, and flaws impact the device's overall mechanical behaviour and performance [18–20]. The technique and implementation of representative cooling systems for indentation devices are examined. To produce accurate nanoindentation curves and computed material properties, several influencing elements, such as thermal drift and temperature-induced influence on the indenter and specimen, enter the picture [21, 22]. Using sintered Si–C and TiB_2 pellets as the targets, the Si–C–N and Ti–B–Si–C(N) coatings were deposited on Silicon (100), stainless steel, and glass substrates using the magnetron sputtering (Hindhivac, Bangalore) process, which was also computationally simulated [23, 24]. Theoretical studies have demonstrated that SiC_2N_2 tetrahedra are the most stable structure of Si-CN [25]. In contrast, superhard and stable carbon isotopes have recently been empirically confirmed in accordance with previous theoretical predictions. A study of Raman and IR active vibrational modes with tetrahedral (sp3) bands suggests surface defect modes [26]. Recent discussions have also focused on the Raman spectra of carbon and its many forms [27]. The stable carbonic Si–C–N phases have been claimed to be attributable to the abundant carbon in the structure [28, 29]. As previously predicted, the Si–C–N coatings deposited revealed the formation of a stable SiC2N2 phase. The additional carbon found in Si–C–N facilitates the formation of these phases. Si–C–N belongs to the class of polymer-derived ceramics, with the option of adding elements such as boron(B) and oxygen [30]. The Si–C–N coatings were also produced via other physical and chemical deposition methods [31].

3 Fracture in Hard Coatings

The study of fracture behaviour is critical, considering the durability of these coatings in adverse conditions or when subjected to external stimuli for film breakdown. Nanoindentation has been widely utilised to characterise the nanomechanical properties of microelectronic coatings. Nanocomposite hard coatings containing phases such as SiC, Si3N4, CNx, TiC, TiN, TiB2, cBN, and others are used to make nano/micro-electro-mechanical systems (N/MEMS) based devices, interconnects, and metallurgical protective coatings. Nanoindentation properties, including hardness, reduced modulus, fracture toughness, adhesion, and wear, require contact between the indenter tip and the material under study [32–35]. In 1881, Hertz began the survey of indentation-based fracture, which later included Frank and Lawn's energy balancing analysis and Griffith's energy-based criterion [36–43]. Researchers have examined and reported on the process underlying the diverse cracking occurrences in micro and nanoindentation using indenters of different shapes for a long time [44–47]. The thickness of the coating and the indenter's geometry affect the fracture mechanism in coated materials. A sharper indenter induces more plastic deformation, and mode I fracture is more common at greater tip radius to sample thickness ratios. The fracture morphology that arises in coated systems depends on the substrate type, contact, and load applied. Interfacial failure is likely [48–52]. Depending on the coating's mechanical and crystallographic characteristics, the presence of any prior defects, internal microcracks, and pre-strain, the indenter's stress impacts the fractured nature. Hertzian (spherical) interactions have been modeled using plate or membrane bending techniques [53–56]. It has been reported on the stress distribution and interfacial fracture measurement related to hard pyramidal indenters based on models by Thouless, den Toonder, and Bull [25, 35, 57, 58].

Using a Berkovich tip, MTS Nano indenter, USA, executed sliding and static indentations at the nanoscale. Increased loads were applied to the indentations to cause cracks in the material. To conduct the fracture analysis. Optical microscopy (Leica, Germany) and electron microscopy (Philips, Netherlands) were used. In the case of films, crack-based stress intensities are transferred to the substrate, causing a range of cracking events, including shallow lateral fractures, half-penny, lateral, secondary palmist, and radial (palmist) fractures [34]. Since radial fractures and fracture toughness are related, crack opening displacement (COD) models have been employed to measure the latter. The lateral cracks, film delamination, and chipping demonstrate the films' adhesive properties and the substrate. Furthermore, established techniques for determining the interfacial toughness include an energy-based method and geometric mapping of the chipped region [4, 34, 35, 45, 58]. Finite element modelling has been used to analyse the relevant stress distributions during fracture. Changes in the geometry and shape of the indenter and a failure mode during indentation in both static and sliding modes were noted [59–64].

4 Interfacial Fracture

The sample experiences plastic deformation when it comes into contact with a sharp point during indentation, which results in the greatest stress area occurring immediately below the indentation region, surrounded by hydrostatic stress regions. This hydrostatic stress can potentially bring on a phase transition in the silicon substrate. Due to variations in the material configuration of the film and substrate, the shear components of the hydrostatic stress at the interface region contribute to the residual stress that already exists, resulting in film breakdown after a specific threshold value. Figure 1a depicts a schematic diagram of the indentation chipped segment, which shows the geometrical characteristics required for estimating interfacial fracture energy using the proposed models. Figure 1b depicts schematically the fracture characteristics found during the indentation procedure in Fig. 1e. The various forms of fractures were seen and marked, including radial (r1), lateral (L1, L2), secondary radial (r2), and edge cracks (e1, e2). The mechanism of various cracking events, as demonstrated in Fig. 1c and d, has been previously reported [34]. Radial and lateral cracks are combined during the indentation process, and edge fractures are visible developing on both surfaces. These are the outcome of an underlying defect that widens fissures when loaded.

Delamination takes place before chipping. Figure 2b shows that the chipped section C1 is surrounded by a delaminated region resulting from edge fractures

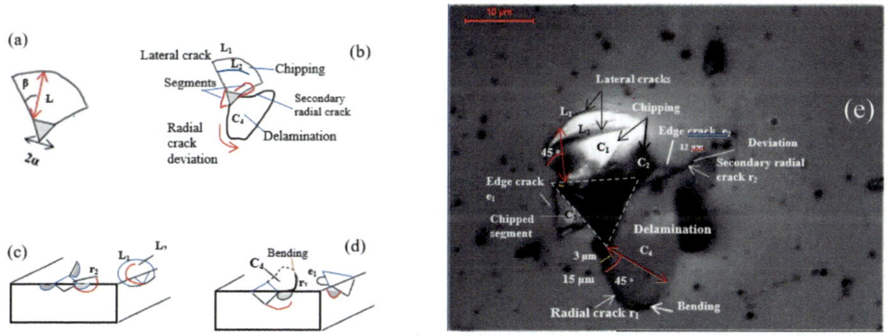

Fig. 1 Diagram illustrating the geometric parameters needed to estimate the interfacial fracture energy according to suggested models for **a** indentation chipped segment **b** The mechanisms, namely **c** radial (Palmqvist) and lateral cracks, and **e** the fracture patterns that emerged throughout the indentation process **d** edge cracks and the way the lateral and radial cracks interact to bend the radial crack (Open Access [58])

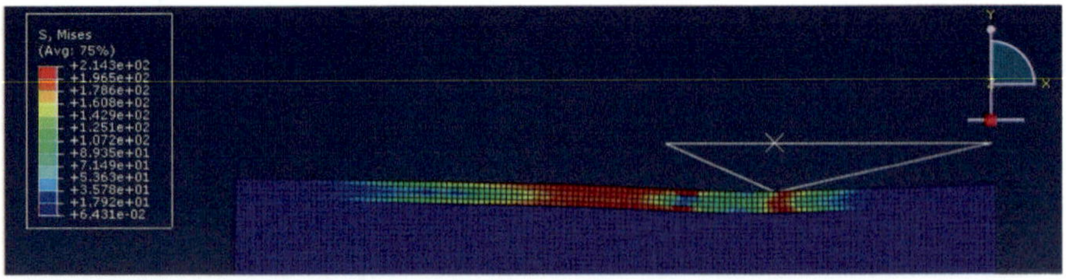

Fig. 2 Von Mises strain for sliding indentation on SiCN/glass by a Berkovich indenter [69]

rather than lateral cracks. Figure 1d schematically depicts the different cracks' interaction. Radial (Palmqvist) cracks form in the coating zone, and the stress distribution is limited to a small area before moving to the surface. The radial crack, a measure of toughness, deviates from its linear route, suggesting high toughness due to the production of nanocrystallites during the film deposition process [65]. The interfacial region becomes particularly significant when lateral cracks occur due to film adhesion and wear. At the film/substrate contact, the passage of hydrostatic stress (σ-hydro) is well defined. It was demonstrated that the pressure on the specimen causes the stress to follow a path along the contact and travel to the substrate, creating lateral fissures. As the stress distribution becomes more uneven, delamination and chipping occur. The geometrical properties of the chipped section, as indicated in Fig. 1b, can be used to compute the interfacial fracture energy expressed as Eq. 1 [34, 35, 45, 57, 58].

$$\Gamma_{\mathrm{i}} = 1.42\frac{Et^5}{L^4}\left(\frac{\frac{\alpha}{L}+\frac{\beta\pi}{2}}{\frac{\alpha}{L}+\beta\pi}\right)^2 + \frac{t(1-\nu)\sigma^2}{E} + \frac{3.36((1-\nu)t^3\sigma}{L^2}\left(\frac{\frac{\alpha}{L}+\frac{\beta\pi}{2}}{\frac{\alpha}{L}+\beta\pi}\right) \quad (1)$$

An approximate estimate of the interfacial fracture energies of the different areas was evaluated as per Eq. 1 using the characteristics of an image for coating with the elastic modulus of 200 GPa, Poisson's ratio of 0.25, and a thickness of 1.5 μm (t). The other terms are disregarded during computations because the first term is the primary term in the result and is quite substantial. The interfacial fracture energy for the chipped segment C1 was estimated to be 59 Jm^{-2}, while for segment C4, it was 68 Jm^{-2}. After dividing Γi by L, the residual stress that leads to film breaking was estimated to be 5.9 and 5.7 GPa for C1 and C4, respectively. Equation 2 [34] can also be used to calculate interfacial modulus (Ei) and toughness (Kic), where Ec and Es are the moduli of the coating and substrate, which were 200 GPa for SiCN and 140 GPa for Silicon, resulting in an interfacial modulus of 167 GPa. C1 and C4 have an interfacial fracture toughness of over 3 MPa $\sqrt{m}$. Use Eq. 3 [34] to get the fracture toughness value. The critical load (Pc = 100 mN) that initiates a lateral crack is represented by the letters H and E in this equation, while θ denotes half of the indenter angle (35° for a Berkovich indenter). A fracture toughness of about 3 MPa/m results from this. While compressive load promotes radial cracking, tensile tension leads to adhesive failure by chipping. Equation 4, a modified version of Stoney's equation, can be applied to substrate systems for coating [66].

$$\frac{1}{E_i} = \left\{ \frac{1}{2}\left(\frac{1}{E_C} + \frac{1}{E_S}\right) \right\}^{-1}; K_{ic} = \sqrt{\Gamma E \mathrm{i}} \quad (2)$$

$$K_c^4 = \frac{P_c(3/4)^2}{1200}(\cot\theta)^{2/3}\left(\frac{H^4}{E}\right) \quad (3)$$

$$\sigma t = pln\left(\frac{KEh^2}{6p(1-\nu)} + 1\right) where\ p = \frac{h^3}{D^2}E\left(\frac{13\cdot 5}{1-\nu^2} - 10.3\right) \quad (4)$$

h is the thickness of the entire film/substrate system, while t is the coating thickness. Parameter D represents the diameter of the coating, which assumes a circular shape and is set at 1 cm. The deposited layer had a thickness of 1 μm on a 1 mm Si substrate. As a result, h is deemed to be 1 mm. The value of k, the spring constant, determines the stiffness. The result is a surface stress value of 300 MPa at 1 mN/nm (MN/m), which rises with indentation. Coating toughness and adherence are determined in a coating/substrate system by thickness fracture and interfacial delamination. The residual stress σres developed in the coating is related to interface toughness as indicated in Eq. 5 [67], where β~0.016 is constant, E and H are modulus and hardness, P (~100mN) is the indentation load, and c is the crack length (15 μm). With 3 MPa$\sqrt{}$m as the fracture toughness (Kc), the residual stress (σ_{res}) is 667 MPaMPa with a shape factor also incorporated into the relation, as seen in Eq. 6 [67].

$$\mathrm{K_c} = \beta(\mathrm{E/H})^{1/2}P/\mathrm{c}^{3/2} + 2\sigma_{\mathrm{res}}(\mathrm{c}/\pi)^{1/2} \quad (5)$$

$$K_c = \beta(E/H)^{1/2}P/c^{3/2} + Z\,\sigma_{res}(c/\pi)^{1/2};\ \ Z = 1.12\sqrt{\pi}\frac{d/c}{\frac{3\pi}{8} + \frac{\pi}{8}\left(\frac{d}{c}\right)^2} \quad (6)$$

The component d is the crack width (3 μm as shown in Fig. 1b, which is mainly the radial crack depth. The Z value comes as 0.33. The energy-based model provided Eq. 7 for fracture toughness at the interface for plane strain mode I fracture, resulting in a residual stress (σ-res) of 4 GPa, where U_f represents the energy dissipated due to fracture and A_f represents the area where the fracture took place. Griffith's criterion considers the total energy release rate (G) for a fracture in a coating subjected to modes I and II loading as G = GI + GII as in Eq. 8. Crack extension occurs when the total energy release rate approaches a critical value (GC), which was 45 J m^{-2} for plane stress (α = 1), as per Eq. 9.

$$\mathrm{K_{IC}} = \left[\mathrm{EU_f}/\left(1-\nu^2\right)\mathrm{A_f}\right]^{0.5} \quad (7)$$

$$\mathrm{G_I} = \alpha\,\mathrm{K_I^2/E} \ \ \&\ \ \mathrm{G_{II}} = \alpha\,\mathrm{K_{II}^2/E} \quad (8)$$

$$Gc = \alpha\, K_{IC}^2 / E \tag{9}$$

5 Sliding Indentations

Scratching or sliding indentations have been shown to be useful in assessing the abrasiveness, wear, and adhesive strength of coatings. Micro and nanoscale scratch studies examined the adhesive strength of Si–C–N coatings created by magnetron sputtering. Because Si–C–N nanocomposites include several challenging phases, they are perfect for use as coatings for protection. The failure sites on the scratch track paralleled the characteristics of the coating/substrate system. Pre-scan, scan, and post-scan are the three phases of nano scratch for elastic recovery. It investigated how load and scratch velocity affected each other. The findings provide insight into the coating's behaviour under various mechanical conditions and its applicability for a particular use [68]. Scratch testing was performed on thermally robust hard Si–C–N coatings applied to steel (SS) 304 using radiofrequency (RF) magnetron sputtering. The abrasion and wear attributes were examined by looking at the shifting failure complexion inside the scratch, which showed a combination of brittle and ductile features. The loading rate variation resulted in notable changes to the failure mode. It was demonstrated that plasticity first exhibited peripheral dominance with an increase in load, but as the loading rate grew, brittle failure replaced it. The adhesion work and wear rate were computed [69].

Strong materials are applied to automotive glass components to protect them from heat stress, abrasion, electromagnetic interference, and UV glare. For this reason, glass was coated with thermal-resistant nanocomposite hard Si–C–N coatings, which were then scratch-tested to verify wearability and damage tolerance. Changes in coating thickness and loading rate caused adhesion strength and wear variations. The failure mechanisms were affected by the different stress distribution beneath the indenter. Coatings with greater thickness and loading rate exhibited a high frequency of elastic recovery. Even though it led to an early failure, the higher loading rate, on the other hand, caused less wear. Finite element modeling (FEM) revealed a coating-specific localised transition between tensile and compressive stress (Fig. 4) [70]. There is uncertainty in the results due to the indentation position and nanocrystalline properties, which call for careful examination. Dislocation movement boosted the toughness of the crystalline sections while making the amorphous portions more brittle. The Weibull distribution was used to calculate the probability of fracture (Pf). The change from ductility to brittleness is shown by sliding indentations. There was evidence of adhesive failure, which occurred significantly sooner and required enhanced coating component inspection. Finite Element Modeling (FEM) was used to analyse the stress and to offer details on the kinematics of dislocation and the influence of indenter shape on fracture. The outcomes are promising for the production of N/MEMS-based devices [71].

6 Probabilistic and Stress Distribution

Scratch tests and high-load nanoindentation were used to evaluate the adhesion properties and film failure. Power law expressions between load (Ph^n) can represent the loading curve during nanoindentation. The first expression, $P \propto h$, exhibits Hertzian elastic behavior for lower loads due to the tip getting blunted, which demands zero-point corrections [72]. Higher orders of h cause the expression of the power law to alter. Elastic–plastic deformation occurs beneath the indenter when the conical sides begin to pierce the sample as the indenter moves deeper under increased stress [73].

The P–h plot of nanoindentation performed on SiCN coatings on Si substrates is shown in Fig. 3a. 50 nm (the maximum depth) was reached with a force of 0.33 mN. After one hour, a residual depth of 20 nm was noted. The ratio hr/hmax is connected to the residual stress (σR), influencing the fracture process under high load indentation [46]. The indenter is pressed by a coating separation beneath it, which results in linear unloading [46]. The adhesive in the coating failed based on the linear expansion in the unloading section of the curve. Pop-in occurrences during the indentation process in the load depth curve [74] indicate a hard coating crack that is through-thickness in the picture frame.

The default discontinuous border of the indentation, depicted in Fig. 3a, is linked to low-intensity lateral cracks in addition to these picture frame cracks. The rectangle shows how the frequency of these tiny cracks rises with increasing loads. The discontinuities were maximum at 0.3 mN, deviating from the plot's typical parabolic shape. One possible explanation for this phenomenon is that indenter penetration creates tiny fissures that eventually coalesce to form larger ones. The fracture energy release was estimated using the variation to be 6.25 pJ (Fig. 3a) [33, 59, 63].

The pop-in event distribution of chances was plotted against contact stress decrease, calculated by the equation in Eq. 10. The corresponding indenter contact areas are A (h) and A (h + h). A (h) and A (h + h) are the loads for the indenter contact areas, and P' is the load at which pop-in occurs for depths h and h + h. The contact stress probability distribution peaks at a given load and then decreases at higher loads, suggesting a mixed-mode failure where brittle failure first predominates and raises the log value before ductile (plastic deformation) failure (Fig. 3b).

$$\Delta\sigma = \frac{P\prime}{A(h)} - \frac{P\prime}{A(h + \Delta h)} \tag{10}$$

The radial cracks, which are markers of toughness, developed during unloading. In contrast, abrasion, wear, or adhesive strength in the case of films are the causes of lateral cracks. The interfacial fracture energy in the failure zone—a chipped area formed by radial fractures can be found using geometry on any side of the indentation imprint [4, 35].

Radial cracks, essentially half-penny-shaped at the cross-section, originate in the corners of depression imprints. As can be seen, the chipping of the indentation only affected one side. Chipping occurs on all three sides when greater loads are applied.

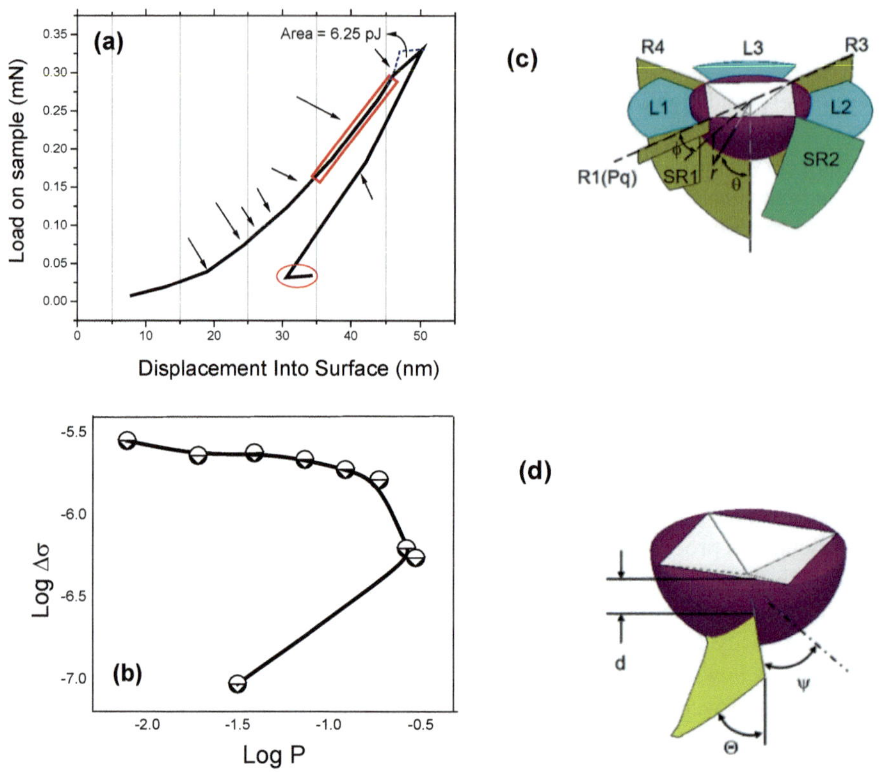

Fig. 3 **a** Nanoindentation load-depth plots for Si–C–N/Si **b** The mixed mode film failure probability distribution (reproduced with permission) [32] **c** schema of cracks by Vickers micro-indents **d** An angled crack's geometry (reproduced under CC BY 4.0) [59, 76]

However, the side with the highest local applied stress is the one that displays the initial chipping phenomenon. This could be related to the geometry of the indenter, which gets ununiformly sharpened at a certain side after a certain number of applications. Chipping happens at lower weights for sharper indenters, such as the 4-sided pyramidal Cube-corner, because the local applied stress is more substantial. In addition, there is little chance of radial cracks in Hertzian (spherical indenter) contacts; most cracking occurs lateral to the indentation imprint [34, 75]. It was found that the anisotropic stress nature at the crack tip controls the formation of radial cracks. The concentration propagates under loading and deviates because of nanocrystalline aggregation during deposition. Prior to the vertex, there are lower-intensity radial fractures, and various stress raisers at the worn-out indenter tip may form many lateral cracks [32].

The chipped region's interfacial energy was discovered, and the radial crack demonstration was also accomplished using a cube corner indenter. The energy generated during the fracture as well as the tip sharpness index (TSI) were computed. Hard coatings like CNx shatter when indented, indicating brittle failure. Lower thickness

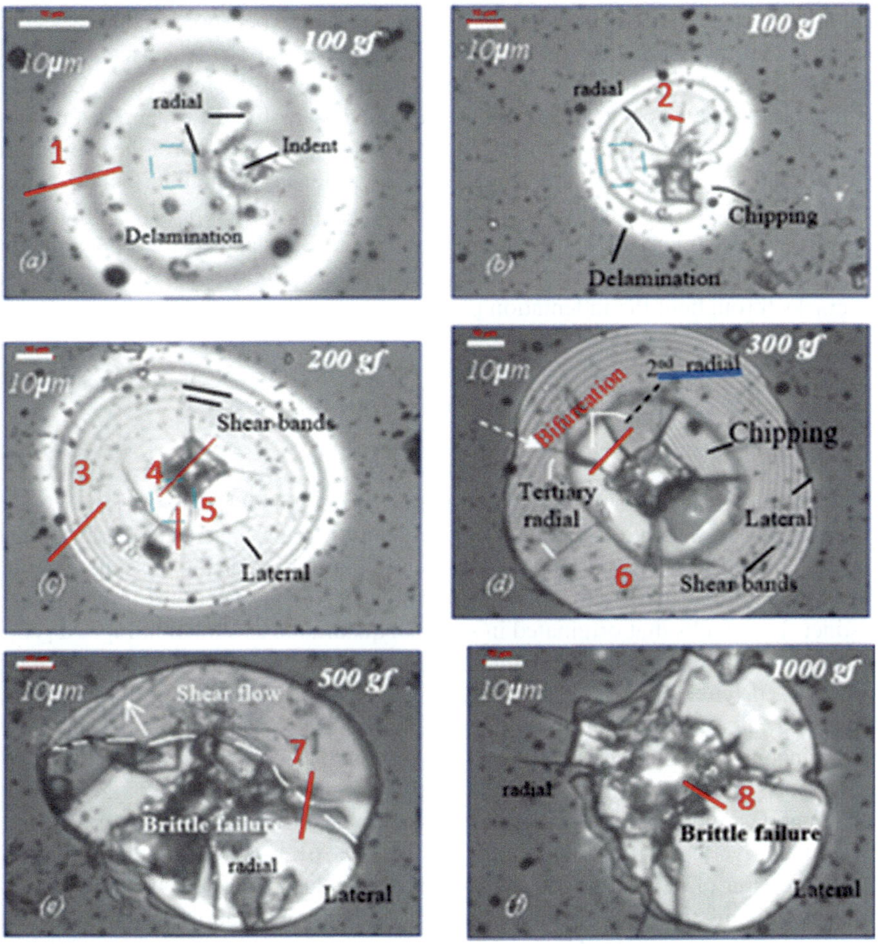

Fig. 4 Optical micrographs of SiCN coatings' fracture mechanisms at different indentation loads **a–f** 100 gf to 1000 gf with silicon substrates. Reproduced with permission [77]

and harder films, on the other hand, cause plastic flow to extend well beyond the indentation area. Strain hardening brought on by crack propagation prevents plastic flow in particular symmetrical linear directions. A bigger plastic flow zone indicates more incredible interfacial fracture energy. To release tension that had built up during hardening, crack deflections happened. There was less strain hardening and more plastic flow because of the Berkovich indenter's increased sharpness [63].

In the coordinate system (r, θ, ϕ), where r is the radial distance, ϕ is the hoop angle about the loading axis, and θ is the angle to the loading axis in spherical polar coordinates (Fig. 3b), cracks created during indentation as a result of the maximum primary stress at the elastic–plastic zone's border are visible. The primary stress, $\sigma_{\phi\phi}$, creates the radial fracture (R), whereas $\sigma_{\theta\theta}$ and σrr are responsible for the median (M) and lateral cracks (L). Shallow lateral (SL) gets its name because lateral cracks can occur at shallow depths and parallel to the surface. Shear stress causes secondary radial (SR) cracks to appear ($\tau r\phi$). After a certain depth, more loading results in inclined lateral cracks that spread into the surrounding material. And are determined by angles Θ and ψ, as shown in Fig. 3c. They are the reason for adhesive failure or coating delamination and are governed by shear tension. The interfacial fracture energy is eventually found by geometrically analysing the chipped segments that arise from these cracks. The blister stress fields and Boussinesq stress describe the $\tau_{r\theta}$ at the indenter contact point [76]. Equation 11 describes the stresses that cause all cracks throughout the indentation process.

$$\sigma = \begin{pmatrix} \sigma_{rr} & \tau_{r\theta} & \tau_{r\phi} \\ \tau_{\theta r} & \sigma_{\theta\theta} & \tau_{\theta\phi} \\ \tau_{\phi r} & \tau_{\phi\theta} & \sigma_{\phi\phi} \end{pmatrix} = \begin{pmatrix} SL & IL & SF, SR \\ IL & M & SF \\ SF, SR & SF & P_R \end{pmatrix} \quad (11)$$

The parallel shear faults (SF) are caused by coupled shear stresses $\tau_{r\phi}$ and $\tau_{\theta\phi}$, resulting in discrete horizontal and vertical coverage that forms parallel lines surrounding the depression. These forces caused the coating on the missing quadrant side to delaminate into two separate delaminated zones, which are linked to two distinct shear faults that originated in separate quadrants, as shown in Fig. 3d [76].

The stress distribution beneath the indentation region varied with the shape of the indenter sed (cube corner and Berkovich), which naturally affected the interfacial toughness calculations and film failure patterns. It has been shown that the sharpness of the indenter encourages complementary stress distribution, reducing the sharpness associated with brittle cracks and boosting toughness at the edge of the pores by inducing plastic deformation surrounding them. Anisotropic fracture will be evaluated qualitatively because of the findings. These discoveries will be helpful for both the fabrication of devices in the microelectronic industry and heavy-duty applications by utilising their load-withstanding or load-bearing capabilities and lifetime to prevent early failures due to hard operating circumstances. The circular circumferential delaminated area may resemble a cardioid due to fracture. This area is equivalent to the residual compressive pulse enclave, which develops when, instead of the elastic zone in the case of fatigue failure, the plastic zone close to the fracture tip expands because of overload and cannot return to its original shape after unloading. When a circle travels around the edge of another lower circle, such as the Mohr circle, it forms the car The plastic zones are estimated using the Tresca and Mises yield criterion for anelastic deformation at the fracture point [59].

7 Fracture Caused by Vicker's Microindenter

Figure 4 depicts a distinct fracture morphology of SiCN coatings placed on silicon substrates. The flow of shear stress was found to accompany coating delamination, Fig. 4a, where the indentation was performed with a 100gf load. Minor cracks propagating radially were also discovered. Figure 4b shows the same indenter imprint 5 s later, with exacerbated radial cracking and chipping. This area is equivalent to the residual compressive pulse enclosure, which develops when, as opposed to the elastic zone in the event of fatigue failure, the plastic zone close to the crack tip has expanded due to stress and cannot return to its original shape after unloading. The increase in stress leads to chipping, which prevents shear flow and delamination. The enclave is forced back into the indented area, forming a cardioid shape. By eliminating two dislocation interaction points at the boundary, the procedure generates a new dislocation inside the cardioid. One may draw a reverse connection between the two scenarios, where the delaminated zone is in concurrence with the dislocation loop and fracture to form a new dislocation. The cardioid around the depression can be used to define the J-integral [77]. At loads greater than 200gf, lateral cracks and shear bands can occur within the delaminated area (Fig. 4c). The enclave has not been drawn into the indented zone because chipping (brittle failure) has not yet occurred. When the load was increased to 300 gf, the chipping segments became visible where the radial and lateral cracks had combined.

A phenomenon of chipped area bifurcation induced by creating a secondary radial fracture can be observed. These secondary radial cracks form because of pre-existing flow along the indentation (indenter) sides, which becomes more intense as the load increases. The shear stress flow persisted outside the chipped zone as shear bands and radial cracks, also known as tertiary radial cracks. The enclave was observed to deform somewhat due to chipping (indicated by the arrow) and transform into a lateral crack [Fig. 4d]. The brittle failure strengthened as the load increased, and as a result, the shear flow zones decreased, as seen for 500 and 1000 gf, and radial cracks were detected to propagate farther [Fig. 4e and f]. This suggests that the sources of radial and, to a lesser extent, lateral cracks continue to regenerate and remain active after brittle failure. Cracking can develop in Vickers micro-indentation on Si–C-N/Si films when the indentation stress increases with time, at 300 gf, radial cracks formed, followed by circumferential delamination, which resulted in uniform 4-blade chipping on both sides. Mixed-mode failure in the indented region and chipped segment bifurcation were observed. The chipped segment model with interfacial fracture toughness produced lower fracture toughness values. Future research should focus on building models that account for the localised film delamination caused by indentation from other perspectives, like crystalline growth [60].

Table 1 The initialisations for the finite element modeling (FEM) [70]

Mesh	The 5057 linear quadrilateral element, type CPS4R, a 2-dimensional, 4-node bilinear plane stress element, is used to mesh the geometry for both glass and SiCN film substrates
Material property	Properties of the isotropic, elasto-plastic material are incorporated. The characteristics of "Material non-linearity" and "Geometric non-linearity" (high strain deformation) are integrated into the model
Contact property	The film's surface that meets the indenter is considered the slave surface since it is a deformable object, and the indenter's surface, which contacts the film, is referred to as the master surface because it is considered a rigid, analytical element. They create a property of surface-to-surface contact with one another. "Single configuration (state)" contact tracking and "surface to surface" discretisation are used. The mechanical normal behaviour, which has a coefficient of friction of 0.25 and is referred to as "hard contact" or "tangential behaviour," is the contact property
Boundary conditions	The substrate's bottom is stationary in the 3 perpendicular directions (x, y, z) and rotates around the Z axis. Using data from the experiment, displacement is provided concurrently along the x and y axes
Control	Hourglass control and the reduced integration approach are used to complete the solution

8 Finite Element Modelling

ABAQUS 6.13 was utilised in conjunction with Finite Element Modelling (FEM) to analyse the indentation-based stress distribution causing film failure. The specific parameters used for the FEM are displayed in Table 1.

The entire film or system experiences a bending moment because of the scratching process, and the sliding process leaves behind residual tension. Unlike with softer substrates like silicon and steel, no plastic zone or radial stress dispersion into the hydrostatic area is visible. The scratched zone contains the shear stress primarily in charge of the plastic deformation. The intriguing finding of a low-stress region (blue) immediately following a high-stress region points to a novel phenomenon in which failure occurs at the underlying regions rather than the surface. In such instances, the coated portion typically experiences chipping. In this aspect, more research is necessary [78].

9 Softer Lubricative Coatings

As was previously reported [80], the CVD-CNx films with higher crystallite sizes displayed variations in the hardness (H) and modulus E magnitudes as well as changes in the P–h plot based on the indenter [79, 80]. A greater mechanical reaction was obtained from indentations made inside the grains (A) as opposed to at the grain borders (B) (Fig. 5a and b). XRD revealed lower H and E values when comparing the soft graphitic C3N4 phase formation to traditional protective coatings. Within the grain region, it was discovered that the hardness increased after 100 nm, suggesting strain hardening. At the GBs, the effect was not all that widespread. It is possible to observe a 50 MPa increase in hardness because of 300 nm penetration (Fig. 5c).

As the sharper sides of the 3-sided *Berkovich* indenter came into touch, the contact shifted from Hertzian (spherical), which was similarly linked to the increased plastic deformation (shown in the inset). On the other hand, the change in elastic modulus is the event that happens before fracture and is connected to the damage during the indentation (Fig. 5d). Pop-ins, or discontinuities in the loading areas, are one way the fracture process shows up. Although the P–h plot did not show any evidence of a cracking phenomenon underneath indentations inside the grains, the derivative of the loading component of the plots did. (A) When the indentation loading rate gradually rises. The indentation speed was significantly more subtle near the grain boundary (B), and a distinct change in contact was seen around 200 nm (Fig. 6a–c). The hardness is in the plastic zone (radius ($\boldsymbol{r_p}$) specified by Eq. 12 below the indented region (where σ_y is the yield stress) [81]. According to Eq. 13 [82], this radius is once more connected to the interfacial fracture energy Γi.

Since **rp** defines the zone in front of the crack tip where new dislocations are formed, a rise in **rp** with penetration depth indicates that more dislocations result in strain hardening. The three-sided pyramidal Berkovich indenter's change in contact with the sample causes the hardness variation; strain hardening brought on by the creation of dislocations increases the hardness. Dislocation-based creep can also be caused by a more considerable penetration depth (load). As a result, grains frequently experience a rise in non-conservative dislocation [83]. These phenomena do not occur in the GB region because it is not crystalline. That explains the differences in hardness and P–h graphs.

$$r_p = \sqrt{\frac{3P}{2\pi\sigma_y}};\ \sigma_y = 2.8\,H \tag{12}$$

$$r_p = \frac{1}{\pi}\left(\frac{\Gamma_i}{H}\right) \tag{13}$$

The rp for the indentations made at the GBs did not exhibit a continuous rise. In some cases, it appeared to reach saturation at 300 nm, but then it rose. When films are produced on a ductile substrate, the substrate's influence begins to show at approximately one-tenth of the coating thickness, which causes the hardness to decrease as penetration depth increases. The effect of the substrate might appear in multiple ways when the coatings, such as in this instance, are less complicated than the substrate material.

One of them is the resumption of the rise in rp, which is now visible in the 304SS substrate rather than only the coating zone (as illustrated in the inset, Fig. 6d). As a result, the strain hardening begins in the substrate itself, independent of the coating. Its plastic flow around the depression is the only impact it might have on the coating.

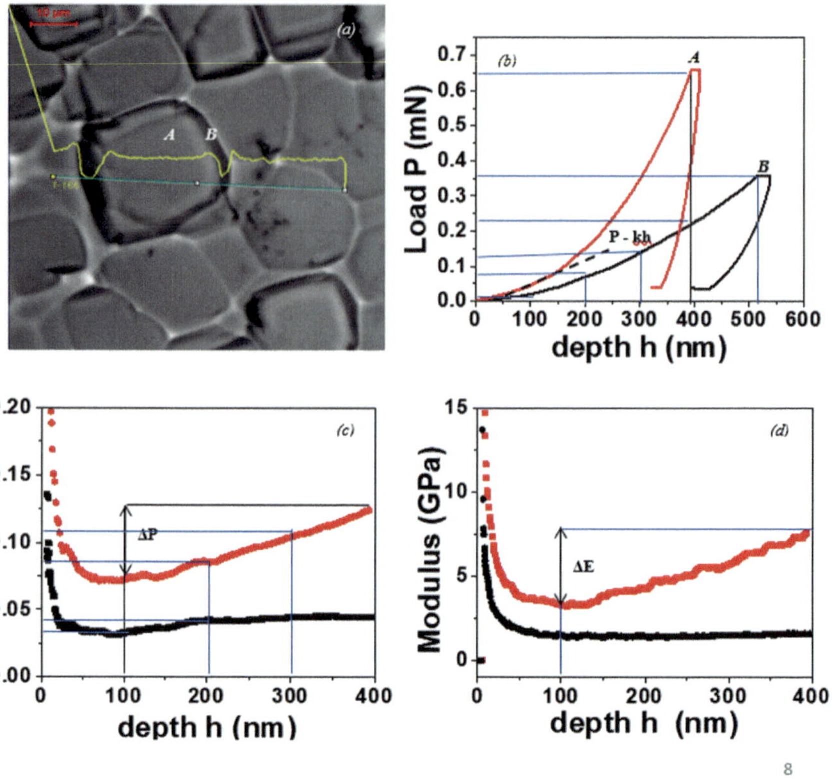

Fig. 5 **a** Nanoindentation and CVD-CNx optical images; **b** P–h plots at various microstructure locations with corresponding changes in **c** hardness and **d** modulus **e** Filling in the nanoindent P–h graphs with data. Differential hardness for grain regions **g** and grain boundaries **f** (Open Access) [79]

At this point, the system looks like a strain-hardened steel surface with a softer, lubricating layer that resists corrosion and wear. The soft-hard interface zone can also accommodate stress concentrations and volumetric expansions. More flexibility and cycle performance are also achieved in terms of device performance. In this case, there is a greater likelihood that any hit will be absorbed than deflected. Since the volume proportion of grain boundaries—the most susceptible to chemical attacks—is comparably higher in larger grains, the softness of g-C3N4 also predicts better substrate adherence and stronger resistance to chemical attack. When dislocations accumulate at the interface during grain center testing, the interfacial fracture energy plot illustrates how the interface becomes tougher (Fig. 6e, plot A). The tougher and more brittle substrate is more likely to fracture via radial and lateral cracks because it cannot support the elastic strain gradient that causes strain hardening.

There is less plastic deformation in the case of grain boundary indents because no crystallographic pattern forms. The area mostly comprises mixed hybridised carbon

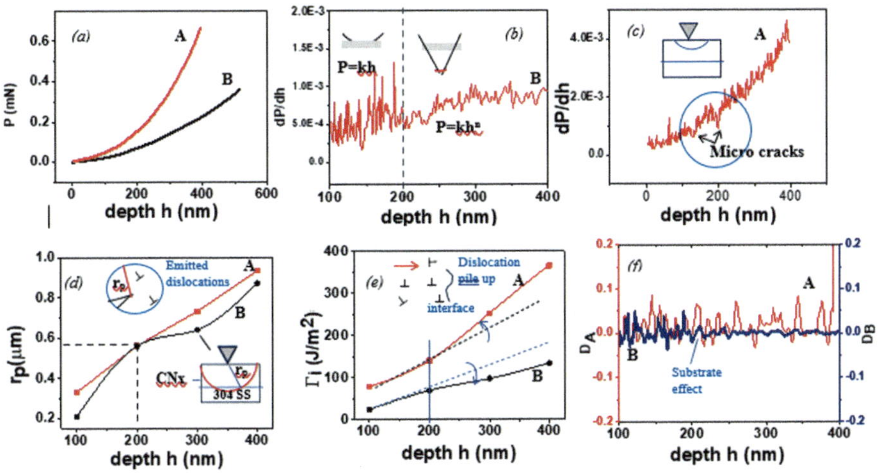

Fig. 6 **a** The P–h plots based on differential hardness for the following: **b** grain boundaries, **c** inside grains, **d** depth variation, **e** interfacial Fracture energies dissipated at the interface, and **f** variations in damage index for both inside-grain and at-grain barriers (Open access) [79]

phases (sp, sp^2, and sp^3) and is vulnerable to brittle failure after damage, as indicated by the decline in Γi (figure B-Fig. 6e). As demonstrated in Fig. 6f, damage is linked to changes in elastic modulus and is represented by a damage index D, which is the derivative of dE/dh. Grain boundary indentation is linked to the DB index. The stainless-steel substrate accommodates the collected strains and helps to prevent brittle failure by lowering the generated damage index. The higher strain rates and lower variations of the damage index parameter for indentations made at the grain boundaries compared to those made in the grains were due to strain hardening due to the presence of the 304SS substrate (Fig. 7a—c) [79].

10 Rheological Coatings

Rheology is the study of flow, and by rheological coatings, we generally refer to polymer coatings having viscoelastic properties [84, 85]. In current times, amorphous conducting polymers are essential in technology due to their lightweight, transparency, and long-lasting mechanical performance. They are finding applications in the aerospace, automotive, and military sectors. Thus, adequate precision-based research is required. The nanoindentation P–h plots of PPy were examined based on load fluctuations observed at specific depth intervals. The observations were used to explain polymeric chain motions and viscoelastic-plastic flow. The ultimate processability of polymers, viscoelastic complex fluids, intimately correlates with their rheological characteristics. Optimising the design of new materials for particular uses, like polymer coatings, requires a thorough understanding of polymeric systems' viscoelastic effects and flow characteristics. Advances in instrumentation, polymer reference materials, material-independent calibration procedures, and new testing and analysis techniques that consider polymers' viscoelastic and viscoelastic behaviour are required to achieve quantitative characterisation of polymer properties through nanoindentation testing [86]. The effects of annealing on the mechanical resistance of photoresist layers were examined using temperature-changing nanoindentation. While thermally-induced polymerisation increased mechanical resistance, it was discovered that raising the annealing temperature decreased the susceptibility to plastic strain. The decrease in strain energy dissipation coefficients with increasing annealing temperature demonstrated an annealing-induced hardening effect [87].

Since amorphous polymers offer excellent mechanical, thermal, optical, and chemical properties and can be produced at reasonable costs, they have become more and more well-known in the scientific and technological communities. Like this, these polymers are widely used in tough industries, including the military, automotive, and aerospace sectors. Military applications, such as machine guards and protective eyewear, specifically use polycarbonate (PC) and poly (methyl methacrylate) (PMMA) due to their lightweight, transparency, and long-lasting mechanical performance.

Nonetheless, loading conditions like temperature, strain rate, and pressure significantly impact the performance of amorphous polymers. A range of deformation mechanisms, such as crazing and shear yielding, further sets them apart. For strain rates ranging from low to high, a visco-elastic-visco-plastic constitutive model has been constructed. We model two molecular mobility mechanisms with the use of recognised rheological components. An efficient implementation follows from the derivation of a fully implicit formulation. A four-phase calibration approach based on optimisation is suggested. Outstanding concordance between two thermoplastics' experimental and numerical results [88]. The non-linear behaviour of pristine and reinforced polymer having amorphous configuration can be studied using a finite strain viscous elastoplastic model, which also includes an amendment of a hardness law [89]. Additionally, conducting polymers such as PPy have been nano-indented at very low stresses (μN), resulting in a substantial viscoelastic deformation, as illustrated in Fig. 8a [90].

The stress–strain response of amorphous polymers under compression has been subdivided into A, B, and C segments for visco-elastic, strain softening, and strain hardening, respectively, and is associated with intermolecular interactions between the polymeric chain segments. The chains align and stretch themselves to reduce the entropy (strain hardening) (Fig. 8b and c). The maximum stress experienced before the requirement of such adjustments is the yield stress for the amorphous polymers. Between yielding and strain hardening, the chains undergo configurational changes to eliminate the residual stress that is causing softening [88]. A softening index (n) has recently been reported in the case of biomaterials, which influences the pace at which the effective elastic modulus varies with depth [91]; due to the viscoelastic –plastic properties of the amorphous polymers, additional molecular motions get triggered on increasing the strain rate. These effects are analogous to getting into the tertiary (accelerating) stage from the steady-state stage [83]. The nanoindentation

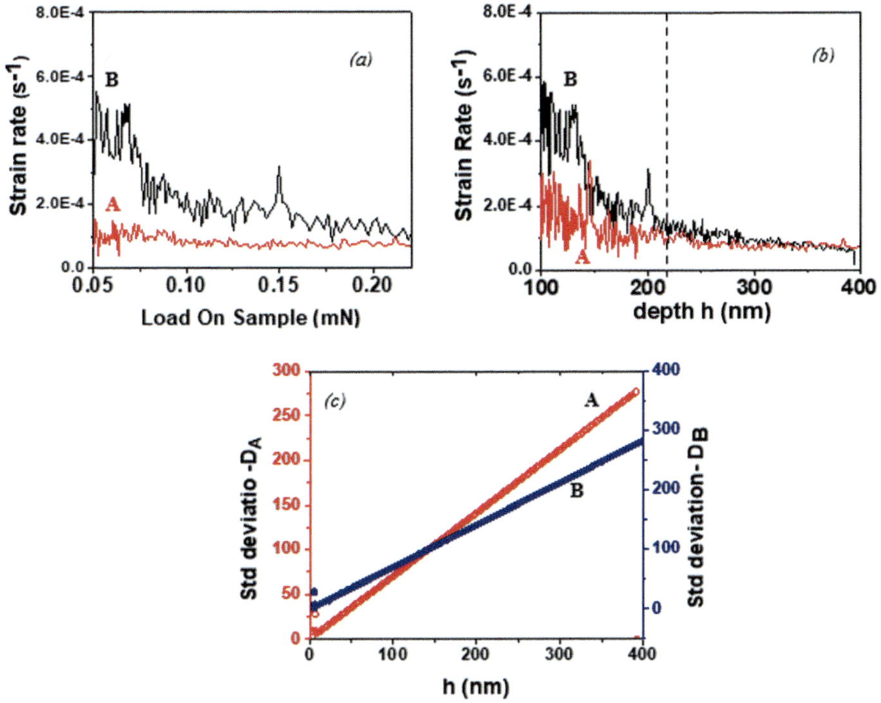

Fig. 7 Strain rate changes with respect to **a** the load and **b** penetration depth and **c** the deviations in the damage index fluctuation for indents made in the grain (D_A) and at the grain boundaries (D_B)-Open Access [78]

P–h plots of PPy showed heterogeneity, which was most prominent between 100–300 nm (Fig. 8a). Test 1 showed a bulging and test 3 showed a contracting nature, like pop-in–out events observed in more complex materials. As mentioned previously, the viscoelastic phenomena followed by chain movements can be linked to this case. However, such events can look random due to a lack of periodicity. Taking the average of a much larger data set shall show any specific trend, and is kept as one of the future scopes of work.

The loads at an interval of 50 nm in the depth mentioned above were determined from the plots for all three tests (Table 2). Taking test 2 as a reference (due to the least variations), the changes in load divided by the corresponding depth were determined. The ΔP/h variations with indentation depth for the two cases were not similar. The one showing an average decreasing trend, i.e., $\Delta P_{23}/h$ variations with depth of indentation, showed the bulging effect, while the one with undulating variations showed the contracting effect. The three tests' unloading nature was similar to their corresponding loading plots. All the tests showed a higher unloading rate, which can be associated with the visco-plastic phenomenon. Further investigation is required before coming to any specific conclusions.

Table 2 Load variations among the tests at specific depth intervals

Depth h (nm)	P_1 (nN)	P_2 (nN)	P_3 (nN)	$\left\|\frac{\Delta P_{21}}{h}\right\|$ (N/m)	$\frac{\|\Delta P_{23}\|}{h}\left(\frac{N}{m}\right)$
100	55	40	70	0.15	0.30
150	75	75	100	0	0.17
200	90	120	140	0.15	0.1
250	130	150	170	0.08	0.08
300	190	220	220	0.1	0

11 AFM Nanoindentation

The AFM-nanoindentation approach has shown promise in studying substrate and horizontal confinement effects and characterising nanomechanical behaviour in thin films and nanostructures. Nanomechanical principles were employed to create multi-component hard coatings for MEMS cantilevers in AFM nanoindentation. It has been investigated how mechanical input affects electrical conduction, and it has been shown how ion beams and mechano-magnetic influences affect Si-based films used in memory and switching devices [91–95]. The strain imposed in such low-precision indentation is more effective in polymers due to their sensitivity with respect to temperature. The morphological changes occurring with parallel load and temperature increase reveal the rheological properties of the photoresists used in lithography.

Lithography based on laser and thermal evaporation methods was used to successfully create arrays of Ag nanodisks on Si (100) substrates with different radii (125–900 nm) and thicknesses (20–150 nm). The impact of thickness and substrate on the mechanical resistance of the [96]. Using the Finite-Difference Time-Domain (FDTD) approach, we investigated Electric-Near Field Enhancement (ENFE) in Au and Ag nanodisk-based photonic crystals. Optimising the visible and near-infrared spectra's ENFE through optimal design parameters was analysed using photonic crystals with squared-lattice characteristics [97]. An efficient method for examining how annealing affects photoresist layers' mechanical resistance is the AFM-assisted DSNI approach. The findings show that the plastic strain susceptibility of the films decreases with increasing annealing temperature, indicating that thermally induced polymerisation enhances mechanical resistance. The decrease in strain energy dissipation coefficients with increasing annealing temperature indicated an annealing-induced hardening effect [98]. The pseudo elasticity brought on by stress influences the freestanding.

The AFM (cantilever bending)-assisted DSNI method was effectively used to measure and analyse Cu-Al-Ni thin films. This enables us to formulate new perspectives about using this technology as a productive tool for next studies and measuring the shape memory effect, also known as pseudo elastic behaviour, in micro/nanostructured SMAs [99]. Atomic force microscopy (AFM) under various liquid conditions using an assembled cantilever probe (ACP) and its size-dependent nonlinear behavior. Demonstrated the system's softer nature [100]. AFM nanoindentation has been applied to a range of biomaterials. This communication focuses on collagen fibrils, which provide mechanical support to the living cells in many animals. Because of its heterogeneity, anisotropy, and water content, collagen fibrils have extended Young's modulus values and are not linearly elastic materials. Three force-indentation curves were shown for a collagen fibril with a radius in Fig. 9a. The Young's modulus values vary greatly. It resembles a softening behaviour as found in the ACPs [100]. In fact, the stiffness(S) of the indented region varies, bringing differences in the E values. For the samples having E = 2.25 GPa, the S was 5 N/m, which rose to 11.11N/m for 1.38 GPa. The stiffness determination was carried out using biomaterials, which have been reported to be beneficial for detecting diseases like organ fibrosis. An effective Young's modulus (E_{ff}) given in Eq. 14 was proposed based on different models. The symbols P, R, and h represent the load-applied radius of the indenter tip at the depth. The change in E_{ff} with respect to depth was again found to be a crossing of the parabolic nature (1. e $1/h^n$ with n varying from 1.5 to 2.5 (Eq. 15). This n can be considered as the softening index in biomaterials and can be related to adhesion force as observed through Fig. 9b and is left as a future scope of work [101].

$$E_{ff} = \frac{3P}{4h^{3/2}\sqrt{R}} \tag{14}$$

$$\left[\frac{dE_{ff}}{dh} \propto \frac{1}{h^{5/2}}\right] \tag{15}$$

12 AI and ML in Mechanical Coatings

In surface coating, artificial intelligence (AI) and machine learning (ML) is quite significant. Successfully implementing and adopting AI and machine learning can lead to more efficient resource utilisation, lower inventory costs, and improved forecasting of any fracture of in-service components [102]. It has been found successful clinically in determining bone fracture, where a process known as meta-regression analysis was adopted [103]. AI refers to any skill in which a machine mimics human intellect. ML is a subfield of artificial intelligence that allows a machine to learn and develop based on experience without the intervention of humans. On the other hand, deep learning (DL) is a highly specialised subset of ML that can analyse larger data sets by translating an algorithm's inputs into outputs using advanced computational models like deep neural networks. Convolutional neural networking CNN is an evolutionary computational technique in deep learning that can potentially alter critical areas of medicine, such as medical imaging [104]. Artificial neural networks (ANN) have been used to model PVD- and CVD-based mechanical films for cutting tool applications. An artificial neural network (ANN) model was developed based on

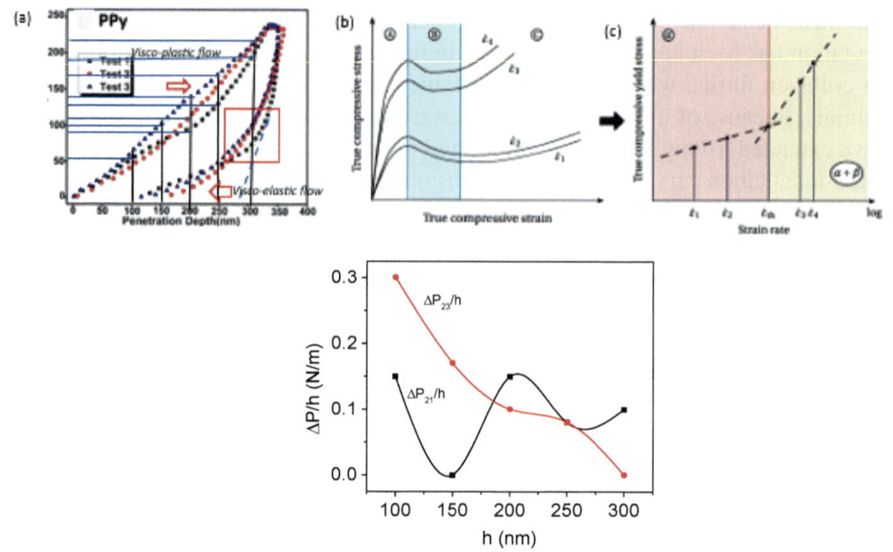

Fig. 8 **a** Nanoindentation load-depth plots on PPY (Open access [90]) **b** stress–strain (compression) of amorphous polymers **c** variation of yield stress with respect to strain rate in low and high strain rate sensitive area of amorphous polymers (Open access [88]) **d** The ΔP/h variations with depth of indentation

experimental data. This model demonstrates how coating parameters, such as hardness, adhesion, grain size, and thickness, impact the cutting durability of tool blades coated with the researched coatings. The computational model was built using an artificial neural network [105, 106]. The image-recognition tool can be calibrated to detect the overall corrosion or coating breakdown area and calculate it pixel-by-pixel, resulting in an accurate and consistent report on the corroded area. This means that it may be used to study any maritime structure, onshore or offshore, if it is trained with enough data to perform the assessment. One of the driving forces behind the development of this technology is that in the future, more and more inspections will be performed remotely, utilising technologies like drones or crawlers for inspecting difficult-to-reach areas such as restricted spaces, at heights, and underwater. In addition to lowering the risks associated with deploying surveyors inside tanks or other restricted locations, these technologies produce a significantly greater volume of data, including images to interpret [107].

13 Conclusions

This chapter provides an overall perspective of the fractographic analysis caused by indentations of coatings having mechanical applications. The coatings discussed were mainly polymer-derived ceramic-based SiCN, TiBSiCN, PVD, and CVD-based

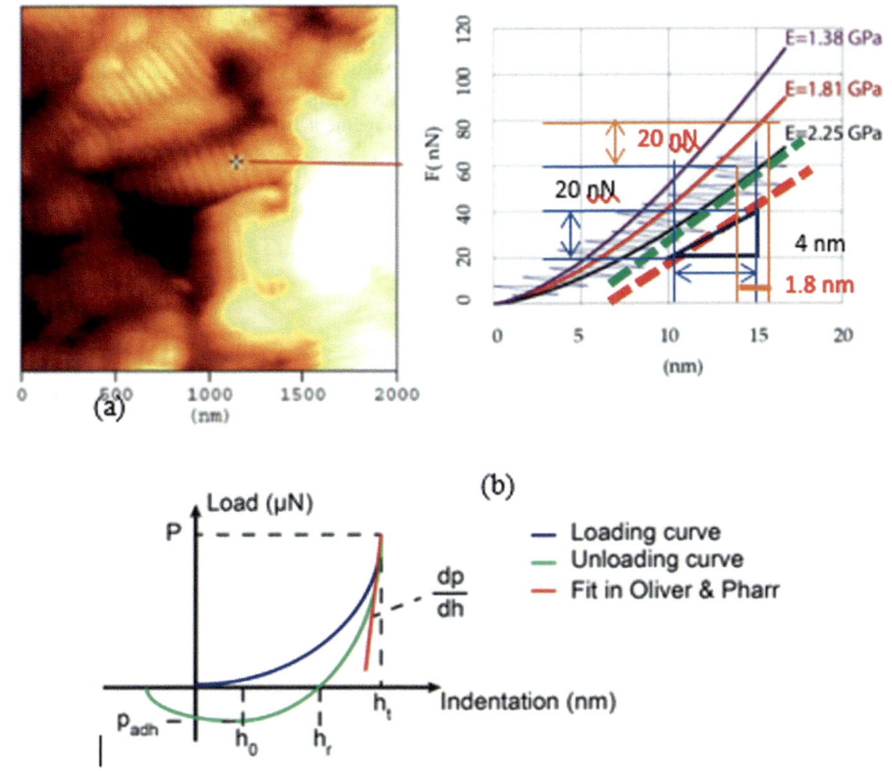

Fig. 9 **a** Load-indentation schematic diagram representing the force of adhesion (P_{adh}) and variation over parabolic nature (Open access [100]). **b** Force–displacement data in AFM nanoindentation on collagen fibrils with varying elastic modulus (Open access [101])

CN coatings. The compositions naturally varied with alterations in initial deposition parameters in phases developed, influencing the mechanical properties. Indentations, both at the micro and nanoscale, were used for the fracture studies. Being of different shapes and sizes, the chronological development of the crack growth with increased load was also different. The cracks developed were analysed pictorially, along with the evaluation of interfacial fracture energy. Sliding indentations (scratch) and finite element analysis were also done. The effect of nanoindentation on rheological (polymer) materials, biomaterials, and AFM nanoindentations was put forward where it was found that the stiffness fluctuations during AFM nanoindentation can qualitatively assess the heterogeneous behaviour of the biomaterial and can also be used for relative quantification; a softening parameter influencing the pace at which the effective elastic modulus varies with depth was determined. The P–h plots of PPy showed bulging and contraction as a response to mechanical stress, affecting the intramolecular interactions of the polymeric chains. Lastly, a short note on using AI and ML to determine coating properties has been given.

Acknowledgements The author hereby shows sincere gratitude towards Prof N.R. Bandyopadhyay, Prof. S. M. Hossain (IIEST, Howrah), Dr S.K. Mishra (CSIR-NML, Jamshedpur), Prof B. N. Dev, Prof G.P., Das (IACS, Kolkata), Prof S. Bhowmik (Amrita University), Prof. S. K. Samdarshi (CUJ-Ranchi), Dean (School of Engineering & Technology, CUJ-Ranchi), and Hon'ble Vice-Chancellor (CUJ-Ranchi) for their support, encouragement, and guidance.

Funding The authors acknowledge the funding by DST-SERB core research grant project (File no CRG/2023/003421).

Data Availability Data shall be made available on request.

Declarations

Conflict of Interest The author(s) declare no conflict of interest.

Compliance with Ethical Standards The submitted work has not been published anywhere else, in any format or language, except for a few preprints and properly attributed open-access content.

References

Here are the references in RSC Advances style:

1. Callister WD, Rethwisch DG (2020) Fundamentals of materials science and engineering: an integrated approach, international adaptation. John Wiley & Sons
2. Dieter G, Dieter G (1986) Mechanical metallurgy. McGraw-Hill
3. Pharr GM, Oliver WC (1992) MRS Bull 17:28–33
4. Bhattacharyya AS, Kumar RP, Priyadarshi S, Shivam SS, Anshu S (2018) J Mater Eng Perform 27:2719–2726
5. Fischer-Cripps AC, Lawn BR, Pajares A, Wei L (2005) J Am Ceram Soc 79:2619–2625
6. Soare S, Bull SJ, Oila A, O'Neill AG, Wright NG, Horsfall A, dos Santos JMM (2022) Int J Mat Res (Z. Metallkd.) 96:1262–1266
7. Cao M, Zhao L J Vac Sci Technol A. https://doi.org/10.1116/6.0002360
8. Jain AK, Kumar M, Thakre GD (2021) Tribol–Mater Surf Interfaces 15:278–291
9. Cellini F, Gao Y, Riedo E Sci Rep. https://doi.org/10.1038/s41598-019-40636-0
10. Rahimian Koloor SS, Karimzadeh A, Tamin MN, Abd Shukor MH (2018) Metals (Basel) 8:421
11. Yan Q, Chen S, Shi H, Wang X, Meng S, Li J (2023) J Eur Ceram Soc 43:7373–7380
12. Aldinger F (1998) Pure & Appl Chem 70:439–448
13. Wang B, Song Y, Zhang X, Chen K, Liu M, Hu X, He L, Huang Q (2022) Ceram Int 48:10280–10287
14. Duan J, Li M, Wang W, Huang Z, Jiang H, Ma Y (2023) Mater (Basel) 16:1665
15. Rajak A, Dash R, Kumari A, Bhattacharyya AS (2024) Mater Today Commun 38:107891
16. Bhattacharyya AS (2024) Mater Today Commun 38:107697
17. Bhattacharyya AS (2024). https://doi.org/10.13140/RG.2.2.10116.33920
18. Fan X, Smith AD, Forsberg F, Wagner S, Schröder S, Akbari SSA, Fischer AC, Villanueva LG, Östling M, Lemme MC, Niklaus F Microsyst Nanoeng. https://doi.org/10.1038/s41378-019-0128-4
19. Chang Y, Wei J, Lee C (2020) Nanophotonics 9:3049–3070
20. Zara JM, Yazdanfar S, Rao KD, Izatt JA, Smith SW (2003) Opt Lett 28:628
21. Wang S, Zhao H (2020) Micromachines (Basel) 11:407

22. Wang S, Xu H, Wang Y, Kong L, Wang Z, Liu S, Zhang J, Zhao H Rev Sci Instrum. https://doi.org/10.1063/1.5054628
23. Bhattacharyya AS (2024) Chem Phys Impact 8:100454
24. Bhattacharyya AS (2024). https://doi.org/10.13140/RG.2.2.15837.63201
25. Matar SF, Etourneau J, Solozhenko VL (2023) SILICON 15:511–520
26. Wang H, Zhang Z-Y, Cai X-W, Liu Z-H, Zhang Y-X, Lv Z-L, Ju W-W, Liu H-H, Li T-W, Liu G, Li H-S, Yan H-T, Feng M (2020) Chin Physics B 29:093601
27. Li Z, Deng L, Kinloch IA, Young RJ (2023) Prog Mater Sci 135:101089
28. Wen Q, Qu F, Yu Z, Graczyk-Zajac M, Xiong X, Riedel R (2022) J Adv Ceram 11:197–246
29. Wen Q, Yu Z, Riedel R (2020) Prog Mater Sci 109:100623
30. Bhattacharyya AS (2024)
31. Kumar D, Das S, Swain BP, Guha S (2024) Ceram Int 50:6526–6542
32. Dash R, Bhattacharyya K, Bhattacharyya AS (2023) Fatigue Fract Eng Mater Struct 46:1641–1645
33. Dash R, Bhattacharyya K, Kumar RP, Bhattacharyya AS (2022) Eng Res Express 4:045012
34. Chen J (2012) J Phys D: Appl Phys 45:203001
35. Toonder J, Malzbender J, With G, Balkenende R (2002) J Mater Res 17:224
36. Hertz H (1881) Reine. Angew Math 92:156
37. Auerbach F (1891) Ann Phys Leipzig 43:61–100
38. Griffith AA (1920) The phenomena of rupture and flow in solids
39. Irwin GR (1958) Fracture, Handbuch der Physik. Springer-Verlag, Berlin 6:551
40. Frank FC, Lawn BR (1967) Proc Roy Soc A 229
41. Oliver WC, Pharr GM (1992) J Mater Res 7:1564–1583
42. Fischer-Cripps AC (1995) J Mater Sci 32:1311
43. Tanaka K (1997) J Mater Sci 22:1501
44. Kruzic JJ, McNaney JM, Cannon RM, Ritchie RO (2009) J Mech Behavior Biomed Mater 2:384–399
45. Bull SJ (2005) J Phys D: Appl Phys 38:R393–R413
46. Holmberg K, Laukkanen A, Ronkainen H, Wallin K, Varjus S (2009) Wear 267:2142–2156
47. Thornton JA, Hoffman DW (1989) Thin Solid Films 171:5–31
48. Mercier D (2013) PhD thesis, Univ. Grenoble
49. Mercier D, Barthel E, Marliere C (2020) J Sol-Gel Sc Technol 93:229–243
50. Pajares A, Garcia C, Guiberteau F, Dominguez-Rodriguez A (1907) J Am Ceram Soc 1996:79
51. Li XD, Diao DF, Ogawa K (1997) Acta Mater 45:4453
52. Malzbender J, de With G, den Toonder J (2000) Thin Solid Films 37:1
53. Bahr DF, Hoehn JW, Moody NR, Gerberich WW (2003) Int J Frac 119:339
54. Kramer D, Huang H, Talke FE (1998) Acta Mater 47:33
55. Timoshenko S, Woinowsky-Krieger S (1959) Theory of plates and shells. McGraw-Hill
56. Hainsworth SV, McGurk MR, Page TF (1987) J Mater Res 1996:11
57. Thouless MD (1998) Eng Frac Mech 61:75
58. Dash R, Bhattacharyya K, Bhattacharyya AS. Nami Jishu Yu Jingmi Gongcheng (Nanotechnology Precis Eng). https://doi.org/10.1063/10.0022175
59. Bhattacharyya AS (2024) Fatigue Fract Eng Mater Struct 47:603–608
60. Bhattacharyya AS, Bhattacharyya K (2024) Next Mater 3:100038
61. Dash R, Bhattacharyya K, Bhattacharyya AS (2023) Int J Refract Hard Met 116:106373
62. Dash R, Bhattacharyya K, Bhattacharyya AS (2023) Eng Fail Anal 150:107353
63. Dash R, Bhattacharyya AS (2023) Fatigue Fract Eng Mater Struct 46:2714–2719
64. Char M, Chakraborty AK, Bhattacharyya AS, Kar A Adv Eng Mater. https://doi.org/10.1002/adem.202100679
65. Bhattacharyya AS, Mishra SK (2011) J Micromech Microeng 21:015011
66. Liu H, Dai M, Tian X, Chen S, Dong F (2021) AIP Adv 11:125009
67. Bull SJ (2011) C R Mecanique 339:518–531
68. Bhattacharyya AS (2021) Surf Topogr 9:035052

69. Bhattacharyya AS, Bhattacharyya K Surf Rev Lett. https://doi.org/10.1142/s0218625x24501233
70. Bhattacharyya AS, Bhattacharyya K (2024) Results Surf Interfaces 14:100196
71. Bhattacharyya AS, Rajak AK, Bhattacharyya K (2024) Surf Eng 40:359–369
72. Geng D, Yu H, Okuno Y, Kondo S, Kasada R (2022) Sci Rep 12:6391
73. Zhu L, Xu B, Wang H, Wang C (2015) Crit Rev Solid State Mater Sci 40:77–89
74. Sato Y, Shinzato S, Ohmura T, Hatano T, Ogata S (2020) Nat Commun 11:4177
75. Michel MD, Muhlen LV, Achete CA, Lepienski CM (2006) Thin Solid Films 496:481–488
76. Okuma G, Maeda K, Yoshida S, Takeuchi A, Wakai F (2022) Sci Rep 12:6994
77. Mishra SK, Bhattacharyya AS (2013) In Silicon-based nanomaterials. Springer, New York, New York, NY, pp 215–241
78. Bhattacharyya AS, Bhattacharyya K (2024) Preprint. (link unavailable)
79. Bhattacharyya AS (2024) Chem Phys Imp 100551
80. Sha Z, Wan Q, Pei Q et al (2014) Sci Rep 4:7437
81. Zak S, Trost COW, Kreiml P et al (2022) J Mater Res 37:1373–1389
82. Dash R, Bhattacharyya K, Bhattacharyya AS (2023) Int J Ref Met Hard Mater 116:106373
83. Wang J, Yang C, Liu Y, Li Y, Xiong Y (2022) ACS Omega 7(16):14317–14331
84. Bhattacharyya AS (2024) Preprint
85. Bertolo MRV, Rodrigues MÁV, Horn MM et al (2023) In polymer-based nanoscale materials for surface coatings. Elsevier, pp 131–149
86. Mark R, Cowie JMG, Sperling LH (2001) Macromol Symp 167:15–43
87. Avellaneda M, Rodriguez-Manzo JA, Castaneda SI (2023) Surf Interface 41:103181
88. Carvalho A, Machado RAF, Ferreira JAM (2023) Int J Plast 169:103712
89. Carvalho A, Machado RAF, Ferreira JAM, Delgado JMPQ (2023) Int J Solids Struct 283:112488
90. Najafi AR, Ayatollahi MR, Khonsari MM (2023) J Mater Res Tech 26:6823–6841
91. Bhattacharyya AS (2024) Preprint. https://doi.org/10.20944/preprints202404.0969.v1
92. Bhattacharyya AS. Preprints. https://doi.org/10.20944/preprints202404.0969.v1
93. Bhattacharyya AS, Bhattacharyya K. Researchgate. https://doi.org/10.13140/RG.2.2.16302.93769
94. Bhattacharyya AS. Research square (preprint). https://doi.org/10.21203/rs.3.rs-4190750/v1
95. Bhattacharyya AS (2024) Metal Res Technol 121:311
96. Roa S, Rodriguez-Manzo JA, Castaneda SI, Cowie JMG (2022) J Phys Chem Solids 163:110605
97. Roa S, Rodriguez-Manzo JA, Castaneda SI, Cowie JMG, Sperling LH (2023) Surf Interface 39:102948
98. Roa S, Rodriguez-Manzo JA, Castaneda SI, Cowie JMG, Sperling LH (2023) Solid State Commun 361:115071
99. Saeedi B, Khonsari MM, Ferreira JAM (2023) Mech Adv Mater Str 1–18
100. Kontomaris SV, Rodriguez-Manzo JA, Castaneda SI (2022) Mater 15(7):2477
101. Wu G, Cowie JMG, Sperling LH (2020) Sci Rep 10:18784
102. Verma J, Khanna AS (2023) Phys Open 14(100133):100133
103. Kuo RYL, Harrison C, Curran T-A, Jones B, Freethy A, Cussons D, Stewart M, Collins GS, Furniss D (2022) Radiol 304(1):50–62
104. AlGhaithi A, Al Maskari S, Musculoskelet J (2021) Surg Res 5:4
105. Dobrzański LA, Staszuk M, Honysz R (2012) Arch Mater Sci Eng 58(2):152–157
106. Dobrzański LA, Staszuk M, Honysz R (2010) Arch Comput Mater Sci Surf Eng 2(3):141–148
107. Snyder J (2024) Using AI to detect–and predict–coating corrosion. Riviera. https://www.rivieramm.com/news-content-hub/news-content-hub/using-ai-to-detect-ndash-and-predict-ndash-coating-corrosion-55480

Chapter 8
Mathematical Modelling of Mechanical Coating

Armab S. Bhattacharyya, Kushal Bhattacharyya, R. Praveen Kumar, Ritambhara Dash, and A. K. Rajak

Abstract This article provides a review of the modelling done in the case of coatings used in mechanical applications, like protective metallurgical coatings having properties of high hardness and wear and abrasion resistance. It discusses the models established during the atomic layer-wise film deposition, as well as static indentation and scratch-based mechanical property determination. Finite element modelling to study the stress distribution, cantilever N/MEMS and magnetic coatings subjected to ion beam irradiation (using SRIM) has also been discussed. The discussions presented are beneficial for the determination of fracture toughness and the use of coatings in sensing devices.

Keywords Modelling · Atomic layer deposition · Nanoindentation · Scratch · Finite element modelling

---

A. S. Bhattacharyya (✉) · R. Dash · A. K. Rajak
Department of Metallurgical and Materials Engineering, Central University of Jharkhand, Ranchi, India
e-mail: 2006asb@gmail.com; arnab.bhattacharya@cuj.ac.in

A. S. Bhattacharyya
Centre of Excellence in Green and Efficient Energy Technology, Central University of Jharkhand, Ranchi, India

K. Bhattacharyya
Department of Mechanical Engineering, Netaji Subhash Engineering College, Kolkata, India

R. Praveen Kumar
Department of Computer Science and Engineering, Nalla Malla Reddy Engineering College Divyanagar, Medchal, Hyderabad, India

B. P. Swain (ed.), *Advances in Mechanical Coating*, Materials Horizons: From Nature to Nanomaterials, https://doi.org/10.1007/978-981-96-7484-8_8

1 Introduction

Mathematical modelling has had an enormous boost in recent times due to improved computing techniques. They are being used in the medical industry to detect fractures in bone or failure in a bioimplant beforehand, as well as the quality and capability of PVD/CVD-based mechanical coatings [1–6]. Mechanical coatings are used mainly as protective coatings applicable to aerospace, marine, automation and tool industries [7–10]. High-precision coatings also have applications in microelectronic packaging, N/MEMS-based sensors and devices [11, 12]. Mathematical modelling saves much time, money, and energy in this regard, as the applicability and performance of any coating system can be assessed without practically putting it to the test. With the development of AI, modelling is becoming more and more accurate and economical in terms of time and money. This chapter focuses on some of the mathematical modelling techniques used in mechanical coatings and the type of physical interpretation they use.

2 Sputtering-Based Atomic Layer Deposition

For thin films with stoichiometry control, sputtering is a useful physical deposition technique. It is employed in the technology of depositing different coatings of technological relevance to produce surgical equipment used in a variety of medical settings, protective coatings for tools and aeroplanes made of metallurgical materials, and microelectronic gadgets. To obtain the desired coating properties, the process parameters and the composition of the target are adjusted. Trials must be conducted using this approach until the targeted properties are obtained. Therefore, computational modelling is significant as it saves time and money. Therefore, the computational model might be quicker and less expensive. We aimed to reproduce the method of deposition via mathematical indices that characterise physical properties using a sputter rate equation and a publicly available database [13, 14]. The model has been applied to coatings of materials having technological significance [13]. It is based on the number of atomic layers computationally visualised by sputtering, depending on the time for which the deposition was carried out and the current density maintained during the process. The sputtering rate or the quantity of target material sputtered in a specific length of time is determined by Eq. 1.

$$zt = M/(r N_A e) S j_p \tag{1}$$

where N_A is the Avogadro number; e is the electron charge; M is the target's molar weight [kg/mol]; r is the material's density; jp is the primary ion current density, and S is the sputtering yield (atom/ion). The sputtering rate of atomic layers (AL) per second is obtained by dividing this rate by the atomic diameter [14, 15].

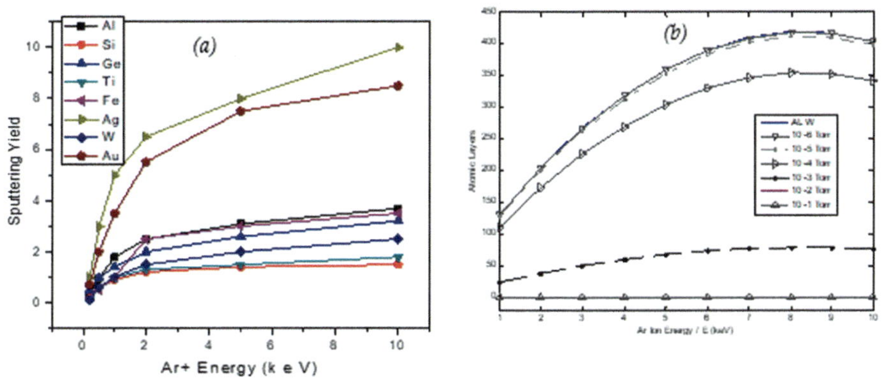

Fig. 1 **a** Sutter-yield versus Ar+ energy. **b** sputtered and deposited atomic layers of W (reproduced with permission) [13]

Octave, a free online programme, was utilised for the computational research. The atomic layer deposition with incident argon energy at varied deposition pressures is studied using the model presented in the preceding section. As demonstrated in the Fig. 1a, the S is observed to rise with ion energy [16]. This is evident since higher energy will have a more significant impact and cause more atoms to be ejected from the target. In an evacuated chamber, an inert gas—typically argon—is added and ionised. Ionisation results in the formation of plasma. In plasma, the atomic density is around $10^{18}/m^3$. Once more, the energy of the incident Ar ion determines the S parameters. We begin with the metal tungsten (W), which has the following properties: radius of 0.137 nm, 12.96 g/cm^3 as density, and molar weight of 184 g/mol. To determine the relationship between S and E, a quadratic polynomial was obtained by fitting the plot involving S and E (Eq. 2)

$$S = -0.033E^2 + 0.55E + 0.3 \tag{2}$$

where the value of 0.3 representing the intercept has been disregarded. Thus, in the sputtering rate calculation, we substitute E for S [16]. The argon ion energy in the model was adjusted between 1 and 10 keV. In terms of 10^{-n}, where the pressure index, denoted by n, ranged from 1 to 6, the deposition pressure P varied from 10^{-1} Torr to 10^{-6} Torr. Figure 1b shows the plotting of the sputtered and deposited atomic layers of tungsten (AL W and ALd W) with an increase in Ar ion energy. The difference between the sputtered and deposited ALs at lower pressures (10^{-6} and 10^{-5} Torr) is not very great. There is a more noticeable change at 10^{-3} and 10^{-4} Torr of pressure drop. The discrepancy grows as the pressure drops. This discovery can be explained by the fact that the adatoms' mean free path is shorter at lower pressure, causing fewer collisions with gas atoms. Naturally, the film thickness exhibits the same pattern as displayed in Fig. 2a. The variation in atomic layers sputtering and depositing at different energies, indicated by D in relation to the pressure index, is provided in Fig. 2b. Although the difference increased with the increase in energy, the

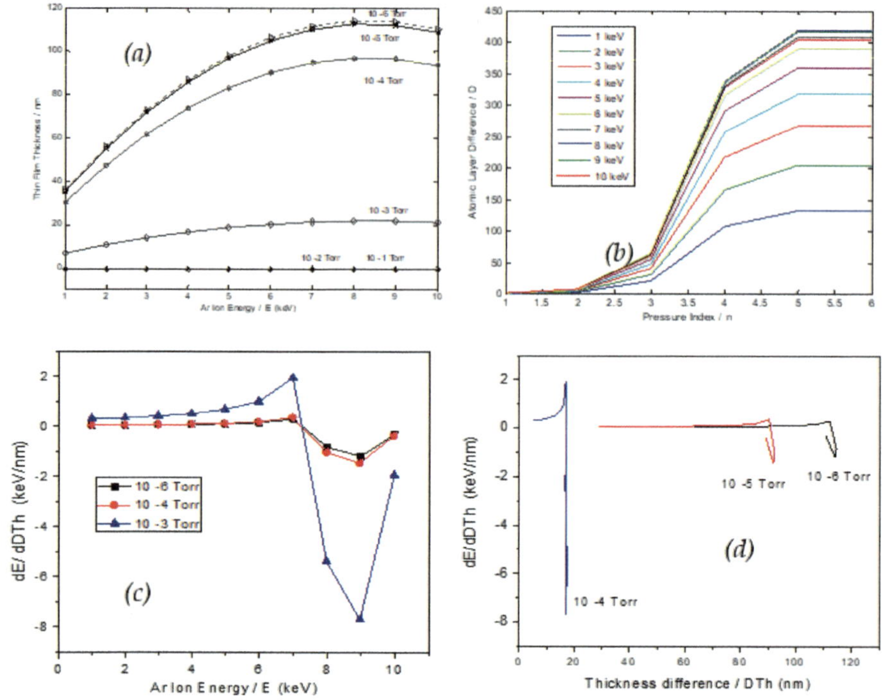

Fig. 2 **a** Thickness-W films thickness with pressures and Ar+ energies **b** Sputtered and deposited atomic layers with Ar+ energies **c** in relation to deposition pressure (index n). The rate at which thickness changes at various deposition pressures in relation to Ar Ion energy **d** An equivalent curve to the Bragg curve, which displays Ar+ energy change rate as a function of film thickness with pressures and stopping potentials (reproduced with permission) [13]

differential rate decreased. At various ion energy ranges, the rate of energy change with thickness is provided in Fig. 2c.

On the other hand, the energy change rate in relation to the thickness difference is shown in the Fig. 2d is like the ion radiation Bragg curve. Through a Bragg curve analogy, we may handle (dE/dD_{Th}) by treating S (D_{Th}) as the stopping potential. The incident ion effectiveness was found to increase with a decrease in pressure [17].

3 Nanoindentation Load-Depth Plots

For thinner coatings (also known as thin films), depth-sensing indentations at the nanoscale (also known as nanoindentation) are made using a continuous stiffness measurement method. We have employed nanoindentation over the past few years, specifically for the purpose of determining mechanical parameters on a very tiny scale that apply to thin films and nanostructures. The load-depth (P–h) plots, which numerous academics attempted to generalise, serve as its foundation. Oliver and Pharr provided the most well-known equation among them, $P = nh^m$. Upon which numerous other unloading curves are suggested [18–22]. However, considering the sample's elastic–plastic nature, we are taking into consideration the loading and unloading curves that were previously proposed [23]. These are provided by Eq. 3a for the loading (which is the same in every case) and Eqs. 3b, 3c and 3d for unloading (which varies depending on the sample type) sections.

$$\text{For loading}: P = nh^m \tag{3a}$$

$$\text{For unloading}: P_1 = nh^m - n(h_{\max} - h)^{1/2} \tag{3b}$$

$$P_2 = nh^m - n.^2(h_{max} - h)^{1/2} \tag{3c}$$

$$P_3 = nh^m - n(h_{max} - h)^{5/2} \tag{3d}$$

$$P_4 = nh^m - n^2(h_{max} - h)^{3/2} \tag{3e}$$

It is not a revolutionary approach to substitute the original value. The algorithm's flow chart diagram is displayed below (Fig. 3). Studies have also been conducted using the value substitution algorithm [24].

4 Computational Studies-Vickers Indentation

Vickers hardness (HV) typically decreases with thickness for hard coatings on relatively softer substrates because of higher substrate influence. The HV also decreases with increased indentation load. In each scenario, the hardness reaches saturation. The case of hard nanocomposite SiCN coatings has also shown this trend. Vicker's Hardness Number (VHN), which represents hardness, can be converted to GPa using the formula (100 VHN = 1GPa). The cubic fit was used for Si substrates, whereas for glass and 304 SS substrates, a quadratic fit of the hardness profile was done (Fig. 4a–c). Equations 4 and 5 provide the fitting relations.

$$H = P_1d^4 + P_2d3 + P_3d^2 + P_4d + P_5 \text{ (4th order fit)} \tag{4}$$

$$H = C_1d^3 + C_2d^2 + C_3d + C_5 \text{ (Cubic fit)} \tag{5}$$

The parameters P and C were varied to measure the substrate effect and thin film hardness as a function of thickness, i.e., H(d) (Fig. 5a). Since the hardness values are proportional, only their changes may be considered [24].

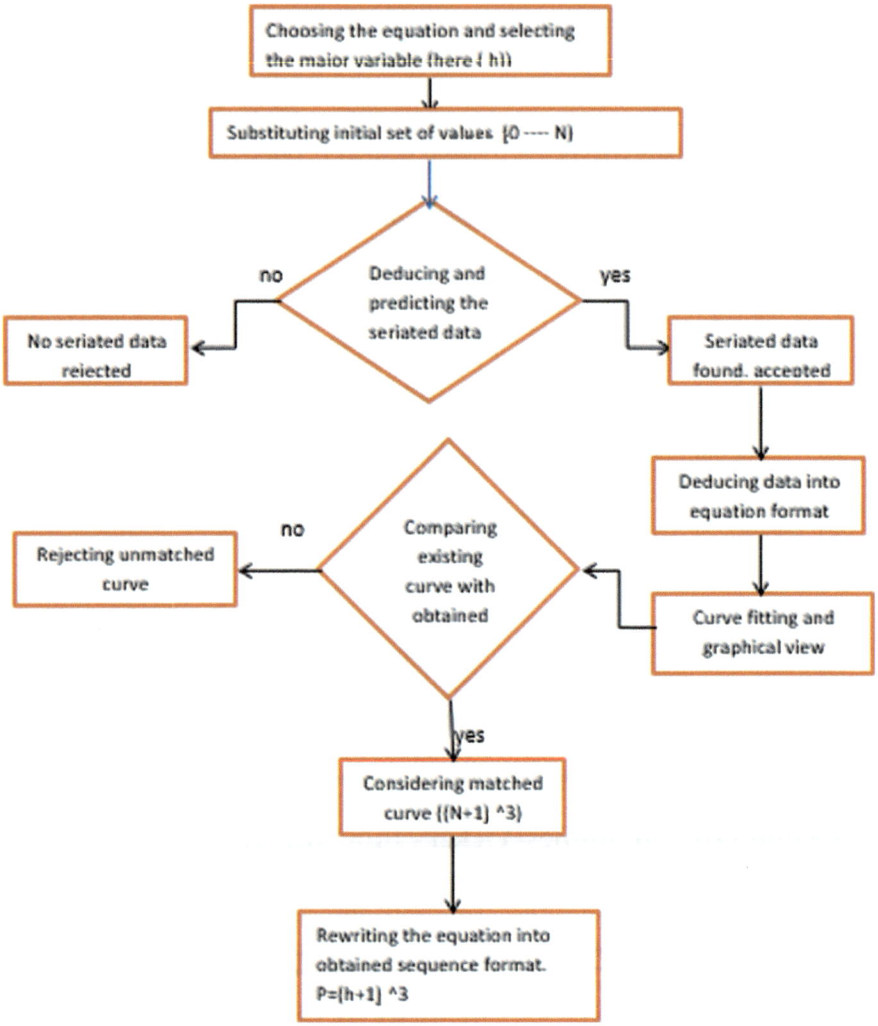

Fig. 3 The flowchart for the algorithm–open access [18]

Hardness, as we all know, is the material's limitation on dislocation movement. The decreased substrate effect is the cause of the hardness increase with thickness. On the other hand, poor adhesion because of increased residual stress can account for a decrease in thickness. Usually, compressive stress causes better film adhesion than tensile stress. The variation of hardness with respect to the change in P_1 for different film thicknesses is given in Fig. 5b. Because the hardness is nearly constant, there is little evidence of stress variation for thicknesses smaller than 2 μm.

On the other hand, increased stress fluctuation in the material has been found to cause instability. The variation of hardness for different thicknesses (1−5 μm) is

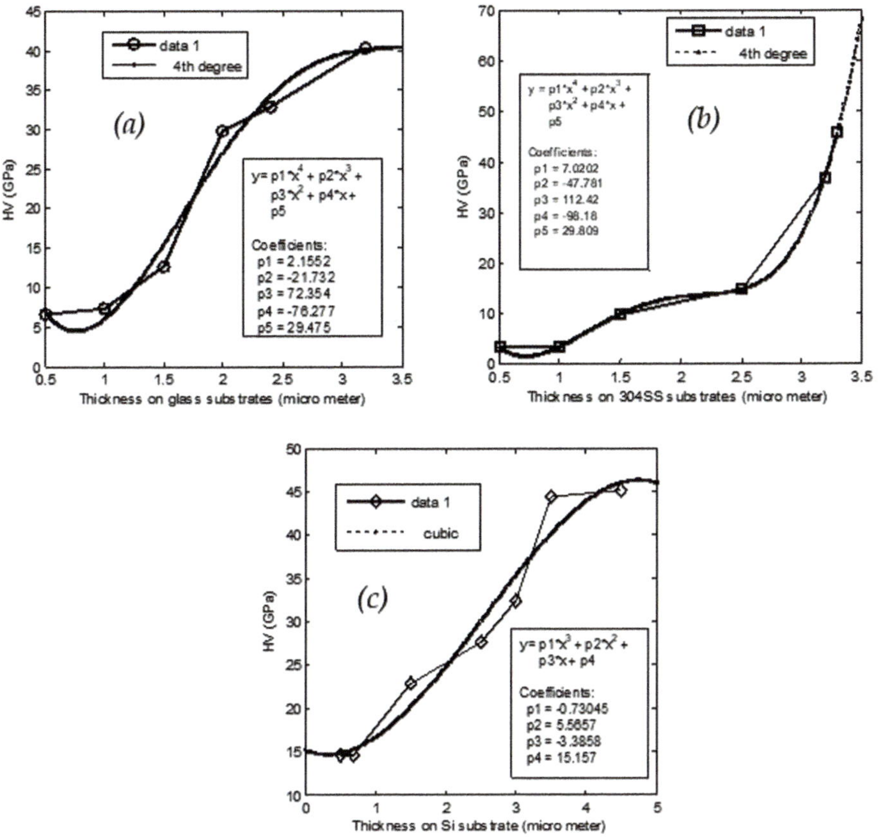

Fig. 4 Polynomial (4th order) fit for the thin film on different substrates **a**, **b**, and **c** glass, 304 SS, and Si (100) substrates-open access [24]

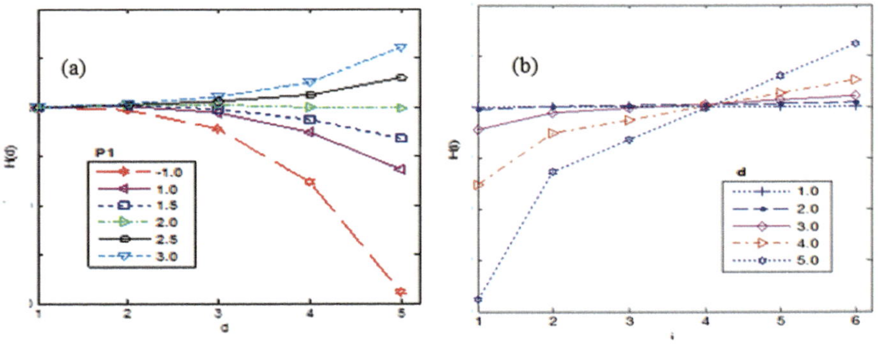

Fig. 5 Hardness variations with **a** thickness and **b** values of P_1-open access [24]

shown in Fig. 6a–d. It is evident that hardness decreases when P2 increases, and this decrease rate increases with thickness. Remarkably, a reversal of the hardness rise is shown for P3, and the rate of increase does not increase or decrease monotonically with thickness; it peaks at 4 μm. For P4, the difference in the reduction rate is substantially smaller, and for P5, it is nearly constant. The fluctuations in hardness with thickness were different for the different parameters and can be correlated with tip sharpness, since a sharper tip results in a lower hardness. P2 and pile-up have a similar relationship in that the harder something is, the more pile-up there is. Conversely, P3 is linked to sink-in effect, work hardening, and blunt tips. Hardness consequently rises with P3 [24].

As P5 is a constant; hence, its value does not change as it increases. The significant 4 μm shift in thickness suggests an ideal thickness with the best characteristics. We can also observe from Fig. 7a that P3 is the most prominent parameter in this instance since the fluctuations in Fig. 7b are more striking than those in the other figures. We can, therefore, assume that SiCN has been significantly sinking and tip-blunting. But for nanoindentation, tip blunting or sharpening is much more significant; thus, we will concentrate on the fitting P parameters, which characterise the material characteristics instead of the indentation procedure.

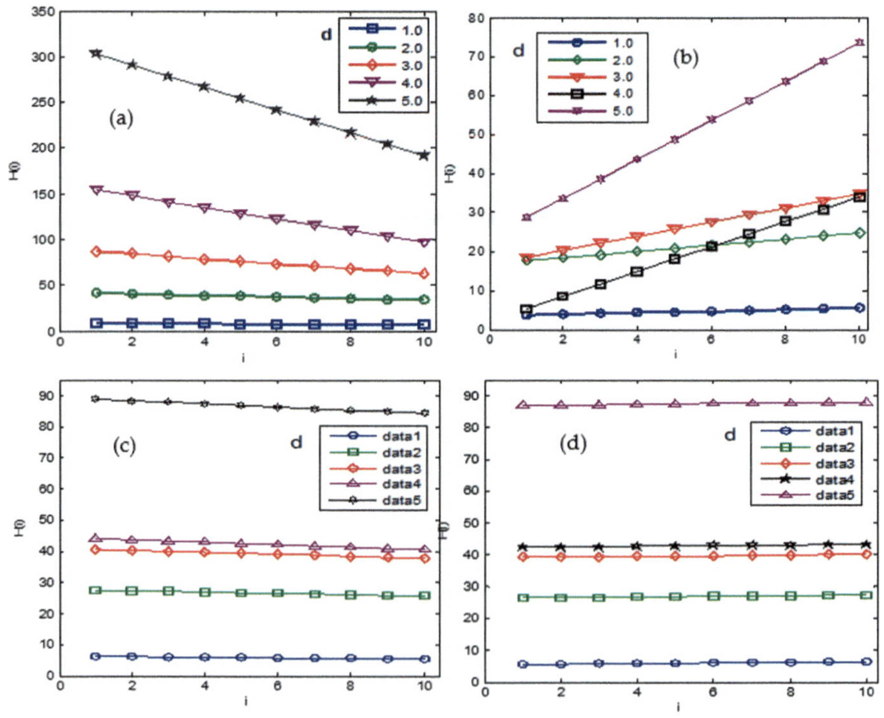

Fig. 6 Hardness variations with fitting coefficients at different thicknesses-open access [24]

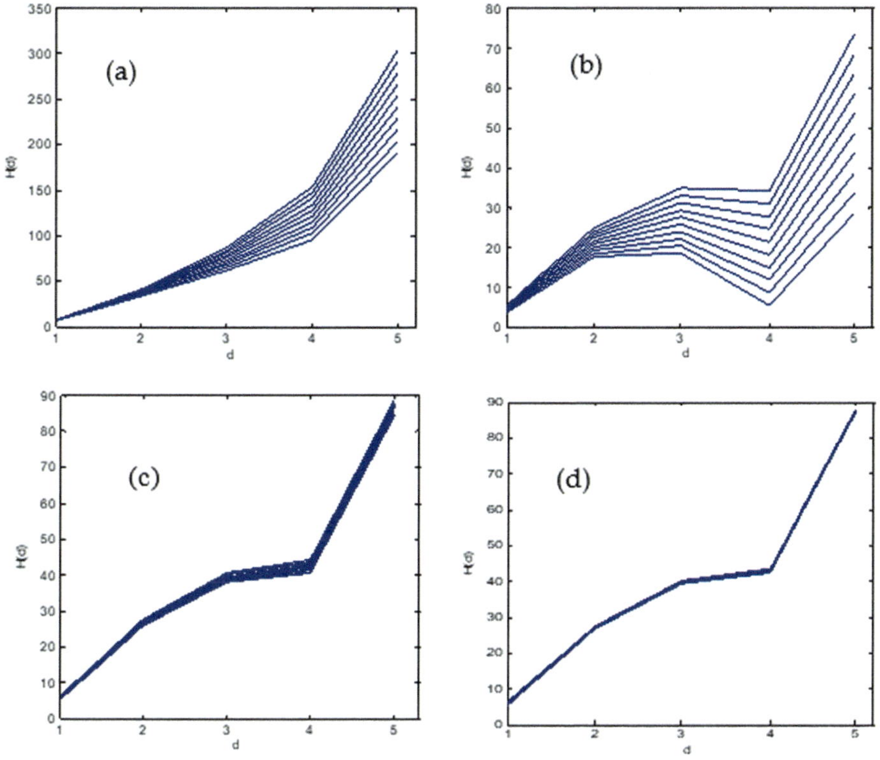

Fig. 7 Hardness changes with thickness for different coefficients–open access [24]

5 Nanoindentation Hardness and Modulus Variation

As seen in Fig. 8a and b, the nanoindentation on Si–C–N/SS304 gave a modulus of 160 GPa and a hardness of 14 GPa, with the substrate effect getting initiated from 250 nm which is one-tenth of the film thickness (2.5 μm). The modulus and hardness graphs changed as penetration depth increased. MATLAB techniques were used to modify the coefficients to generate different modulus and hardness responses. As Fig. 10 shows, we found that coefficient variation affected peak hardness and modulus values [20].

In order to ascertain and document the fluctuation in hardness and modulus values resulting from changes in deposition parameters, our group carried out experiments. Depending on the substrate type and treatment conditions, the results varied from 1.4 GPa hardness and 50 GPa elastic modulus to high values of 25 and 250 GPa (Fig. 9a and b).

Similar findings have been achieved when the temperature has increased. Thus, one may see the coefficient p8 as a measure of the influence of temperature. Since the differences in Fig. 9c are much larger, more research is needed to determine how

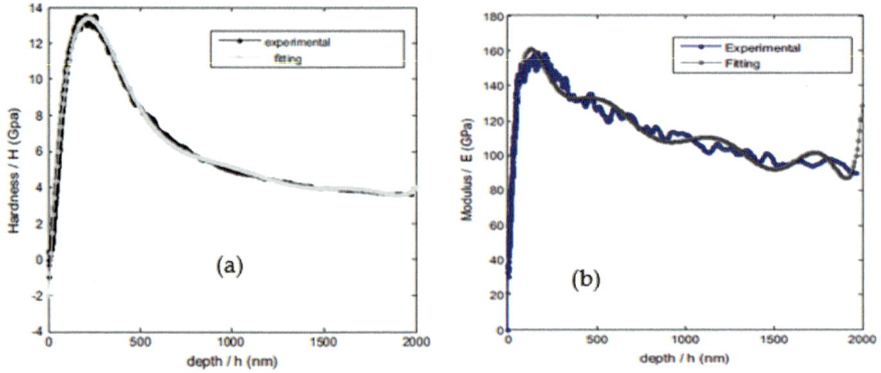

Fig. 8 **a** Hardness and **b** modulus of Si–C–N/SS304 substrate–comparative experimental and fitted plots [20]

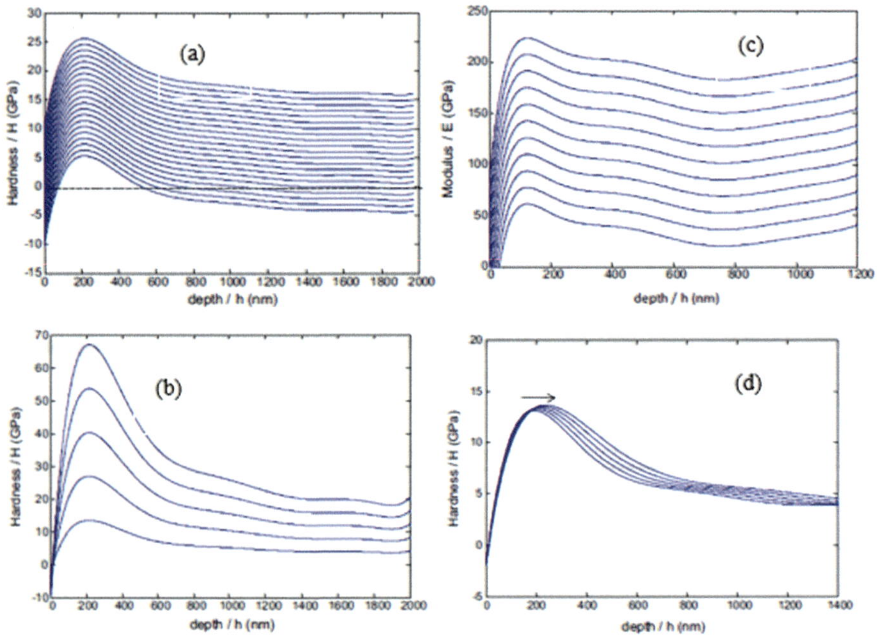

Fig. 9 Hardness **a–c** with different fitting coefficients resembling **d** change in thickness or tip-blunting [20]

additional coefficients can be introduced. The depth at which maximal hardness is reached got shifted in Fig. 9d. This happens when the tip blunts or the coating gets thicker [20].

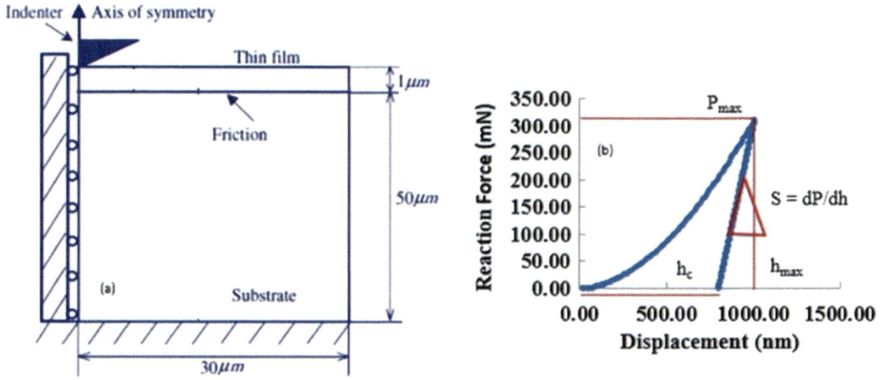

Fig. 10 **a** Finite element mesh model **b** Force versus displacement curve resulted from the finite element simulation (reproduced with permission) [25]

6 Finite Element Modelling

ABAQUS was used to do FEM simulations in order to study the stress distribution that resulted from the indentation, which led to delamination and fracture. The model is simulated in axis-symmetric 2D due to the axial symmetry of the specimen and the indenter. An axis-symmetric discrete stiff element with a 70.3° half-included tip angle is used in the nanoindentation. The thin film and substrate materials were represented by four nodes of axisymmetric CAX4R elements27, respectively. For the model depicted in Fig. 11a, 68,764 elements with a side measurement of 150 by 150 nm were employed. This was necessary because accurate simulation results need a finer mesh size of the element used for analysis [25].

The model's regions—between the film and substrate and between the stiff indenter and the thin film, respectively—are subjected to contact analysis. The first scenario creates surface-to-surface contact, with the thin film surface acting as the

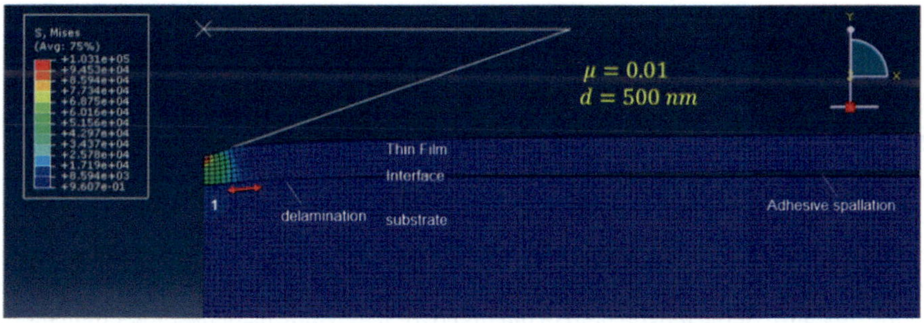

Fig. 11 FEM simulation of stress distribution during nanoindentation with different coefficients of friction and indentation depths as $\mu = 0.01$, 500 nm (reproduced with permission) [25]

slave surface and the stiff indenter surface acting as the master surface. The contact characteristic in this scenario is Normal behaviour with hard contact. The substrate surface is referred to as the slave surface and the film surface as the master surface in the second scenario. For the several simulations that are run, tangential behaviour with different friction coefficients is provided as contact attributes.

There are two steps in completing the analysis. Using a firm indenter, the load is delivered as 1000 nm displacement in steps of no more than 1/100000 to a maximum of 1/100 of the recommended displacement in Step 1. In the subsequent phase, the indenter is removed to a depth of 800 nm in order to simulate elastic recovery, as shown in Fig. 10a. The map of the load-depth created by FEM is displayed in Fig. 10b. Measurements are made of the maximum load (Pmax) of 300 mN and the corresponding maximum load (h max) of 100 nm. Elastic modulus is computed using stiffness (S = dP/dh), which is determined by the unloading component. The projected area $A = 24.5\ hc^2$, which is used to calculate hardness, is calculated using the contact depth (hc).

7 The Effect of Friction

In order to investigate the function of the interface in fracture, the coefficient of friction (μ) was changed in order to use FEM to examine the function of contact mechanics. We utilised μ values of 0.01, 0.02, and 0.08. Each example employed two penetration depths (500 nm and 1000 nm) to assess the impact of indenter penetration. As the indenter pierces the substrate, the stress distribution can be observed. Shear stress from indentation acts in the slip plane that contains the dislocation. For a 500 nm penetration, the dislocation bulges out and induces slip and annihilation to form a large loop (#1); as the penetration depth increases to 1000 nm, a new source of dislocation emerges and creates a smaller loop inside (#2) (Fig. 11c). The second loop maintains a significant distance from the contact by being almost totally contained within the first loop. Similar to this, adhesive spallation increased with a penetration of 1000 nm [25]. The plastic zone forms a cardioid and adheres to the Von-Mises yield requirement [26].

8 The Effect of Indenter Shape

An FEA analysis of a silicon substrate coated with a 1 μm thick layer of TiB_2, using a cube corner indenter to display the stress distribution, is shown in Fig. 12a. The strain zones that appear during nanoindentation are caused by the tensile and compressive fields. Because of the tensile field in the first part, it expands in-plane rather than more deeply horizontally. Its lower part experiences compressive stress, which impacts the coating substrate contact and causes more profound expansion than lateral expansion. Plotting the compressive and shear stress distributions against the tip's displacement from the contact point to the film (Fig. 12b) showed that the compressive stress varied more significantly along the axes of indentation, with a slope of 80 MPa/nm, than the shear stress, which varied by 40 MPa/nm. There is a greater chance of coating detachment since the shear force was often parallel to the film surface contact [27].

When indenting, different cracking events might happen, especially if you use a sharp pyramidal indenter like the Berkovich or Cube Corner. With the sharper cube corner indenter (Fig. 12c), the corresponding cone angle for Berkovich is 70.3° (half angle 35.15°), which is lowered to 42.23°. The two scenarios have different stress distributions. Increased sharpness somewhat counteracts the impact of preexisting microcracks in the sample, reducing the through-thickness lateral cracks in the coating. While lateral cracks are associated with coating adhesion and wear resistance, radial cracking is connected to film toughness. The risk of the film breaking from the substrate increases with greater sharpness because it creates more shear

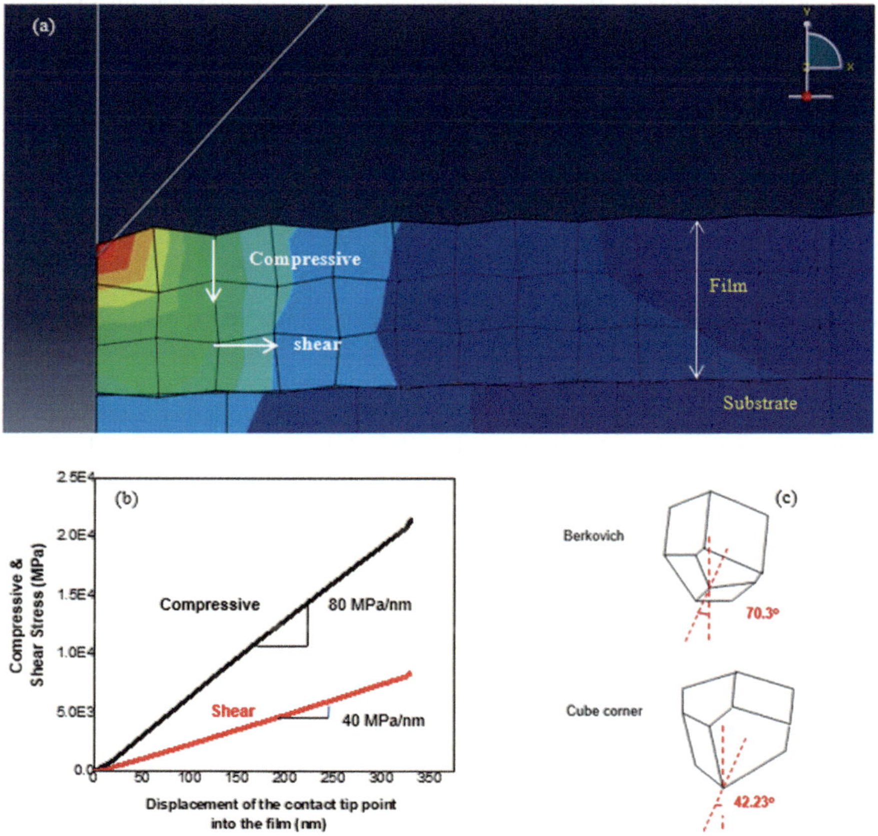

Fig. 12 **a** Finite element modeling of nanoindentation of TiB_2 coatings on Si substrates **b** Compressive and shear stress distribution and **c** Geometry of Berkovich and cube corner indenter (open access) [27]

stress parallel to the film substrate contact. Lateral cracking removes the coating, whereas radial cracking happens on the film and has no effect on the substrate. The radial fissures that cause film failure become more familiar with finer indenters [27].

The Cohesive Zone Model (CZM), in conjunction with Finite Element Modelling (FEM), has been used to study thin film failure mechanisms, including interfacial fracture and delamination, that are related to coating adhesion, a crucial parameter for microelectronic device performance. As mentioned previously, with regard to interfacial adhesion, the stress distribution that develops beneath the indentation affects the interfacial fracture energy release rate [28].

SAC (Sn–Ag–Cu) lead-free solder connections, which are employed in electronics packaging, have also been subjected to nanoindentation and stress measurement [29]. Since the broken or highly plastically deformed regions suggest substantial strain fields surrounding the indentations, which should interfere with free electron flow motion, they also have an impact on electrical characteristics. Due to a conflict between the mechanical load and the electric field during the domain-switching process, it has been discovered that the electric field influences nanoindentation [30, 31]. For Li_xSn alloys, the effect of charge states on yield and elastic modulus was revealed by stress–strain plots based on nanoindentation using the power law hardening model [32].

9 Coating Failure Analysis

To ascertain the stress distribution during indentation, Finite Element Analysis (FEA) was carried out for four different indentation depths: 240, 336, 432, and 480 nm. Since Si–C–N is a hard material, the experimental curve deviation was corrected using FEA using the brittle fracture approach. For SiCN, the yield stress—which establishes the compressive stress property—was fixed at 1000 MPa. As shown in Fig. 13 over a range of load indentation depths, the area where the Maximum Principal Stress equals twice the Yield Stress is known as the fracture process zone (FPZ). The formula for FPZ is to increase the volume of each element by the number of elements whose maximum shear stress is greater than twice the yield stress [33]. Figure 14 shows a linear relationship (Eq. 6) between the indentation depth and FPZ

$$y = 7E + 06Ex + 2E + 08 \tag{6}$$

where "x" represents the indentation depth in nm and "y" represents the FPZ expressed in volume (nm3) as the number of crossing elements whose Max-Principal stress is twice that of the Yield stress × 4 (as a quarter specimen) × volume of each element 200 × 200 × 200 nm.

As seen in Fig. 15a, the experimental hardness is also contrasted with the findings of the FEA. Hardness is calculated in FEA by dividing the reaction force by 24.5 h2, which is the contact area for each indentation depth and is projected as an area in

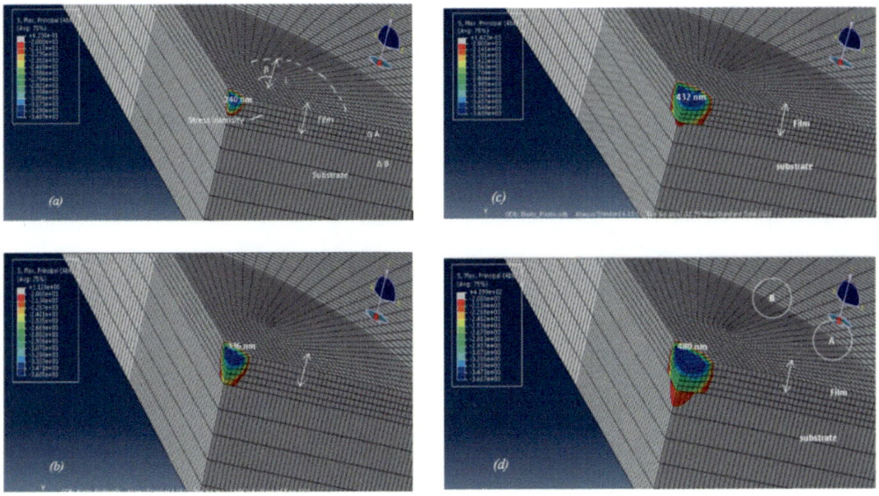

Fig. 13 The stress distribution using finite element modelling during nanoindentation on a SiCN/Si system at different indentation depths **a** 240 nm **b** 336 nm **c** 432 nm and **d** 480 nm (reproduced with permission) [33]

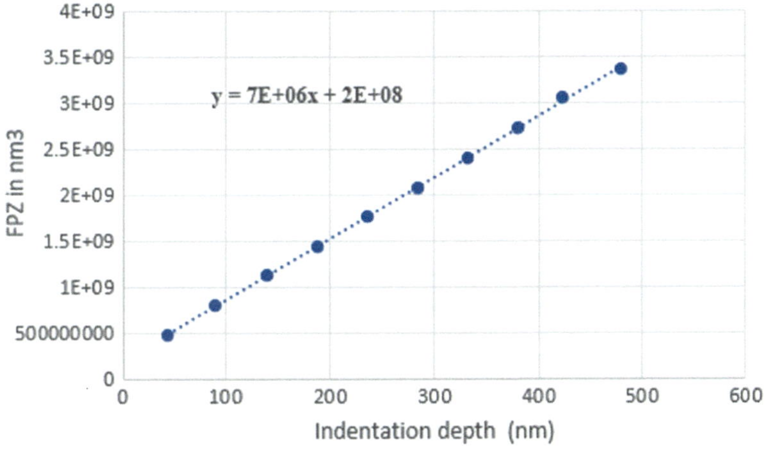

Fig. 14 Fracture process zone variation with indentation depth (reproduced with permission) [33]

the case of a Berkovich indenter for indentation displacement h. Between 300 and 400 nm in the experiment, there is a localised failure in the film that causes a decrease in hardness. However, contact is restored because of the presence of the plastic zone surrounding the crack, which experiences strain hardening and increases hardness, as seen in both experimental and FEA approaches.

The dislocation distribution area with dimensions is the plastic zone $r_c = \frac{1}{\pi}\left(\frac{\Gamma_i}{H}\right)$ next to the fracture point [34], where Γi is the previously determined interfacial

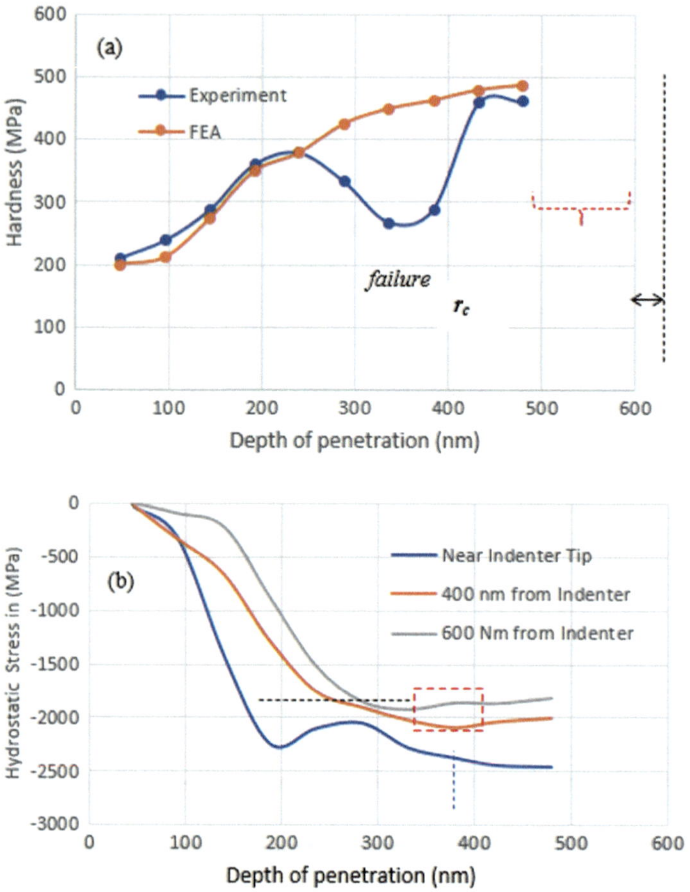

Fig. 15 **a** Comparison of hardness obtained from experiment and FEA **b** Hydrostatic stress versus depth of indentation (reproduced with permission) [33]

fracture energy of 30 Jm^{-2}. The region's hardness (H) is approximately 300 MPa, and the plastic zone's radius (r_c), which measures the distance across which the hardness rise occurred, was found to be 32 nm. The reduction in response value and hardness is a localised event that doesn't add to the material's overall failure. The hydrostatic stress versus indentation depth is displayed in Fig. 15b.

$$\frac{\sigma_{xx} + \sigma_{yy} + \sigma_{zz}}{3} \tag{7}$$

Equation 7 shows that this is the direct component of the stress's mean. The element layer directly beneath the indenter exhibits no change, subsequently increases as the indentation depth increases, and approaches double the yield stress (2000 MPa) value of about 200 nm. This finding is not displayed in the figure and was replicated at a distance of 200 nm below the indentation point. The element did not, however, reach the 2000 MPa threshold value at 400 nm from the indenter, and the value declined further from the threshold value as we went away from the indenter tip.

As Eqs. 8 and 9 suggest, Fig. 15b demonstrates that the shear stresses σxy and σzy did not reach the ideal stress to cause fracture. Less than 1333.3 MPa, the highest shear stress measured was 911 MPa in the xy direction and 910 MPa in the zy direction [35, 36]. This implies that shear stress was not the source of the film's breakdown, but rather that direct compressive force governs the entire failure process.

$$\tau_{yield} = \frac{2}{3}\sigma_{yield} \tag{8}$$

$$\tau_{fracture} = 2\left(\frac{2}{3}\right)\sigma_{yield} = 2\left(\frac{2}{3}\right)1000 = 1333.33 \text{ MPa} \tag{9}$$

A region similar to the chipped zone emerged despite the lack of coating disintegration at these depths (shown by the dotted line with parameters L and in This suggested that the beginning of coating failure originates from far earlier stages of indentation and that fracture formation is the only thing that makes them apparent. Where the mesh sizes in area A are comparable to those surrounding the indent, one can observe the difference in mesh form in the regions designated as A and B and marked by circles.

Due to the reliance of FE analysis on nodes, a larger mesh size implies an inaccurate stress distribution. Lateral fractures emerge in Region B, whereas radial (palmqvist) cracks mainly originate in Region A. This leads to the conclusion that stress variations are more significant for lateral cracks than for radial cracks, suggesting that the coating is more vulnerable to adhesive failure that causes chipping than to the formation of radial cracks. This is consistent with experimental observations that showed no signs of linear radial cracks [33]. The four strain zones that appear during nanoindentation are caused by compressive and tensile fields. The first region has a tensile field, which causes it to stretch in the plane without horizontally deepening. It comes into contact with the coating substrate interface in the region beneath it, which expands more deeply than laterally due to compressive stress. The substrate makes a discernible difference even at small indentation depths. Stress is generated below the compressive area by the film's resistance, which progressively disappears with additional penetration.

The highest stress is located at the fracture tip and is split into an axial and a lateral component at the film/substrate interface, which may cause lateral cracks to form. Previously, extended FEM, the Cohesive Zone model, and the Virtual Crack Closure technique have been utilised in FEA to evaluate fracture toughness and crack trajectory [37]. Determining the fracture toughness and crack pattern depends on the stress intensity component at the crack tip (K_I). The data available for the materials were used to determine the mesh size employed in the analysis.

The SiCN film's meshes were not the same as those in the substrate area. With mesh areas of 0.09 μm2 for ΔA (film) and 0.45 μm^2 for ΔB (substrate), the coating had a thickness of 1.5 μm. The regions of two adjacent meshes, ΔA and ΔB, represent the strain energy release rate. The applied force is denoted by F, while the displacement is represented by u, which has a displacement of 1.5 μm in the vertical and 0.3 μm in the horizontal directions. Using Eq. 10, the strain energy release rate at the crack tip was 300 kJ/m^2 and at the interface was 60 kJ/m^2. This resulted in a mode 1 stress intensity factor of 230 MPa$\sqrt{m}$ in the vertical direction ($K_{IC}{}^{v}$) and 103 MPa $\sqrt{m}$ ($K_{IC}{}^{i}$) at the interface (Eq. 11); the force applied was 4 mN [38].

$$G = \frac{2F\,u}{\Delta A + \Delta B} \tag{10}$$

$$K_I = \sqrt{\frac{G \cdot E}{1 - \nu^2}} \tag{11}$$

Previous studies have examined mixed-mode film failure by utilising FEM to illustrate the stress distribution beneath the depression and radial and lateral cracks. It was found that anisotropic stress concentration at the fracture tip controls the onset of radial cracks. The cracks propagate and are subject to loading and diverging because of the aggregation of nanocrystallites formed during the deposition process [38, 39].

When the indenter descends in Finite Element Modeling, it encounters a new layer of material, reestablishing contact and raising the reaction force through the loading phase. As seen in Fig. 16, the area is known as a fracture process zone (FPZ). Lateral fissures and adhesive failure of the film are caused by a considerable amount of stress created in the FPZ, which reaches up to the interface between the film and the substrate. Comparable loading at depths of up to 400 nm is displayed in the inset. It took a little bit more than 10 mN of load to get to this depth. In order to demonstrate good adhesion and interfacial toughness, loading to deeper penetration depths is required. The tension generated by the load indicated above depth was inadequate to cause adhesive failure in the film [39].

10 FEM of Scratching Event

Figures 17 and 18 display FEM models of the stress distributions during the scratching event. For a total scratch length of 5 mm, a cohesive coating breakdown happens at 1 mm, which is in line with the previously mentioned experimental data. The cohesive failure zone experiences an increase in stress from 20 to 160 MPa. As the slide progresses and reaches a peak tension of about 217 MPa, adhesive failure happens. In this case, the usage of a glass substrate prevented the stress from transferring to the substrates, which is a feature observed in substrates like SS304. Instead, it was contained within the coating.

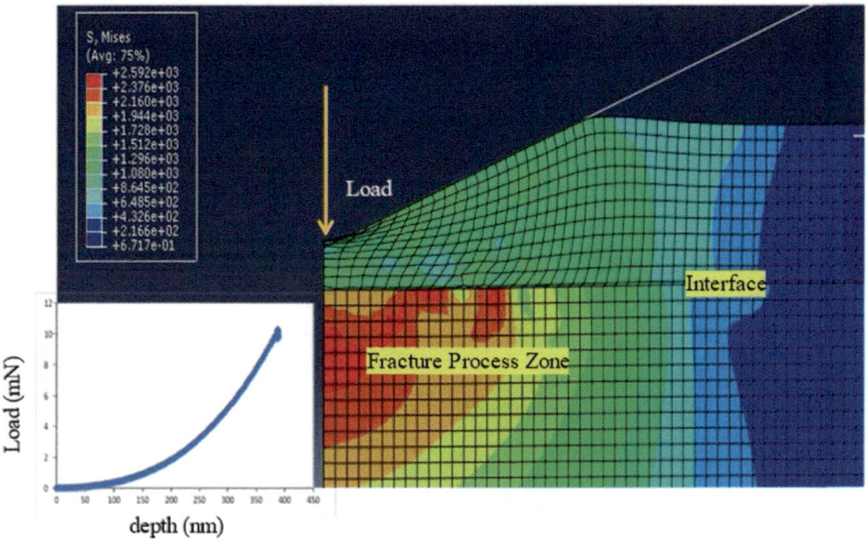

Fig. 16 Finite element modelling of nanoindentation on SiCN/Si-stress distribution (inset P–h plots)-reproduced with permission [38]

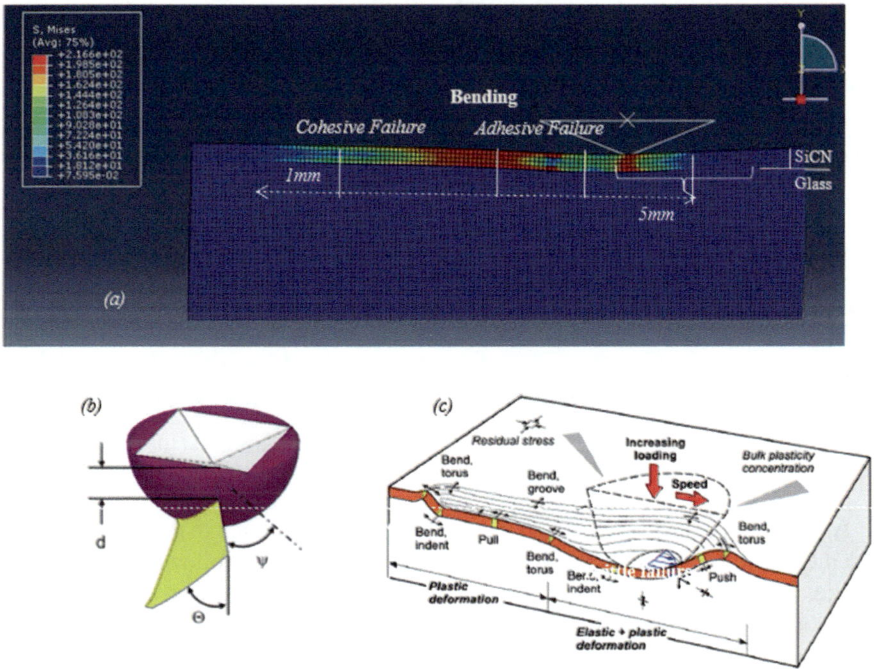

Fig. 17 **a** FEM stress distribution **b** coordinate system **c** schema of scratch track under the influence of bending-Open Access [40–42]

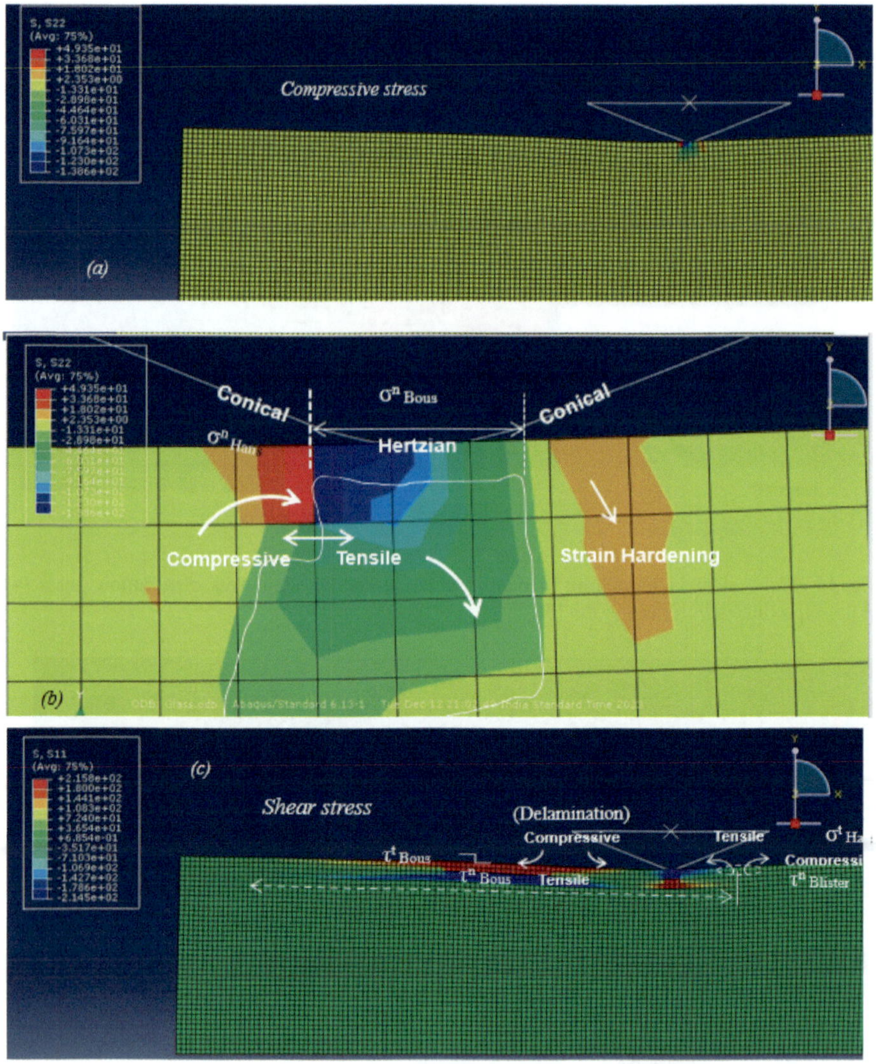

Fig. 18 **a** The **a**, **b** compressive and **c** shear stress distributions-open access [40]

A Hertzian contact results from the high stress being limited to the near-spherical indenter tip, as indicated by the Von-Mises stress distribution. Following the adhesive failure, the glass substrate now exerts a more significant influence. Partial delamination is shown by buckling, whereas entire delamination is represented by spallation. On the other hand, chipping has been observed along the edges of the scratch track. Conformal cracks are brought on by a tensile bending moment within the coating, whereas tensile cracking is brought on by frictional stress behind the stylus. In certain cases, like when employing spherical or Rockwell C indenters, circular

Hertzian cracks can be observed. Two symmetrical splits that come together close to the scratch line's edge are the cause of these cracks.

Symmetric radial cracks from the center, sometimes referred to as Chevron cracks, are another kind of coating failure observed during scratching [43]. The combination of Boussinesq (σBous) from indenter loading (normal and tangential) and Blister (σBlister) from residual stress forms the stress field in the scratch test. Radial fractures are the result of replacing the typical Bous component in conical indenters with Hanso's field (σHans). Equations 12a and b show how these fundamental stresses relate to shear components; the superscripts n, t, and r stand for "normal," "tangential," and "residual," respectively [44, 45].

$$\sigma = \sigma^{\mathrm{n}}\mathrm{Hans} + \sigma^{\mathrm{t}}\mathrm{Bous} + \sigma^{\mathrm{r}}\mathrm{Blister} \tag{12a}$$

$$\tau = \tau^{\mathrm{n}}\mathrm{Hans} + \tau^{\mathrm{t}}\mathrm{Bous} + \tau^{\mathrm{r}}\mathrm{Blister} \tag{12b}$$

The distribution of stress and elastoplastic response is caused by the contact between the indenter tip and the sample. Equation 13 describes the coordinate system (r, θ, ϕ) that represents the radial distance (Fig. 17f), the hoop angle about the loading axis, and the angle to the loading axis in spherical polar coordinates. The value for coating thickness is d. Shear stresses primarily cause adhesive failure and coating delamination. The Boussinesq stress is a representation of the residual stress-induced Blister stress fields and the shear stress $\tau_{r\theta}$ at the indenter contact site as a result of indenter loading [41, 44–46].

$$\sigma = \left(\sigma_{rr}\tau_{r\theta}\tau_{r\phi}\tau_{\theta r}\sigma_{\theta\theta}\tau_{\theta\phi}\tau_{\phi r}\tau_{\phi\theta}\sigma_{\phi\phi}\right) \tag{13}$$

Coating failure has been associated with surface roughness, coating thickness, indenter form, and test parameters like sliding velocity and loading rate. The elastoplastic reaction caused by the indenter tip and sample contact, as well as stress distribution, are the ultimate concerns of all these components. Tensile tension at the rear of the indenter tip causes a through-thickness cohesive failure, which is the first stage of coating failure. The shear components ($\tau_{\phi\theta}$, $\tau_{\phi r}$) of the principal compressive stress ($\sigma_{\phi\phi}$) or Boussinesq stress are imposed.

The compressive stress distribution is shown in Fig. 17a, and an enlarged view is shown in Fig. 17b. At this point, when the load increases, the compressive stress increases until the contact becomes Hertzian. Film collapse and localised tensile stress are caused by the dominance of the conical sides in the sphere-conical-shaped Rockwell C indenter under higher loads. The change from Boussinesq to Hanso's field occurs at this level. One can see a limited zone of elevated stress as a result of strain hardening.

Nonetheless, the film exhibits remarkable mechanical stability because of its reduced exposure to compressive forces. The failure was noticeably more brittle and the coating showed no elastic rebound, indicating the impact of the glass below [40, 41, 47].

Figure 17e clearly shows the influence of the bending indent ($\sigma_{\theta\theta}$). The shear component ($\tau_{\theta\phi}$) is more intense for coatings with smaller thicknesses, resulting in delamination and fragmentation. The radial shear component ($\tau_{\theta r}$) interacts with the Blister field, resulting in the bend torus [42], as shown schematically in Fig. 17g. A shift from compressive to tensile stress, which implies coating delamination, is also revealed by the shear stress close to the adhesive failure spot.

Below the indenter region, an opposing shift that suggests a strain-hardened zone can be seen. On non-ductile substrates, scratching causes Hertzian cracks and complete coating delamination, which leads to chipping. The interfacial fracture energy is then determined using the chipped segment's form [33]. Scratch tests were performed on Si–C–N coatings applied to SS 304 steel. By analysing the changing failure complexion within the scratch, which showed a mix of ductile and brittle characteristics, the abrasion and wear attributes were examined.

There were noticeable changes in the failure mode as a result of the loading rate fluctuation. With increasing load, it was discovered that plasticity had a peripheral dominance initially, but as the loading rate grew, brittle failure took its place. Both the wear rate and the adhesion effort were computed [48]. Recent work has concentrated on a novel approach that takes into account changes in dislocation kinetics, sliding modes, indenter sharpness, and crystallisation to calculate the probability of fracture (Pf) as well as the transitions from brittle to ductile and ductile to brittle in the same coating/substrate system. Equation 14 illustrates how the Pf was determined using the Weibull distribution. The changes from ductility to brittleness are shown by sliding indentations.

$$P_f = 1 - exp\left[-\left(\frac{h_{max} - h_{min}}{h - h_{min}}\right)^6\right] \tag{14}$$

The deepest and lowest penetration depths in a sequence of indentations made on the same material are denoted by the terms hmax and hmin. Finite-element modeling was used to assess the stress and to get data on the kinematics of dislocation and the influence of indenter shape on fracture. The results point to the possibility of creating nano- and micro-electromechanical system-based devices [49].

11 Computational Studies on Nanoindentation

The films' mechanical properties were examined using nanoindentation XP (MTS, USA). Using great spatial and depth resolution, nanoindentation is a depth-sensing indentation technique that may be used to characterise the mechanical characteristics of materials at very small scales. It is advantageous to thin films and establishes characteristics like adhesion strength, toughness, hardness, and modulus. Taking into account that the substrate effect starts at one-tenth of the film thickness [50–52]. Although indenters of different shapes are available, the 3-sided pyramidal

Berkovich indenter is the most common tool used for nanoindentation. Figure 19a and b shows the visual and experimental measurements of the indentation load and different stages of displacement associated with residual (hr), elastic (he), total (ht), contact (hc), and sink-in (ha) depths [53]. The displacement of 50 nm required a load of about 0.7 mN. An arrow indicates the horizontal displacement that results from afterload withdrawal, which points to film delamination [52].

After every 2 nm penetration, the H and E values were determined. Using the indenter's shape and the empirical relation for projected area A = 24.5hc2, the hardness is computed by dividing the maximum load (Pt) by the predicted location. Elastic modulus is calculated from unloading by the stiffness S = dP/dh at maximum penetration; good mechanical qualities were revealed by the deposited films with a modulus of 170 GPa and a hardness of 15 GPa (Fig. 19c and d). Apart from hardness (H = Pt/A), bulk modulus (E), which is related to reduced modulus (Er), is determined in Eq. 15.

$$\frac{1}{E_r} = \frac{(1 - v^2)}{E} + \frac{(1 - v_i^2)}{E_i} \tag{15}$$

The modulus of the diamond indenter is 1140 GPa, and its Poisson ratio is 0.07. Numerous facets of nanoindentation have been the subject of prior computer studies.

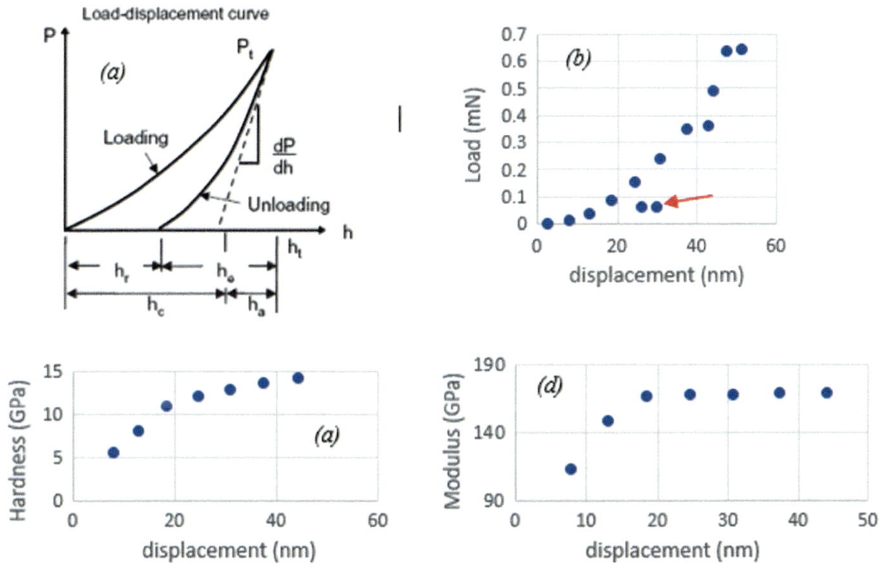

Fig. 19 **a** Schema of the different depths (h_r-residual, h_e-elastic, h_c-contact and h_t-total) in the nanoindentation load-depth plot (*P–h*) [21]; experimental **b** P–h **c** H **d** E variations with indent depth (reproduced with permission) [50]

Studies have been done on elasto-plastic responses as temperature changes, micro-crack formation during loading or unloading, instabilities during nanoindentation, and indenter selection [54, 55].

Using the formulae, the load-unload curves for nanoindentation were simulated using $P = kh^2$ for the loading curve and $P = kh^n - k\ (ht\text{-}h)^N$ for the unload portion (Fig. 20a). MATLAB (octave-online) was used for the programming [56]. The indenter's geometry affects the P–h curve; a spherical indenter, for instance, has a linear P–h relation and a Hertzian contact. Since there hasn't been any significant plastic deformation, the initial contacts—even with a sharp 3-sided pyramidal indenter—can be regarded as spherical for shallow depths.

As seen, plastic deformation results from the indenter entering the material, increasing the exponent value n in P = khn.in Fig. 20b. Thus, as the indenter digs deeper into the substrate, the value of n shows the type of contact. Since the shallow depth, contact established is more spherical, the Hertzian value n is equal to 1, and further, as the sharper sides get into contact with the sample. Simulated load depth curves for various elastic–plastic reactions were performed. There are no differences

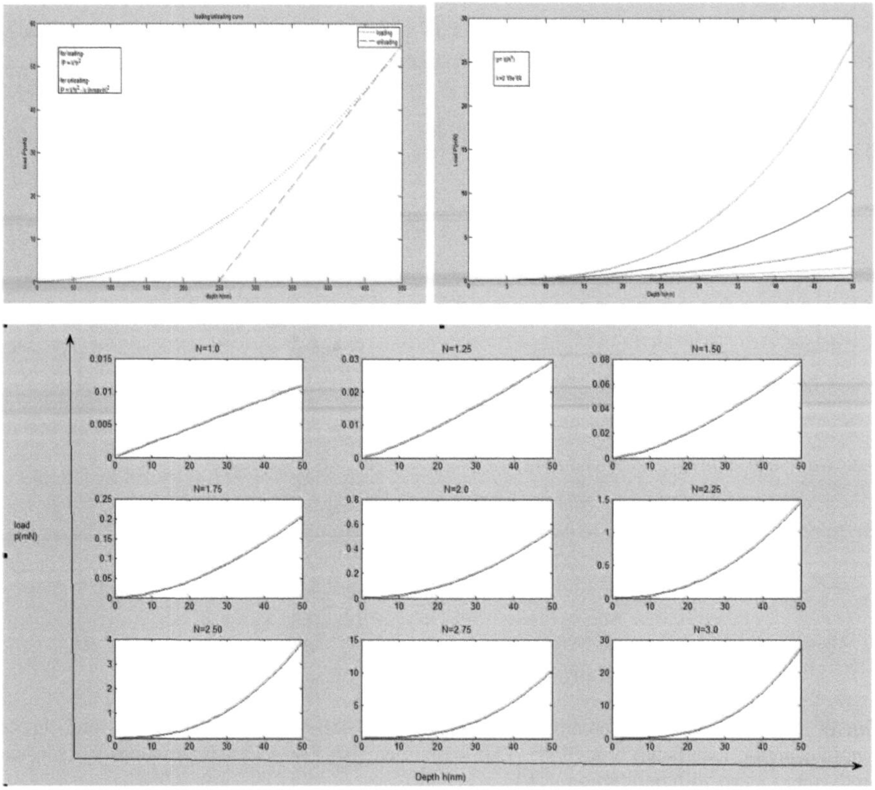

Fig. 20 Graphical representations of load-depth P with varied N displayed separately as obtained computationally (reproduced with permission) [50]

between the P–h curves in a fully elastic response. As plastic deformation increases, the divergence starts to rise. The area beneath the plot, which is inversely correlated with hardness, reflects the plastic work required to produce the deformation [23]. Figure 20c displays the P–h plots that were created for a range of N values. The sample's elastic–plastic reaction determines how N is valued. As demonstrated here, the value of N changes for various elastic–plastic reactions from ½ to 5/2. A helpful technique for analysing the mechanics involved in cantilever MEMS switches is nanoindentation [57, 58].

12 Cantilever N/MEMS

Nano/micro-electro-mechanical systems (N/MEMS)-based devices employ novel materials such as carbon nanotubes, piezoelectric materials, Si-nanowires, and two-dimensional materials. Actuators, sensors, and a variety of other systems like biosensors, magnetometers, gyroscopes, accelerometers, and pressure sensors make up these devices [59, 60]. These devices are made of silicon because of its reproducible and controllable properties as well as its established manufacturing processes. Because of soft clamping's ability to reduce dissipation, a quality factor of 10 billion and more has been demonstrated recently for silicon-based nanomechanical resonators [61]. Si-based nanomechanical resonators used in 3D printers have very high precision mass sensing capabilities [62].

On silicon substrates, nanomechanical devices are also being fabricated for the semiconductor industry [62]. As Si, as well as Ge, can be grown as single crystals or as polycrystalline using CVD methods. Low-dimensional layers of Si-based compounds with robust properties are used to produce devices. Si–C–N, a ceramic generated from polymers, is used in N/MEMS-based devices [63, 64]. The overall mechanical behaviour and performance of the device are affected by the defects, internal and residual stress, and variable microstructure [65].

Si-based resonators in the form of cantilevers are advised for making sensors due to their excellent quality and sensitivity. The Si-based structures can't operate at temperatures more than 150 °C, which limits it's use in high-temperature conditions. Hence, the composite material Si–C–N is utilised in high-temperature pressure sensors because of its superior temperature stability [66, 67]. The problem with MEMS resonator sensors based on cantilevers is that they cannot achieve consistent mass sensitivity. The cantilever's free end experiences the highest vibration intensity, which also produces the most significant mass sensitivity. As one approaches the fixed end, the vibration intensity diminishes. The cantilever's mass sensitivity is highest at its free end due to the highest vibration intensity, which, however, gets lowered as one moves towards the clamped end.

Impulse is essential to these devices because the resonance frequency of operation is inversely related to mass. Figure 21a illustrates the spatially non-uniform mass sensitivity of greater than 100% from the cantilever's free end to its middle. A unique design with a pedestal shape has been made to solve this issue [68]. Figure 21b shows

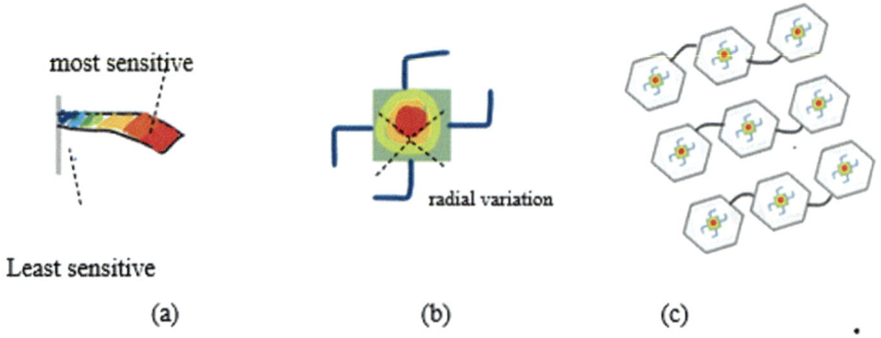

Fig. 21 Schema of **a** pedestal design with less than 4% non-uniformity in mass sensing from the platform's centre to its edge, and **b** high spatial non-uniformity from the cantilever's free end to its middle. Red is the most sensitive colour, while blue is the least sensitive, as shown by the colour distribution **c** linear configuration of the pedestal sensors-reproduced with permission [69]

the resonator, which is made out of a square pedestal that is held up by four beam springs. The pedestal shakes vertically because the ends of four beam springs are fastened to the substrate.

The pedestal design is also shown in Fig. 21b, where the mass sensitivity's spatial non-uniformity decreases by less than 4% as the platform moves from its centre to its edges. Instead of linear variation like in the previous scenario, the difference is caused by radial variation. Sensitivity is represented by the colour distribution, where blue is the least sensitive and red is the most sensitive. The pedestal cantilever configured as an array for use as a sensor is shown in Fig. 21c [68]. This creative design was subjected to finite element analysis (ANSYS 2022 R12, ANSYS Inc.) utilising the geometry and mesh distribution. The results showed the values for directional deformation, total deformation, and minimal primary strain (Fig. 22). Figure 22a displays the dispersion of the mesh. The pedestal structure's responses to the stresses put on it by the capture of biological molecules are shown by the directional deformation in Fig. 22b and the overall deformation in Fig. 22c, which is diagonal for the second case and linear for the first. However, as the above illustration shows, the pedestal structure is arranged in an array for sensing, which eliminates the effect of one pedestal cantilever compensating for another. As can be seen from the red zone over the surface and the uniform minimum principal strain, these structures fared better than standard cantilevers with a single fixed end. Figure 22e illustrates the geometry of the resonator, which consisted of a square pedestal measuring 14 $\times$ 14 mm, suspended by four beam springs with a length of 9 mm and a separation of 4 mm from the pedestal edge. The ends of four beam springs are fixed to the substrate, and the pedestal vibrates vertically. Based on the variation of mass with coordinates (Fig. 22f) [69], sensitivity evaluations are conducted.

13 Magnetic Coatings

Because of its enormous magnetoresistance (GMR), cobalt(Co) on cobalt oxide(CoO) is a combination representing a ferromagnetic (FM) layer on an antiferromagnetic (AFM) layer that is crucial to technology. Cobalt (Co)/CoO thin films were deposited by magnetron sputtering over Si (111) substrates that were exposed to Si ion beam radiation. A good technique for analysing morphological changes in thin films is ion beam irradiation. Localised heating can result in intriguing phenomena such as surface energy shifts and atom displacement, which could have applications in low-scale design during device fabrication in the future [70].

Because of the higher surface energy that the incoming atoms got from the ion beam radiation, the films showed a shift in the magnetisation curve along the x-axis (known as exchange bias) and coagulations due to high surface energy (dewetting) [71]. The M-H plot, which was produced by fitting the experimental plot to a mathematical one and assessing the impact of the different parameters on the plot, was the subject of a computational study [72]. Figure 23a displays the experimental M-H curve, which demonstrates the superparamagnetic behaviour of the system with almost negligible coercivity. The magnetisation in SPM can reverse directions, as ferromagnetic nanoparticles demonstrate. For the SPM state, the M-H plot displays a Langevin function that fits better over a smaller range of H values. A polynomial was used to fit the experimental plot, and by varying the coefficients, other simulated M-H plots could be created. Figure 23b shows the one approximating exchange bias. The results have applications in magnetic MEMS devices [73–76].

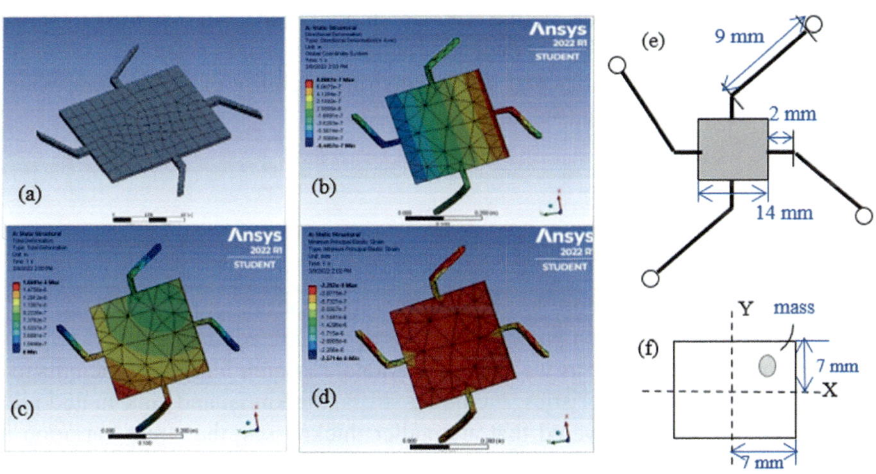

Fig. 22 **a** New pedestal design's geometry and mesh distribution of the resonator's geometry; **b**, **c** directional and total **d** lowest strain at the primary level and **f** fluctuation in mass sensing efficiency-reproduced with permission [69]

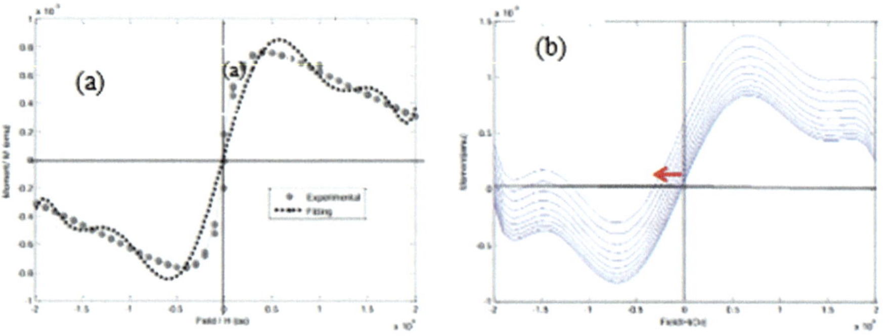

Fig. 23 **a** Experimental and fitted **b** simulated magnetization plot of Co/CoO thin film-open access [76]

14 Ion Beam Implantation Modelling

Si has previously undergone ion beam implantation to change its shape and prepare it for the manufacture of devices, especially small sensors [77–82]. Analogous memristors were recently shown to be able to benefit from ion beam-assisted SiGe solid phase epitaxy [83]. Using the SRIM/TRIM program, a simulation-based study of Na+ ion implantation on Si was carried out [84]. Simulated graphs of recoil ionisation and range distribution are shown in Fig. 24. These factors affect the morphological modifications needed for memory structures based on memristive devices to be produced at the wafer-scale in CMOS [85]. These energy recoils result in defects, amorphous pockets, and recrystallisation, which may produce conducting channels that are advantageous to produce devices. Radiation-induced vacancies modify the current–voltage characteristics and state retention capability of memristors, which impacts device performance [86–88].

15 Conclusions

Numerical models for fracture toughness, nanoindentation, and atomic layer-wise coating deposition have been presented. It was demonstrated that as coverage grew, the rate of partial yield rose as well. Computational research linked coefficients to a variety of physical characteristics, including stress, thickness, and sink-in and pile-up processes. It was discovered that at smaller thicknesses, the stress variation in the films is reduced, resulting in nearly constant hardness. The load-unload plot, which represents the various mechanical responses, was the subject of computational investigations. A key system made of materials based on SiC is the MEMS-based cantilever utilised in pressure sensors. Aside from the materials perspective, though, the cantilever sensors' geometry must be adjusted in order to account for the impacts of mass sensing capabilities. A FEM model for a novel pedestal design has

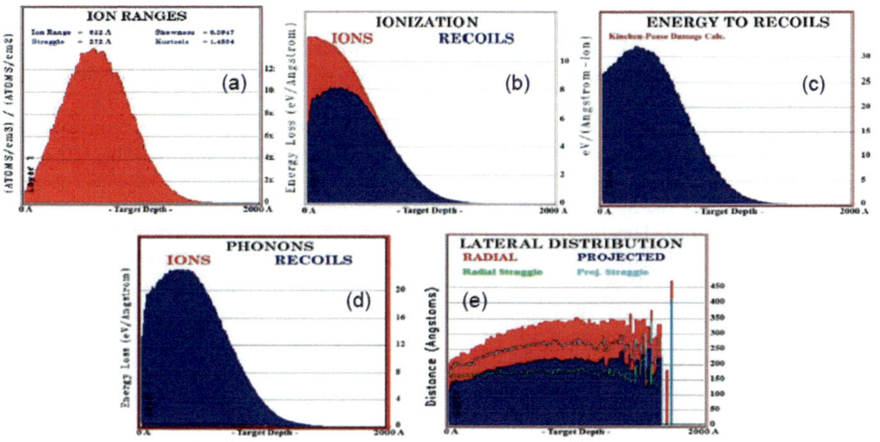

Fig. 24 **a – c** The distribution of ions in range, ion and recoil ionisation, energy to recoil, **d**, **e** energy loss associated with phonon recoils, and the lateral distribution curve for the ions of Na on Silicon–open access [89]

been presented, beneficial for the fabrication of MEMS-based cantilevers. The mass sensing was associated with mass-absorbing locations through resonant frequency shift.

Acknowledgements The author hereby shows sincere gratitude towards Prof N.R. Bandyopadhyay, Prof. S. M. Hossain (IIEST, Howrah), Dr S. K. Mishra (CSIR-NML, Jamshedpur), Prof B. N. Dev, Prof G. P., Das (IACS, Kolkata), Prof S. Bhowmik (Amrita University), Prof. S. K. Samdarshi (CUJ-Ranchi), Dean (School of Engineering & Technology, CUJ-Ranchi), and Hon'ble Vice-Chancellor (CUJ-Ranchi) for their support, encouragement, and guidance.

Funding The authors acknowledge the funding by DST-SERB core research grant project (File no CRG/2023/003421).

Data Availability Data shall be made available on request.

Declarations

Conflict of Interest The author(s) declare no conflict of interest.

Compliance with Ethical Standards The submitted work has not been published anywhere else, in any format or language, except for a few preprints and properly attributed open-access content.

References

1. Verma J, Khanna AS (2023) Physics Open 14:100133
2. Kuo RYL, Harrison C, Curran T-A, Jones B, Freethy A, Cussons D, Stewart M, Collins GS, Furniss D (2022) Radiol 304:50–62

3. AlGhaithi, A, Al Maskari S (2021) J Musculoskelet Surg Res 5(4)
4. Dobrzański LA, Staszuk M, Honysz R (2012) Arch Mater Sci Eng 58:152–157
5. Dobrzański LA, Staszuk M, Honysz R (2010) Archives of computational materials science and surface. Engi 2:141–148
6. Snyder J Using AI to detect–and predict–coating corrosion. https://www.rivieramm.com/news-content-hub/news-content-hub/using-ai-to-detect-ndash-and-predict-ndash-coating-corrosion-55480. Accessed 10 August 2024
7. Cao M, Zhao L J Vac Sci Technol A. https://doi.org/10.1116/6.0002360
8. Jain AK, Kumar M, Thakre GD (2021) Tribol-Mater Surf Interfaces 15:278−291
9. Cellini F, Gao Y, Riedo E Sci Rep. https://doi.org/10.1038/s41598-019-40636-0
10. Rahimian Koloor SS, Karimzadeh A, Tamin MN, Abd Shukor MH (2018) Metals (Basel) 8:421
11. Char M, Chakraborty AK, Bhattacharyya AS, Kar A Adv Eng Mater. https://doi.org/10.1002/adem.202100679
12. Bhattacharyya AS, Mishra SK (2011) J Micromech Microeng 21:015011
13. Bhattacharyya AS (2024) Mater Today Commun 38:107697
14. Bhattacharyya AS, Kumar S, Jana S, Kommu PY, Gaurav K, Prabha S, Kujur VS, Bharadwaj P Int J thin Film Sc Tech. https://doi.org/10.48550/ARXIV.1508.04920
15. Bhattacharyya AS, Prabha S, Vidya S, Kujur P J Nano Sci Eng & Trends 2:1–5
16. http://www.specs.de/cms/upload/PDFs/IQE11-35/sputter-info.pdf. Accessed 10 August 2024
17. Bhattacharyya AS (2015) Researchgate. https://doi.org/10.13140/RG.2.1.2735.5366
18. Praveen Kumar R, Bhattacharyya A (2024) Preprints. https://doi.org/10.20944/preprints202401.0848.v1
19. Oliver WC, Pharr GM (1992) J Mater Res 7:1564–1583
20. Bhattacharyya AS, Priyadarshi S, Sonu S, Shivam S, Anshu S. https://doi.org/10.48550/ARXIV.1704.04525
21. Fischer-Cripps AC (2011) In mechanical engineering series. Springer, New York, NY, pp 21–37
22. Soare S, Bull SJ, Oila A, O'Neill AG, Wright NG, Horsfall A, dos Santos JMM (2022) Int J Mat Res (Z. Metallkd.) 96:1262–1266
23. Bhattacharyya AS (2015) J Sci Ind Res 74(4):223−224
24. Kumar RP, Mitra S SciFed J Met Sci
25. Bhattacharyya AS (2024)
26. Dash R, Bhattacharyya K, Kumar RP, Bhattacharyya AS (2022) Eng Res Express 4:045012
27. Mishra SK, Bhattacharyya A (2013) Silicon-based nanomaterials. Springer, New York
28. Dash R, Bhattacharyya K, Bhattacharyya AS Nami Jishu Yu Jingmi Gongcheng (Nanotechnology Precis. Eng.). https://doi.org/10.1063/10.0022175
29. Reuther GM, Albrecht J, Dudek R, Rzepka S (2022) Microelectron Reliab 133:114530
30. Khodabakhshi F, Zareghomsheh M, Khatibi G (2020) Mater Sci Eng A Struct Mater 797:140203
31. Wang F, Zhou C, Zhang C, Dong C, Yang C, Jiang C, Jia C, Xue D (2014) Appl Phys Lett 105:062407
32. Gao X, Ma Z, Jiang W, Zhang P, Wang Y, Pan Y, Lu C (2016) J Power Sources 311:21–28
33. Soshnikov AI, Kravchuk KS, Maslenikov II, Ovchinnikov DV, Reshetov VN (2013) Instrum Exp Tech 56:233–239
34. Dash R, Bhattacharyya KAS Engineering failure analysis
35. Ly TH, Zhao J, Cichocka MO, Li L-J, Lee YH (2017) Nat Commun 8:14116
36. Xu A, Vodenitcharova T, Kabir K, Flores-Johnson EA, Hoffman M (2014) Mater Sci Eng A Struct Mater 599:125−133
37. Burke B (2011) Graduate theses and dissertations
38. Haddad M, Sepehrnoori K (2016) Rock Mech Rock Eng 49:4731–4748
39. Dash R, Bhattacharyya K, Bhattacharyya AS (2023) Fatigue Fract Eng Mater Struct 46:1641–1645
40. Dash R, Bhattacharyya AS (2023) Fatigue Fract Eng Mater Struct 46:2714–2719
41. Qian H, Chen M, Qi Z, Teng Q, Qi H, Zhang L, Shan X (2023) Crystals (Basel) 13:428
42. Bhattacharyya AS (2024) Fatigue Fract Eng Mater Struct 47:603–608

43. Bhattacharyya AS, Bhattacharyya K (2024) Results Surf Interfaces 14:100196
44. Bhattacharyya AS (2021) Surf Topogr 9:035052
45. Wang J-W, Yu K, Li M, Wu J, Wang J, Wan C-W, Xiao C-L, Xia B, Huang J (2022) Sci Rep 12:8717
46. Karimzadeh A, Koloor SSR, Ayatollahi MR, Bushroa AR, Yahya MY (2019) Sci Rep 9:15763
47. Okuma G, Maeda K, Yoshida S, Takeuchi A, Wakai F (2022) Sci Rep 12:6994
48. Holmberg K, Laukkanen A, Ronkainen H, Wallin K, Varjus S, Koskinen J (2006) Surf Coat Technol 200:3793–3809
49. Bhattacharyya AS, Bhattacharyya K Surf Rev Lett. https://doi.org/10.1142/s0218625x24501233
50. Bhattacharyya AS, Rajak AK, Bhattacharyya K (2024) Surf Eng 40:359–369
51. Dash R, Bhattacharyya AS (2022) Mater Today 66:3233–3237
52. Das S, Kumar D, Borah R, Dutta A, Guha S (2022) SILICON 14:9643–9657
53. Bull SJ (2005) J Phys D Appl Phys 38:R393–R413
54. Ananthakrishna G, Krishnamoorthy S (2018) Crystals (Basel) 8:200
55. Goel S, Cross G, Stukowski A, Gamsjäger E, Beake B, Agrawal A (2018) Comput Mater Sci 152:196–210
56. Dash R, Bhattacharyya K, Bhattacharyya AS (2023) Int J Refract Hard Met 116:106373
57. Octave Online. Cloud IDE compatible with MATLAB. https://octave-online.net/. Accessed 10 August 2024
58. Lee H, Coutu RA, Mall S, Leedy KD (2006) J Micromech Microeng 16:557–563
59. Bhattacharyya AS (2024) Metall Res Technol 121:311
60. Yadav S, Tripathy S, Sarkar D (2022) In advanced MEMS/NEMS fabrication and sensors. Springer International Publishing, Cham, pp 133–185
61. Pu D, Hu H (2022) In advanced MEMS/NEMS fabrication and sensors. Springer International Publishing, Cham, pp 1–18
62. Beccari A, Visani DA, Fedorov SA, Bereyhi MJ, Boureau V, Engelsen NJ, Kippenberg TJ (2022) Nat Phys 18:436–441
63. Stassi S, Cooperstein I, Tortello M, Pirri CF, Magdassi S, Ricciardi C. Nat Commun. https://doi.org/10.1038/s41467-021-26353-1
64. Hagelüken L, Sasikumar PVW, Lee H-Y, Di Stadio D, Chandorkar Y, Rottmar M, Maniura-Weber K, Blugan G, Brugger J (2022) J Eur Ceram Soc 42:1963–1970
65. Fan X, Smith AD, Forsberg F, Wagner S, Schröder S, Akbari SSA, Fischer AC, Villanueva LG, Östling M, Lemme MC, Niklaus F. Microsyst Nanoeng. https://doi.org/10.1038/s41378-019-0128-4
66. Sadhukhan D, Singh GP (2020) AIP conference proceedings. AIP Publishing
67. Shi S Nanotechnology and precision engineering (NPE)
68. Nili SH, Walia S, Balendhran D, Strukov M, Bhaskaran S (2014) 24:6741–6750
69. Kim N (2018) M/NEM devices and uncertainty quantification
70. Rajak AK, Dash R, Kumari A, Bhattacharyya AS (2024) Mater Today Commun 38:107891
71. Lian J, Wang L, Sun X, Yu Q, Ewing RC (2006) Nano Lett 6:1047–1052
72. Bhattacharyya AS, Kabiraj D, Yusuf SM, Dev BN (2014) Phys Procedia 54:87–89
73. Bhattacharyya AS (2016) arXiv [cond-mat.mtrl-sci]
74. Oh Y-J, Ross CA, Jung YS, Wang Y, Thompson CV (2009) Small 5:860–865
75. Manna PK, Yusuf SM, Shukla R, Tyagi AK (2011) Phys Rev B 83:1844121–1844125
76. Manna PK, Yusuf SM, Shukla R, Tyagi AK (2010) Appl Phys Lett 96:242508
77. Kumar P, Yadav BS, Singh K, Bhattacharyya AS. https://doi.org/10.48550/ARXIV.1705.00525
78. Lu QY, Chu PK (2012) Rev Sci Instrum 83:075116
79. Korol VM (1988) Phys Status Solidi Appl Res 110:9–34
80. Stepanov AL, Vorobev VV, Nuzhdin VI, Valeev VF, Osin YN (2017) J Appl Spectrosc 84:785–789
81. Baccar ZM, Jaffrezic-Renault N, Martelet C, Jaffrezic H, Marest G, Plantier A (1996) Sens Actuators B Chem 32:101–105
82. Fritzsche C, Goetzberger A, Axmann A, Rothemund W, Sixt G (1971) Radiat Eff 7:87–93

83. Nouira W, Haddad R, Barhoumi H, Maaref A, Bausells J, Bessueille F, Léonard D, Jaffrezic-Renault N, Errachid A (2009) Sens Lett 7:689–693
84. Kim K, Kang DC, Jeong Y, Kim J, Lee S, Kwak JY, Park J, Hwang GW, Lee K-S, Ju B-K, Park JK, Kim I (2021) J Alloys Compd 884:161086
85. Ziegler JF, Ziegler MD, Biersack JP (2010) Nucl Instrum Methods Phys Res B 268:1818–1823
86. Pacheco JL, Perry DL, Hughart DR, Marinella M, Bielejec E Appl Phys A Mater Sci Process. https://doi.org/10.1007/s00339-018-2041-3
87. Nord J, Nordlund K, Keinonen J Phys Rev B Condens Matter. https://doi.org/10.1103/physrevb.65.165329
88. Vujisic M, Stankovic K, Marjanovic N, Osmokrovic P (2010) IEEE Trans Nucl Sci 57:1798–1804
89. Bhattacharyya (2024) engrXiv. https://doi.org/10.31224/3664

Chapter 9
Tribological Properties of Coating Materials

Sumitra Dutta, Aishwarya Madhuri, and Bibhu Prasad Swain

Abstract Tribology is the study of lubrication, friction and wear, which are closely related. Interacting surfaces in relative motion are of interest to science and technology. The chapter provides a comprehensive overview of tribology's fundamental principles, exploring the interactions between surfaces in relative motion and the significance of friction, wear, and lubrication in these processes. It then delves into the different types of coating materials, such as hard coatings, soft coatings, and lubricious coatings, highlighting their unique properties and mechanisms of action. Alloying, composite structure and soft or hard film coating have all been created to reduce wear and friction. This is accomplished by enhancing surfaces and materials with properties that increase their resistance to wear and friction. Many innovative solid lubricants and contemporary lubrication ideas have been developed in recent years to obtain excellent lubricity and longer wear life in demanding tribological applications. Most conventional solid lubricants are made of ceramic, metal, or polymer-matrix composites, which have proven effective in various technical applications. Through multiplex surface treatments, recent advancements in thin-film deposition methods have created new generations of self-lubricating coatings with composite or multilayered structures. This chapter discusses the usual wear behaviours of representative ceramic, polymeric, metallic alloy and composite materials and their friction behaviours. Furthermore, wear prediction modelling is described. Coating materials are extensively used to reduce friction, wear, and corrosion, improving the durability and efficiency of moving parts in various applications, including automotive, aerospace, and industrial machinery.

Keywords Friction · Wear materials · Mechanical properties · Surface roughness · Hardness

---

S. Dutta · A. Madhuri · B. P. Swain (✉)
Department of Physics, National Institute of Technology Manipur, Langol, Imphal, Manipur, India
e-mail: bibhuprasad.swain@gmail.com

B. P. Swain (ed.), *Advances in Mechanical Coating*, Materials Horizons: From Nature to Nanomaterials, https://doi.org/10.1007/978-981-96-7484-8_9

1 Introduction

Tribology encompasses the investigation of wear, lubrication and friction on surfaces that interact and move relative to each other. An explanation of tribology was given in 1967. The word "tribology" comes from the Greek word "tribos," which means "detriments" or "gliding". Following the first phase of uncertainty, the term "tribology" became widely used. Because tribology is a relatively new concept, some people are even confused by what it means and whenever the term is used, amusing parallels with tribes frequently arise.

Because the term 'tribology' is relatively recent, some community members find its meaning unclear. As a result, humorous comparisons to tribes often come up when discussing tribology [1]. Friction is the resistance that separates two bodies in contact while they move. The Latin word "fricare", which means "to rub", gives us the phrase. The bodies of the substance might be liquid, gas, or solid. Alternatively, the friction could be brought on by internal energy indulgence processes in a single body.

On the other hand, wear is a phenomenon that affects every piece of equipment that comes into contact with objects. Wear can result from three different situations: erosion by corrosive elements (corrosive wear), surface contact with foreign substances (abrasive wear) and surface-to-surface contact (frictional wear). Abrasive wear can be clogged by creating a filtering instrument to eliminate the irritating particles. Additives that counteract the imprudent variety that impacts the surface can help reduce corrosive wear. To minimise wear and friction and initiate lubrication between two sliding solids, a gaseous, liquid or solid lubricant is accumulated at the sliding contact to eliminate the heat and debris the sliding track produces. Depending on the general geometry of the contacting bodies, the roughness and texture of the sliding surfaces, the contacting load, the pressure and temperature, the rolling and sliding speeds, the environmental conditions, the material composition, physical and chemical properties of the lubricant and properties of the near-surface deposit, lubrication processes can take on a wide variety of forms. Different types of lubrication processes can be used, depending on various factors such as the shape of the surfaces in contact, their roughness and texture, the amount of pressure and temperature involved, the speed of rolling and sliding, the environmental conditions, the characteristics of the lubricant, and the material properties at the surface [2]. Figure 1 shows the schematic presentation for a concise history of tribology.

The tribological properties of materials—those relating to friction, wear, and lubrication—are paramount in the design and performance of mechanical systems. In automotive, aerospace, manufacturing, and energy industries, components' efficiency, reliability, and longevity are heavily influenced by their tribological behavior. As machines and devices continue to evolve, operating under increasingly demanding conditions, the need for advanced materials that can withstand harsh environments and reduce energy losses due to friction becomes ever more critical. Coating materials have emerged as a vital solution to these challenges, providing an effective means of enhancing the tribological performance of surfaces. By applying thin films or layers

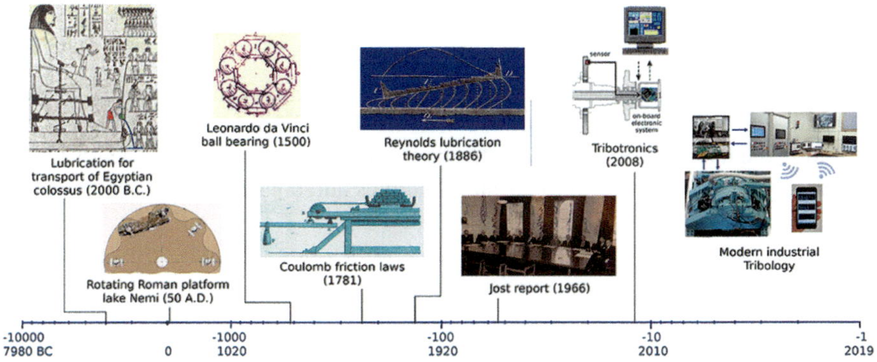

Fig. 1 Concise history of tribology

of specialised materials onto the surfaces of components, engineers can significantly reduce wear and friction, thus extending the life of the components and improving overall system efficiency. Coatings are essential in applications where direct material modifications are impractical or the base material lacks the necessary tribological properties. Figure 2 shows tribology as a multi-disciplinary subject.

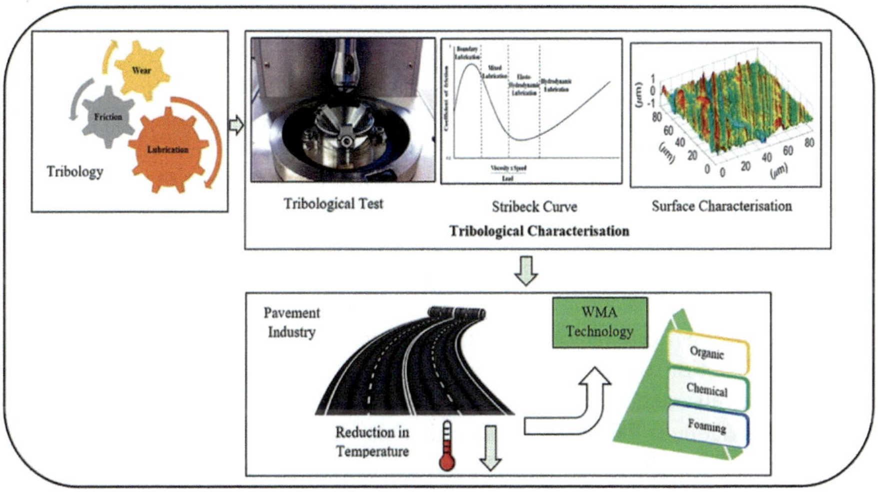

Fig. 2 Tribology as a multi-disciplinary subject

2 Basic Theory of Solid Friction

The theory of solid friction, often called dry friction, deals with the forces that resist the relative motion of solid surfaces in contact. The motion can occur through sliding or rolling; the terms used are "sliding friction" and "rolling friction" respectively [2]. Here is a concise overview of the basic principles and concepts.

2.1 *Static Friction*

Static friction is the frictional force resisting sliding motion between two surfaces. It must be overcome to start moving an object. The frictional force that opposes the relative motion of two surfaces already in motion is called kinetic (dynamic) friction. A combination of surface roughness, adhesive forces, deformation, plowing, mechanical interlocking and energy dissipation governs friction at the microscopic level. The real contact area between two bodies is usually smaller than visible because the surfaces are not smooth at the atomic level. This depends on various factors such as the nature of the surfaces, the materials, the surrounding environment, the conditions under which they operate and the vibrations and securing of the equipment being used [2].

2.2 *Laws of Friction*

Amontons's first law, also known as Amontons's friction law, states that the frictional force is directly proportional to the normal force between the surfaces [3].

$$F_{friction} = \mu F_{normal} \tag{1}$$

where $F_{friction}$ is the frictional force resisting the relative motion of the surfaces, μ is the coefficient of friction, which is a dimensionless constant that depends on the materials of the surfaces in contact and F_{normal} is the normal force or normal load, pressing the surfaces together. This law implies that the frictional force resisting motion increases proportionally as the normal force increases. Amontons's first law provides a fundamental basis for understanding and predicting frictional behaviour between solid surfaces in contact. Amontons's second law is also known as Amontons's law of temperature dependence of friction, which states that the frictional force is independent of the apparent contact area between the surfaces. In other words, the coefficient of friction (μ) remains constant regardless of the size or shape of the contacting surface. The second law implies that the macroscopic area of contact does not significantly influence the microscopic interactions between the surfaces, which determine the frictional resistance. Therefore, even if the apparent contact area changes due to surface roughness, texture, or geometry, the ratio of frictional force to normal force remains constant under the same conditions, such as temperature, humidity, and surface conditions. Guillaume Amontons established these laws in the seventeenth century and has since been confirmed and widely accepted in the field of tribology.

Coulomb's law of friction was formulated by Charles-Augustin de Coulomb in the eighteenth century and remains fundamental in understanding and predicting frictional behaviour in various mechanical systems and everyday scenarios. The kinetic frictional force is independent of the sliding velocity. This means that the kinetic friction force remains constant regardless of the speed of motion. Coulomb's law of friction states that the magnitude of the frictional force ($F_{friction}$) between two surfaces is directly proportional to the normal force (F_{normal}), pressing the surfaces together [4]. This law applies to static and kinetic friction, meaning it describes the frictional force when the surfaces are either at rest relative to each other (static friction) or sliding against each other (kinetic friction). The coefficient of friction (μ) is typically determined experimentally for specific pairs of materials under particular conditions, such as dry or lubricated surfaces.

2.3 Friction Models

Classical (Coulomb) friction model is the simplest model, which assumes that frictional forces are proportional to the normal force and independent of the contact area and velocity (for kinetic friction). It describes frictional force as proportional to the normal force, pressing the surfaces together with a coefficient of friction, determining the proportionality. It distinguishes between static and kinetic friction and assumes independence of frictional force from velocity [5]. Modern friction models consider factors like surface roughness, material properties, temperature and lubrication, providing a more detailed and accurate description of frictional behaviour. These models have evolved from simple empirical formulations to sophisticated simulations integrating advanced physics, material science and computational techniques.

2.4 Factors Affecting Friction

The strength of materials is influenced by three main factors in friction conditions [6]. These factors include: (a) Material properties dictate internal reasons, (b) the type of friction (slipping, rolling) and working conditions such as relative movement speed, load, application type and temperature and (c) the working environment and lubricants. The microstructure of the surfaces in contact affects the frictional force. Rougher surfaces typically exhibit higher friction [7]. Rough surfaces have more contact points, leading to higher frictional forces due to increased adhesion and mechanical interlocking. Different materials have different intrinsic frictional properties. The hardness, stiffness and ductility of materials influence how they deform and interact under contact, affecting frictional behaviour on rough surfaces. An increase in the normal force increases the frictional force proportionally. The force pressing two surfaces together (normal force) directly affects friction. Higher normal forces typically result in higher frictional forces. Lubricants, dust, or other contaminants can significantly alter frictional characteristics. Foreign materials on surfaces, such as dust, dirt, oils or moisture, can alter friction by filling in surface irregularities, changing surface properties or introducing additional lubrication or adhesion effects. Temperature changes can affect both the material properties and the frictional force. Friction can change with temperature due to alterations in material properties (e.g., thermal expansion), viscosity of lubricants if present and surface oxidation. Increasing temperature can sometimes reduce friction by softening materials or improving lubricant flow. Hard metals show less friction than softer metals because they have stronger interatomic connections [8].

2.5 Lubrication

Lubrication is the application of a substance (lubricant) to minimise friction and wear. The presence of lubricants between surfaces can drastically reduce friction by forming a protective film that separates and cushions the surfaces, minimising direct contact and frictional forces. Plastohydrodynamic lubrication (PHL) is often called thick-film lubrication [9]. Types of lubrication include (a) Hydrodynamic lubrication: An entire lubricant film separates the surfaces. A surface with tiny asperities will contact if two plates are placed together. If one of the plates were to slide over the other, friction would increase, the asperities would break, and the surfaces would wear. In hydrodynamic lubrication, a fluid film separates the surfaces, prevents wear and reduces friction. The pressure generated in a narrowing gap demonstrates the vital role of hydrodynamic lubrication in creating a wedging layer. (b) Boundary lubrication occurs when the thin lubricant film and asperities contact each other. In this regime, the surfaces are often in direct contact, and the lubricant molecules act primarily to reduce friction and wear at these contact points. Boundary lubrication is crucial when operating conditions or surface roughness prevent full fluid film lubrication, such as in some engine components, gears and sliding contacts. (c) Mixed lubrication: A combination of hydrodynamic and boundary lubrication. In fluid film contribution, a lubricant film is still present, but it is not thick enough to completely prevent asperity contact. Asperities of the surfaces come into contact intermittently, causing localised increases in pressure and temperature. Friction and wear in mixed lubrication regimes can be influenced by both fluid film viscosity and surface roughness, with friction often higher than fluid film lubrication but lower than boundary lubrication. Mixed lubrication is commonly found in applications such as heavily loaded machine components, automotive engines and certain types of bearings. Figure 3 shows the lubrication regimes [10]. Muratore et al. [11] examined

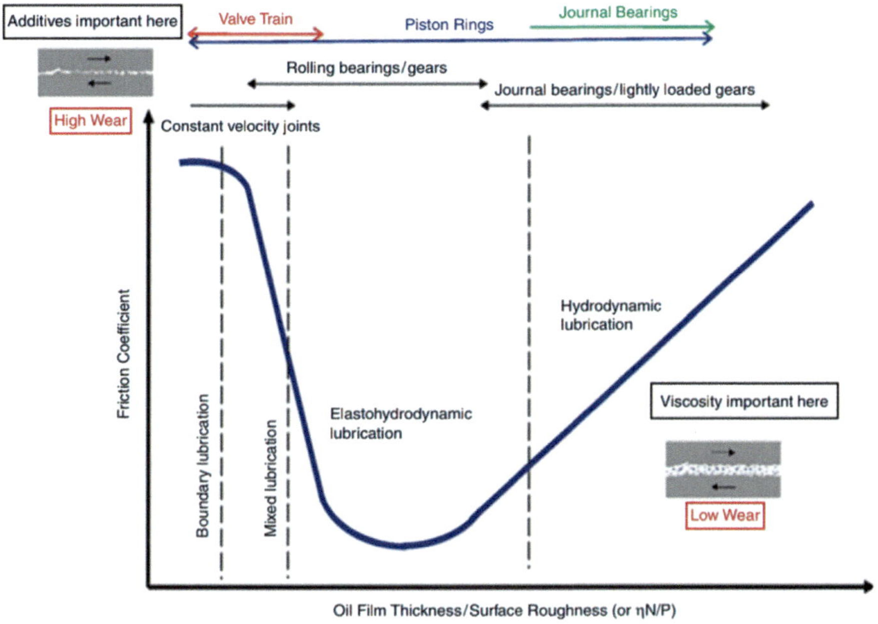

Fig. 3 Lubrication regimes [10]

different types of tribological coatings, including solid lubricants and adaptive coatings, designed to reduce friction and wear under extreme conditions, such as high temperatures, while maintaining their effectiveness over multiple thermal cycles. Solid lubricants like molybdenum disulfide (MoS_2) and graphite have limitations, particularly at high temperatures where oxidation can degrade their performance. Muratore introduced the concept of adaptive coatings, which can form lubricious phases in response to heat, maintaining low friction and wear characteristics even after undergoing significant thermal cycling. An example of an adaptive coating is a combination of transition metal dichalcogenides (like MoS_2) with hard phases such as titanium nitride (TiN) or diamond-like carbon (DLC). These coatings can adjust their structure or composition to retain lubricity across a wide temperature range.

2.6 Wear

Wear properties of coating materials refer to their ability to resist wear, which is the gradual removal or deformation of material at solid surfaces due to mechanical action. Over time, friction can cause wear, leading to material degradation. Wear mechanisms involve complex interactions of mechanical, chemical and environmental factors. If not managed, they can lead to surface roughening, dimensional changes, material loss, and eventual component failure. Strategies to mitigate wear include selecting appropriate materials based on wear resistance, applying surface treatments like coatings or platings, optimising lubrication and maintenance practices, improving design to reduce stress concentrations and controlling operating conditions.

2.6.1 Adhesive Wear

Adhesive wear occurs when two surfaces in contact slide against each other, transferring material from one surface to the other. This transfer happens due to the formation and subsequent breaking of micro-welds between the surfaces. It commonly occurs in metal-to-metal contact under insufficient lubrication or high load conditions. Adhesive wear typically results in smooth wear tracks, galling, and material transfer (smearing) between surfaces. Studies have shown that rough surfaces in contact during adhesive wear are more likely to produce wear particles than abrasive wear. The chances of wear particle production vary from one type of contact to another, with a lower likelihood for smoother surfaces. Adhesive wear is a common cause of breakdown in metal sliding contacts, highlighting the importance of preventing it for optimal machinery performance. Low-friction coatings such as diamond-like carbon (DLC) or solid lubricants like MoS_2 reduce adhesive wear. Sui et al. [12] synthesised CrN/DLC/Cr-DLC multilayer coatings by PECVD and effectively enhanced their adhesive wear resistance. The low friction provided by the DLC layers reduced the likelihood of material transfer between contacting surfaces.

2.6.2 Abrasive Wear

Abrasive wear occurs when hard particles or rough asperities on one surface slide against a softer coating, leading to material removal through micro-cutting or plowing. The wear tracks typically have grooves or scratches corresponding to the abrasive particles. This type of wear is common in dust, sand, or other hard particulate matter environments. Strategies to mitigate abrasive wear include using harder materials, surface treatments and lubrication. Hard coatings like titanium nitride (TiN), chromium nitride (CrN), or alumina (Al_2O_3) are commonly used to resist abrasive wear. Particles or grit can remove material through micro-cutting, micro-fracture, pulling out individual grains or accelerated fatigue from frequent deformations. Typically, the material being abraded should be less than 80% of the particle's hardness for fast abrasion. Even if a material's yield stress is higher than the abrasive particle's, some wear and damage can still occur. Materials with softer phases or insufficient strain hardening may experience wear until their hardness is 1.2–1.4 times that of the abrasive. The abrasion caused by coal is not usually due to carbonaceous minerals like vitrinite, but more by pollutant minerals such as pyrite and hematite [13]. On the other hand, minerals that are too soft to abrade, such as calcite, can still cause wear on a material, albeit through different mechanisms like thermal fatigue. The wear resistance of various metals and alloys after heat treatment is compared, as shown in Fig. 4.

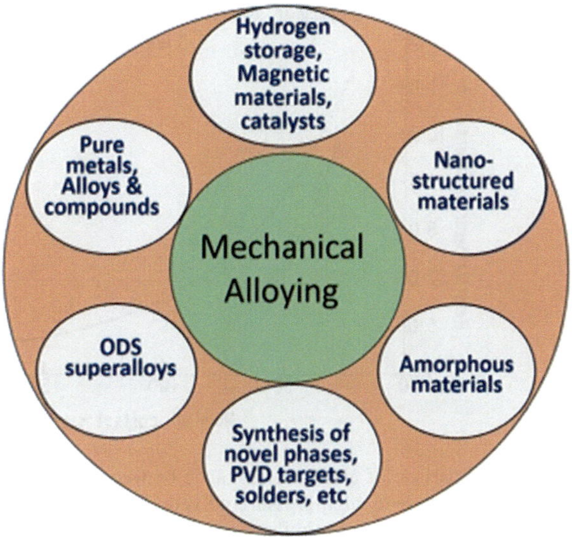

Fig. 4 Comparison of how different pure metals and alloys handle wear and hardness

Abrasive wear typically involves micro-cutting with chip formation in most cases. Polymers have remarkably high resistance to abrasive wear compared to metals of similar hardness. Figure 5 displays the relative resistance to abrasive wear of plastics and soft metals [14]. This is due to their high resistance to abrasion by dull grits and their inability to break grits and create new sharp edges. Field tests on agricultural ploughs with alumina surfaces showed significantly less wear than traditional steel ploughs [15]. Combining alumina with zirconia can increase toughness and improve wear resistance through transformation toughening, where the zirconia particles undergo a phase transformation under stress, helping to absorb energy and inhibit crack propagation [16]. Higher zirconia content generally improves toughness but may also decrease hardness, which can influence the wear rate depending on the specific application. Ceramic matrix composites with metal fibres to enhance toughness have been found to have better abrasive wear resistance than pure ceramics [17]. Ultra-hard ceramic composite coatings, like alumina (Al_2O_3), obtained through microarc oxidation (MAO), demonstrated excellent wear resistance, particularly in high-stress environments where conventional coatings might fail [18]. MAO involves applying a high voltage to the metal substrate in an electrolyte solution, forming plasma discharges on the surface. These discharges facilitate the growth of a ceramic oxide layer that is both hard and wear-resistant. The high hardness of the oxide layers allowed the coatings to resist abrasive and adhesive wear effectively. Laser-clad tool steel coatings showed significant improvement in wear resistance compared to untreated or conventionally coated materials studied by Wang et al. [19]. The high hardness of the clad layer, combined with the presence of hard carbides, makes the coatings particularly effective against abrasive wear.

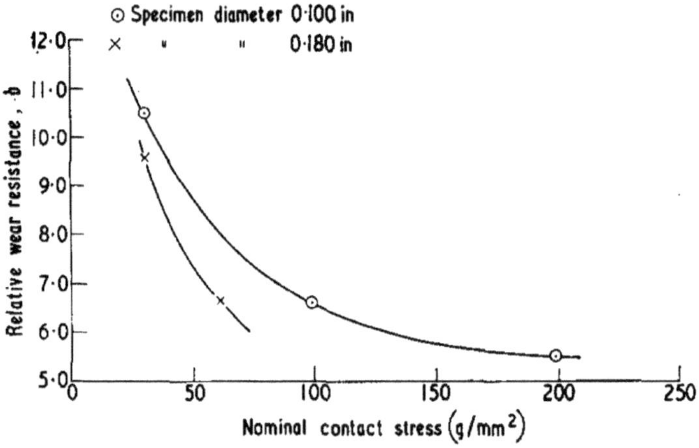

Fig. 5 Comparison of abrasion resistance between (I) plastics such as L54, L68, nylon 6, low-pressure polyethylene, high-pressure polyethylene, polyfluoroethylene and (II) metals like silver, zinc, cadmium, and lead of similar hardness [15]

2.6.3 Erosive Wear

Erosive wear occurs when a stream of particles or fluid droplets impacts the coating surface at high velocities, leading to material removal. Erosion often leads to pitting, cratering, or rounded edges on the surface. The severity of erosion depends on the velocity, angle of impact, and hardness of the impacting particles. Erosion-resistant coatings like tungsten carbide (WC), cermets, or certain ceramic coatings are used in high-speed environments, such as in turbines or fluid-handling equipment. The review by Krella et al. [20] highlights that PVD coatings, such as Titanium Nitride (TiN), Chromium Nitride (CrN), and Diamond-Like Carbon (DLC), exhibit high hardness, which is a key factor in their ability to resist wear. The balance between hardness and toughness is crucial, as overly hard coatings may become brittle and prone to cracking under high stress. PVD coatings generally perform well under erosive conditions due to their high hardness and ability to resist particle impact. However, the performance can vary depending on the impact angle, particle size, and coating material. For example, TiN and CrN coatings are known for their excellent resistance to erosive wear. Singh et al. examined the behavior of atmospheric plasma-sprayed (APS), both conventional and nanostructured alumina coatings, under erosive conditions, where high-velocity particles impact the coating surface. Nanostructured alumina coatings demonstrate superior erosion resistance compared to conventional alumina due to their ability to dissipate the energy of impacting particles more effectively and their higher toughness, which allows them to better withstand the repeated impacts of abrasive particles [21].

2.6.4 Corrosive Wear

Corrosive wear is also known as tribocorrosion, which occurs due to the combined action of mechanical wear and chemical or electrochemical reactions with the environment. This leads to accelerated material degradation. Corrosive wear can result in pitting, rough surfaces, and the formation of oxides or other reaction products that can further promote wear. Coatings like chromium nitride (CrN) or titanium nitride (TiN) resist wear and corrosion. Additionally, corrosion-resistant alloys or polymer coatings can be used in corrosive environments. It typically consists of the presence of corrosive agents such as acids, alkalis, salts or other chemicals, which attack the surface of a material, while mechanical forces such as sliding or rubbing further exacerbate the degradation. In erosive corrosion, abrasive particles in a corrosive environment mechanically abrade the surface. Cavitation corrosion occurs in liquids under conditions, causing the rapid formation and collapse of vapour bubbles near a surface. Fretting corrosion results from repeated small-scale relative motion between surfaces in contact under pressure. Material selection, protective coatings, and environmental controls are crucial strategies to mitigate corrosive wear. These help reduce the chemical attack and mechanical abrasion contributing to surface degradation. This kind of wear is known as 'corrosive wear' with atmospheric oxygen acting as the corroding agent, leading to possible 'oxidative wear' as a subsequent result.

If a metal is corroded to form a film on its surface while also experiencing sliding contact, one of these four outcomes may occur: (a) a lubricating solid film that prevents both corrosion and wear might be developed; (b) a weak film with a short lifespan under the sliding contact may be produced. The friction coefficient could be low or high in this case; (c) the protective surface films may be worn away and a galvanic coupling between the remaining films and the base material may lead to rapid corrosion of the worn region on the surface; (d) the corrosion and wear processes may progress independently to result in material failure, which is essentially the combination of these two processes. The oxides of iron are very fragile at high temperatures. It is doubtful that wear and corrosion can occur independently in the same system, as sliding contact's heat and mechanical friction will inevitably accelerate corrosion. Figure 6 illustrates the process of corrosive wear through the repeated removal of protective films. The study by Ji et al. emphasises that multilayer Fe-based amorphous coatings offer dual protection against both corrosion and wear. The amorphous structure is particularly effective in resisting corrosive attacks, as it lacks the grain boundaries and dislocations that are typical sites for corrosion initiation in crystalline materials. The primary wear mechanisms in these coatings are abrasive wear and oxidative wear. The amorphous structure, combined with the high hardness of the Fe-based material, helps to minimise material loss due to these mechanisms [22]. Adding corrosion inhibitors to lubricating oil or process liquid can effectively reduce corrosive wear. In situations where lubrication is not a concern, such as when the contact is immersed in the process liquid, adding corrosion inhibitors to the wetting liquid can significantly reduce wear. Lee et al. used a nickel matrix reinforced with aluminium oxide particulates to create a composite coating on Ti-6Al-4 V alloy to enhance their surface properties [23]. The Ni/Al_2O_3 coatings provide

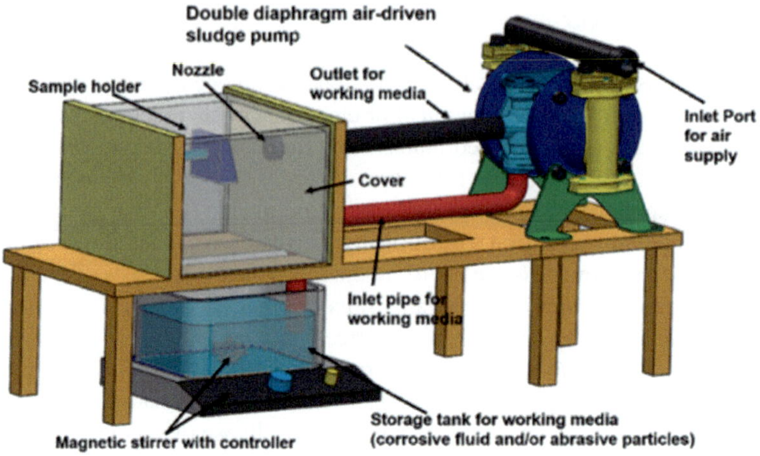

Fig. 6 A theory explaining how corrosive wear occurs due to the repeated removal of protective films [1]

dual protection against both wear and corrosion. The nickel matrix offers corrosion resistance, while the Al_2O_3 particulates enhance wear resistance. The study finds that coatings reinforced with nano-sized Al_2O_3 particles perform better in corrosive wear tests than those with micro-sized particles.

2.6.5 Fatigue Wear

Fatigue wear results from cyclic loading and unloading, which causes the coating material to crack, spall, or delaminate over time due to the accumulation of damage. The formation of cracks, material spalling, and the eventual detachment of the coating from the substrate often characterises this wear type. These cracks initiate and propagate gradually, eventually leading to material failure. It usually manifests as progressive cracking, pitting, spalling or delamination on the surface of the material. These signs indicate localised damage due to the cyclic nature of the applied loads. Factors influencing fatigue wear include the magnitude and frequency of the cyclic stresses, the material's fatigue strength and endurance limit, surface roughness and environmental conditions. Fatigue wear is critical in components subjected to repetitive stresses, such as aircraft components, automotive parts, bridges and machine elements like gears and bearings. To resist fatigue wear, coatings need to have good toughness and adhesion. Multilayered or composite coatings, which distribute the stress more evenly, are often used to combat fatigue wear. Yonekura et al. investigated how the application of arc ion plated Cr/CrN multilayer coating influences the mechanical performance of the Ti–6Al–4 V alloy, particularly in terms of fatigue resistance and wear behavior [24]. The Cr/CrN multilayer coating has a positive impact on the fatigue life of the Ti–6Al–4 V alloy. The coating reduces the initiation and propagation of fatigue cracks by providing a protective barrier that absorbs and redistributes stress more effectively than the uncoated alloy.

3 Microscopic Roughness

3.1 Asperities

Even the smoothest surfaces have microscopic irregularities, known as asperities. These asperities make contact and interlock with each other, leading to resistance against motion. Asperities are tiny protuberances or peaks that can exist on surfaces at a microscopic level. The surface roughness of coatings is directly related to the height and distribution of asperities and significantly influences wear behaviour. Smoother surfaces with lower asperities tend to have lower friction and wear rates because there is less stress concentration and fewer points of high contact pressure. During sliding contact, asperities can deform elastically or plastically. Elastic deformation may lead to temporary wear, while plastic deformation results in permanent wear and material loss. The degree of deformation depends on the hardness and toughness of the coating material. Xie et al. investigated the wear of plasma-sprayed alumina (Al_2O_3) coatings when in sliding contact with harder asperities, such as those found on metal or ceramic counterparts [25]. The harder asperities on the opposing surface can plow through the alumina coating, removing material and creating wear tracks. The presence of sharp, hard particles or asperities exacerbates this abrasive wear.

3.2 Adhesion

The thin oxide film, typically only a few nanometres in thickness, prevents metals from making proper contact and reduces severe wear unless intentionally removed. The effectiveness of hard coatings in protecting substrates from wear, corrosion, and other forms of degradation largely depends on their adhesion to the substrate. Strong adhesion ensures that the coating remains intact under mechanical stress, thermal cycling, and other challenging conditions. The study by Drobný et al. focuses on the methods used to assess the adhesion strength of hard coatings, mainly through indentation testing and acoustic emission monitoring [26]. By correlating the acoustic emission signals with the load applied during indentation, the study assesses the onset of coating failure. Sudden spikes in acoustic emission signals typically indicate crack initiation or delamination. Figure 7 shows the erosion wear mechanism and performance of Cr/CrSiN multi-layer coatings fabricated on Ti alloy (TC4) by ion-plating technology [27].

Markov et al. highlight the direct relationship between adhesion and wear, emphasising that high adhesive forces can lead to increased material transfer between

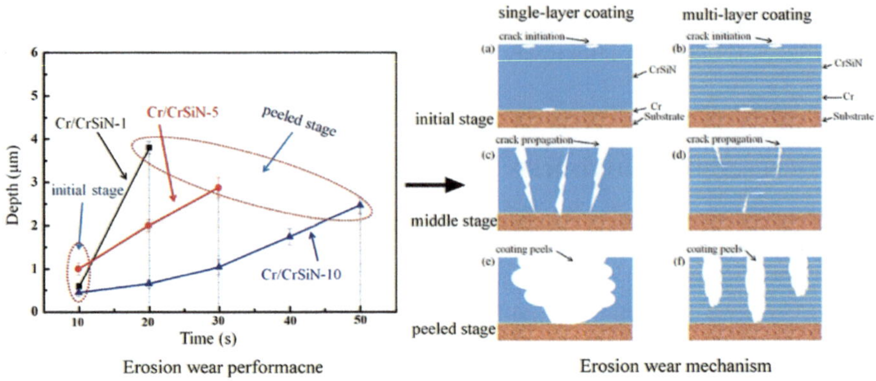

Fig. 7 Erosion wear mechanism and performance of Cr/CrSiN multi-layer coatings [27]

surfaces, resulting in adhesive wear [28]. This study explores the role of adhesion in frictional contact, where the force required to initiate and maintain sliding between surfaces is influenced by the adhesive forces at the interface. Atoms and molecules at the contact points experience intermolecular forces such as Van der Waals forces, metallic bonds or chemical bonds. These attractions create an adhesive force that contributes to friction. Van der Waals forces play a significant role in adhesion, especially in cases where polar interactions like hydrogen bonding are minimal. Hydrogen bonds are stronger than van der Waals forces and can contribute significantly to adhesion, especially in materials like water, specific polymers and biological molecules. Electrostatic interactions involve attractions between positively and negatively charged ions or molecules. Electrostatic forces can contribute to adhesion when surfaces carry opposite charges or when ions in a solution interact with surface charges. Covalent solid or ionic bonds can form between atoms or molecules at the interface of materials. Chemical bonding leads to extreme adhesion but typically requires specific conditions, such as chemical reactions or surface treatments, to achieve. Capillary forces arise due to the surface tension of liquids like water and can cause adhesion between a liquid and a solid surface. Schall et al. highlight how new bonds are formed during sliding and how these bonds contribute to the adhesive forces between the Si surfaces [29]. The research also addresses the reversibility of these bonds, noting that sliding can either enhance or weaken the adhesion depending on the sliding history.

3.3 Surface Energy

The energy required to create new surfaces during sliding or to break existing bonds adds to the frictional force. Rough surfaces have more contact points or asperities that increase the effective contact area between materials. This increased contact area can enhance adhesion and affect surface energy interactions. Surface roughness can alter the wetting behaviour of a material. For example, hydrophobic materials may become more hydrophilic with increased roughness, leading to changes in surface energy and wetting characteristics. Rough surfaces may have different chemical compositions or concentrations of functional groups at peaks and valleys. These variations can influence surface energy by affecting interactions with other materials. Microscopic features can influence surface tension and surface free energy at specific points on the surface of the material. This can impact how materials interact with liquids or other surfaces. The roughness-induced changes in surface energy can affect adhesive strength. Higher roughness can promote mechanical interlocking or increase the effective contact area, enhancing adhesion. Packham et al. provide a comprehensive exploration of how surface characteristics, such as energy and topography, influence adhesive interactions [30]. Surface energy plays a crucial role in adhesion, as higher surface energy typically leads to better wetting of adhesives or coatings, enhancing adhesive strength.

3.4 Elastic Deformation

When asperities come into contact, they deform elastically. This deformation stores energy and part of this energy is dissipated as heat during sliding, contributing to friction. When a material undergoes elastic deformation, especially under load, its surface can exhibit micro-scale changes. These may include surface undulations, ridges or localised deformations, depending on the nature of the material and the applied stress. Microscopic roughness can also result from non-uniform strain distribution during elastic deformation. Areas with higher stress concentrations or material inhomogeneities may exhibit more pronounced surface roughness. Even though elastic deformation typically does not lead to permanent changes in shape, it can cause temporary alterations in surface texture due to localised stretching or compression. This can affect surface finish and, consequently, surface interactions such as friction and wear. Different materials respond to elastic deformation in distinct ways, influencing the nature and extent of microscopic roughness. For instance, metals and polymers may exhibit different patterns of roughness under similar loading conditions. Tribology influences the contact mechanisms between surfaces, affecting friction and wear characteristics. Zhang et al. investigate the impact of squamous (scale-like) textured silicon carbide (SiC) surfaces on the tribological behaviour of graphite asperities during dry-sliding conditions [31]. The friction coefficient is reduced due to the presence of the textured pattern, which likely promotes the formation of a stable transfer film of graphite on the SiC surface. The wear rate of graphite is also decreased, attributed to the reduced contact area and the ability of the texture to trap wear debris, preventing it from further abrading the graphite surface.

3.5 Plastic Deformation

If the contact pressure is high enough, asperities may deform plastically, permanently changing shape and increasing resistance to motion. Plastic deformation can cause visible changes in the surface topography of a material. This includes the formation of slip bands, dislocation pile-ups, surface cracks and other irregularities that are typically smaller in scale than those observed with macroscopic deformation. In crystalline materials, plastic deformation can alter the grain structure, leading to changes in surface roughness due to grain boundary interactions and dislocation movements. The surface texture of a material can change significantly after plastic deformation, depending on factors such as strain rate, temperature and material composition. This altered texture can affect surface properties like friction, wear resistance and adhesion. The extent of plastic deformation and resulting roughness can vary based on the mechanical properties of the material, including its hardness, ductility and strain-hardening behaviour. Different materials exhibit distinct behaviours during plastic deformation, influencing the nature and magnitude of surface roughness. For instance, ductile materials like metals tend to deform plastically through mechanisms such as dislocation motion, while brittle materials may undergo more catastrophic failure. Tribology impacts the friction, wear and lubrication behaviour of surfaces in contact. Chen et al. [32] employ elastic–plastic contact models to analyse the behaviour of coated rough surfaces. The research shows that the elastic–plastic behaviour of rough surfaces with hard coatings is significantly influenced by the coating's thickness relative to the roughness of the substrate. Thicker coatings generally provide better protection against plastic deformation, as they can absorb more stress before yielding. However, if the coating is too thin, it may not fully protect the substrate from plastic deformation, leading to premature wear and failure.

3.6 Plowing

When a harder surface slides over a softer surface, the harder asperities can dig into or plow through the softer material. This plowing action creates resistance and contributes to the frictional force. Plowing hardness is typically defined as the resistance of a material to plastic deformation or wear caused by a harder indenter or tool sliding across its surface. Unlike traditional hardness tests such as Vickers or Rockwell hardness, which measure resistance to penetration, plowing hardness is assessed by observing how the material behaves under sliding contact. It considers factors like the formation of wear tracks, material displacement and the degree of damage caused by the sliding contact. Plowing hardness is important in understanding wear mechanisms, surface integrity and the performance of materials in applications where sliding or abrasion is a significant factor. It helps engineers and scientists evaluate the durability and suitability of materials for specific uses, such as in machinery components, cutting tools or automotive parts. Several factors influence plowing hardness, including the hardness of the indenter or tool, the surface roughness of both interacting materials, the applied load, sliding speed and environmental conditions like temperature and lubrication. The study by Barletta et al. finds that the mechanical properties of the substrate significantly influence the plowing and scratch behaviour of the aqueous two-component polyurethane (2 K-PUR) coatings [33]. The coating exhibits higher scratch hardness and less plowing on glass substrates, as the hard substrate supports the coating and resists deformation. In contrast, on polycarbonate, a softer and more ductile material, the 2 K-PUR coating shows more pronounced plowing and lower scratch hardness. The softer substrate allows more significant deformation under the same loading conditions, leading to deeper grooves and more extensive material displacement.

3.7 Mechanical Interlocking

The asperities of one surface can become mechanically interlocked with the asperities of the opposing surface. Overcoming this interlocking requires additional force, contributing to friction. Mechanical interlock refers to a connection or joining mechanism where two or more components fit together physically to create a secure bond or connection. In mechanical contexts, interlocking mechanisms often enhance the stability, strength and reliability of assemblies. Such interlocking mechanisms are utilised in manufacturing, construction, aerospace, automotive and consumer products. In engineering and manufacturing, understanding and optimising the interlocking of mechanical interlocks is crucial for ensuring the integrity and performance of assembled structures or devices. It involves considerations of both the design of the interlocking features and the operational conditions under which they will function. Zhang et al. focused on creating a novel Ag and carbon nanotube (CNT) composite film using a two-component spin-coating method [34]. The research emphasises enhancing adhesion through mechanical interlocking and chemical grafting, leveraging a silver heterogeneous nucleation mechanism. As Ag nucleates and grows on the CNT surfaces, it creates an interwoven structure that enhances the mechanical stability and adhesion of the composite film to the substrate. Adhesion tests demonstrate that the combined mechanical interlocking and chemical grafting techniques substantially improve the bonding strength between the composite film and the substrate. The films exhibit excellent mechanical flexibility and resistance to peeling or delamination under various stress conditions. The friction surfacing process is an advanced surface modification technique used to deposit a metallic layer onto a substrate to enhance surface properties such as wear resistance, corrosion resistance, and hardness. Stegmüller et al. focus on the influence of inductive heating during the friction surfacing of stainless steel onto an aluminium substrate, aiming to improve the quality of the deposited layer and the bonding strength between the two materials [35]. The introduction of inductive heating results in a more uniform heat distribution across the interface between the stainless steel and aluminium, reducing the risk of thermal stresses that could lead to defects such as cracks or poor bonding. The additional heat from induction enhances the diffusion of materials at the interface, leading to stronger metallurgical bonds. Microstructural analysis shows a more homogeneous and defect-free interface compared to friction surfacing without inductive heating.

3.8 Vibrations and Acoustic Emission

The interaction of asperities can also produce microscopic vibrations and acoustic emissions, which are other forms of energy dissipation. Mechanical vibrations are oscillatory motions of mechanical components or structures, often caused by forces acting within the system. Vibrations can occur due to rotating machinery, reciprocating components or external disturbances. Acoustic emission (AE) refers to the release of energy in the form of stress waves or sound waves from a material when it undergoes deformation or damage. For example, vibrations can induce stress and strain in materials, leading to acoustic emissions as cracks or defects propagate; vibrations can generate acoustic emissions due to the frictional contact or impacts within the system and monitoring acoustic emissions can help detect and analyse vibration-induced damage or wear in mechanical components. AE is a non-destructive testing technique that captures transient elastic waves generated by the release of energy from within a material, such as during crack initiation and propagation. Aggelis et al. studied the impact of different chemical coatings on steel fibers and how these coatings influence the AE behavior of steel fiber reinforced concrete (SFRC) during mechanical loading [36]. The AE data indicate distinct differences in the damage mechanisms of SFRC with different fiber coatings. Fibers with strong bonding coatings tend to produce higher amplitude AE signals during crack initiation and fiber pull-out, reflecting more energy-intensive failure processes. Conversely, coatings that reduce bond strength result in more frequent but lower amplitude AE signals, indicative of gradual and distributed damage rather than sudden catastrophic failure. Coatings that enhance the fiber-matrix bond generally lead to higher ultimate strength and toughness, as the fibers contribute more effectively to crack bridging and energy absorption. However, coatings that promote more accessible fiber pull-out result in improved post-cracking behavior, with the concrete maintaining its integrity even after initial failure.

3.9 Surface Films

Oxide layers, contamination, and lubrication can significantly alter the frictional properties by modifying the interactions at the contact points. Effective mechanical interlock often begins with proper surface preparation. This can include techniques such as roughening the substrate surface, creating micro- or nanoscale features or applying a primer layer that enhances adhesion by physically interlocking with the subsequent film. Surface roughness plays a significant role in mechanical interlocks. Irregularities on the substrate surface create sites where the film can physically anchor itself, increasing the contact area and improving adhesion strength. Methods such as spray coating, dip coating or spin coating can influence the mechanical interlock of surface films. The choice of application technique can impact the uniformity of film coverage and the extent of interlocking with the substrate. The study by Oblak et al. finds that the nanoscale topography of tribochemical films significantly influences the friction behaviour of DLC coatings [37]. Films with smoother surfaces tend to reduce friction, as they minimise asperity contact and allow for more uniform sliding. In contrast, rougher films with larger asperities can increase friction due to the increased mechanical interlocking between the contacting surfaces. DLC coatings with tribochemical films that exhibit both smooth topography and high hardness generally show lower friction coefficients.

4 Friction

(a) **Fluid Friction**: Also known as viscous friction, fluid friction arises when an object moves through a fluid (liquid or gas). It is caused by the viscosity of the fluid, which is the internal frictional resistance to flow. Fluid friction increases with the speed of the object moving through the fluid and is influenced by the viscosity and density of the fluid.

(b) **Rolling Friction**: Rolling friction occurs when an object rolls over a surface. It is generally lower than sliding friction because the contact points between the object and the surface are in motion and experience less resistance compared to sliding. Rolling friction is significant in applications involving wheels, bearings and rolling objects. Rolling friction, also known as the resistance to rolling, can have very low values as low as 0.001 for hard materials. Lubricants are not very effective in reducing rolling friction, which is influenced by three main mechanisms: (i) Micro-slip and friction at the interface, (ii) Material inelastic properties, (iii) Surface roughness. Most of the energy transferred in this state comes from the interaction between surface bumps, leading to an increase in rolling friction as the rolling speed increases. Through frequent traverses, the rolling resistance can decrease as the surface roughness is gradually smoothed out by the repeated plastic deformation of protruding points.

(c) **Sliding Friction**: Also known as kinetic friction, sliding friction occurs when two surfaces are sliding past each other. It is generally weaker than static friction, meaning less force is needed to keep an object moving than to start it moving. Sliding friction depends on factors like the roughness of the surfaces and the presence of lubricants.

(d) **Static Friction**: This type of friction occurs between two surfaces that are not moving relative to each other. It prevents an object from starting to move when a force is applied to it. Static friction increases with the applied force until it reaches its maximum limit, known as limiting friction.

Chvedov et al. investigated the impact of polymer coatings on the frictional behavior of rolled surfaces [38]. The study considered various types of polymer films, including PTFE, polyurethane, and epoxy resins. PTFE, known for its low friction coefficient, shows the most significant reduction in friction, whereas polyurethane offers a balance between friction reduction and wear resistance. It was observed that higher loads generally increase friction, but the polymer coatings help mitigate this effect. At higher sliding speeds, some polymers demonstrate enhanced lubrication properties, further reducing friction. Wear tests revealed that polymer coatings significantly extend the life of rolled surfaces, reducing direct metal-to-metal contact. Aihua et al. compared the tribological behavior of four different PVD nitride coatings: TiN, TiAlN, AlTiN, and CrAlN [39]. AlTiN demonstrates the highest wear resistance among the coatings studied, making it ideal for applications involving extreme temperatures and aggressive wear conditions. Its superior performance is due to the high aluminium content, which forms a protective and wear-resistant alumina layer. TiN coatings provide good wear resistance but are less effective in high-temperature or oxidative environments. AlTiN and CrAlN stand out for their superior performance in extreme conditions compared to others.

5 Material Properties Affecting Tribology

5.1 *Hardness*

Harder materials generally exhibit better wear resistance. Techniques like Vickers and Rockwell hardness tests are used to assess material hardness. Materials with higher hardness typically exhibit better resistance to wear. This is because harder materials are less prone to deformation and abrasion when in contact with another surface. As a result, they tend to maintain their shape and surface integrity over time, reducing the rate of wear. The hardness of materials can affect the friction coefficient between sliding surfaces. Madhuri et al. examine how varying the tungsten (W) concentration in $Al_{1-x}W_xN$ alloys affects their mechanical properties. The incorporation of Tungsten into the AlN matrix is shown to enhance the hardness and toughness of the alloys, making them promising candidates for hard coating applications [40]. Figure 8 shows the plot between hardness and W at.% in the $Al_{1-x}W_xN$ films. The nano-mechanical tests conducted by Das et al. showed that the hardness of TiCN coatings increases with the N_2 gas flow rate [41]. Higher nitrogen content contributes to a harder coating, which improves its wear resistance and extends the lifespan of coated components. Ghadai et al. [42] observed that the hardness of Si-doped DLC films is generally affected by the silicon content. Moderate levels of silicon doping improve hardness by enhancing the internal structure of the film and reducing the tendency for graphitisation. However, very high silicon concentrations may lead to a reduction in hardness due to increased sp^2 bonding and the formation of softer SiC phases. Si doping improves wear resistance by strengthening the DLC films' structure and reducing friction. TiAlN films deposited under optimal conditions by CVD exhibit high hardness, as shown by Das et al. [43]. The study shows that increasing the aluminum content and optimising deposition parameters enhances the hardness of the films. Hardness is attributed to the strong covalent bonding within the TiAlN structure and the formation of a solid solution. The wear resistance of TiAlN films improves with increased hardness and better crystallinity.

In some cases, harder materials may experience higher friction coefficients due to increased adhesion forces between surfaces. However, in properly lubricated systems, harder materials can reduce frictional losses because they produce smoother and more stable surfaces for sliding. Hardness influences the susceptibility of materials to surface damage such as scratches, grooves and indentations. Softer materials are more likely to suffer from such damage when subjected to abrasive wear or contact stresses. Harder materials are more resistant to such damage and can maintain smoother surface finishes, which in turn can reduce friction and wear. In tribological applications, where two surfaces come into contact (e.g., bearings, seals, gears), the hardness of the materials must be compatible to avoid excessive wear due to mismatches in hardness. Hardness affects the effectiveness of lubrication in tribological systems. Harder materials may require different types or amounts of lubrication compared to softer materials.

The choice of lubricant and its ability to adhere to or interact with surfaces can be influenced by the hardness of the materials involved. Hardness can impact how materials respond to changes in temperature during operation. Yamamoto et al. investigated the relationship between the hardness of hydrogenated DLC coatings and their friction and wear behavior under water lubrication [44]. The study involves varying the hardness of a-C coatings through different deposition parameters and evaluating their performance in tribological tests conducted in water-lubricated environments. Harder coatings tend to form a more effective tribolayer in the presence of water,

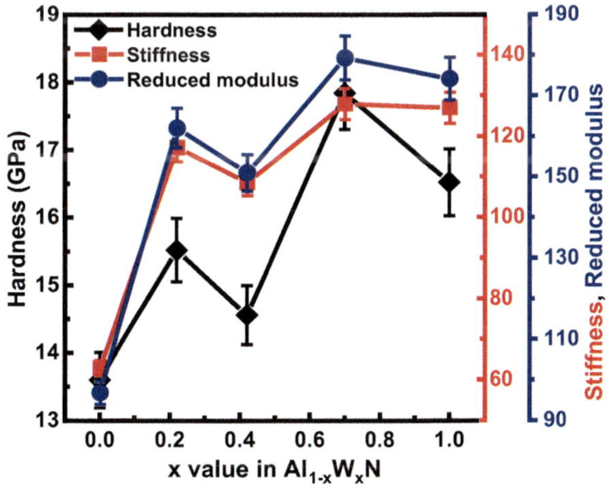

Fig. 8 Hardness, stiffness and reduced elastic modulus with the variation of W at % in the $Al_{1-x}W_xN$ films, including error bars [40]

which helps to maintain low friction and high wear resistance. Additionally, the hydrogen content in the coatings plays a role in preventing corrosion and maintaining coating integrity.

5.2 *Surface Roughness*

Surface topography influences friction and wear. Surface roughness directly affects frictional forces between interacting surfaces. Asperity contacts and adhesion between rough surfaces increase frictional resistance. Smoother surfaces typically exhibit lower friction coefficients due to reduced asperity contact. Surface roughness influences wear mechanisms. Abrasive wear occurs more prominently on rough surfaces, where asperities can break off or plow into softer counter-surfaces. Smoother surfaces generally experience less wear. Roughness affects the performance of lubricants. High roughness surfaces may retain lubricant better due to increased surface area and potential for reservoir formation, leading to improved lubrication. However, rough surfaces can also lead to higher friction if the lubricant cannot adequately separate the contacting surfaces. The study by Tian et al. showed that surface roughness significantly affects the friction coefficient of TiB_2/Al composites [45]. Increased surface roughness generally leads to higher friction coefficients due to greater mechanical interlocking and increased real contact area between the sliding surfaces. This can result in higher friction forces and heat generation during operation. The wear rate of TiB_2/Al composites is influenced by surface roughness. While increased surface roughness can lead to higher wear rates due to more significant abrasive interactions, there is a complex relationship where optimal roughness can potentially improve wear resistance by distributing contact stresses more evenly. Figure 9 shows the 3D AFM images of the steel disc.

Roughness alters the distribution of contact pressure. High points (asperities) bear more load, which can lead to localised stress concentrations and wear. Surface roughness can influence the formation and stability of lubricant films, affecting pressure distribution. Roughness can accelerate surface fatigue by promoting crack initiation at stress concentration points (asperities). This can lead to material degradation and failure over time, especially under cyclic loading conditions. Surface roughness influences adhesive forces between materials. Microscopic irregularities increase contact area and enhance adhesion, which affects both friction and wear characteristics. Goral et al. compared the tribological behaviour of electrodeposited pure Ni coatings with Ni/Al_2O_3 composite coatings and also the relationship between their surface roughness, structural features, and tribological properties [46]. The incorporation of Al_2O_3 particles into the Ni matrix improves the hardness and wear resistance of the coatings. Ni/Al_2O_3 coatings generally exhibit lower wear rates and improved frictional performance compared to pure Ni coatings.

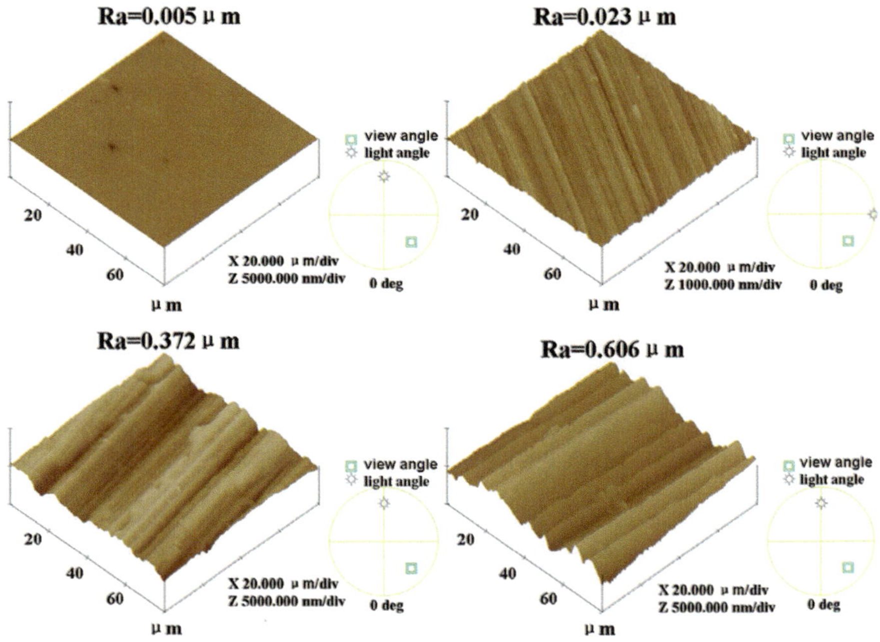

Fig. 9 3D AFM images of the steel disc [45]

5.3 *Chemical Composition*

Material composition affects tribological behaviour, particularly in corrosive environments. Chemical composition determines the hardness of a material, which affects its resistance to wear and deformation under load. Harder materials generally exhibit lower wear rates and reduced plastic deformation during sliding contact. The strength and toughness of a material dictate its ability to withstand applied loads and resist fracture under sliding and impact conditions. Tougher materials may withstand higher stresses before failure, but could exhibit higher wear rates if they lack hardness.

Certain chemical compositions may react differently to environmental factors such as oxygen, moisture, and contaminants. This can lead to surface oxidation, corrosion or chemical wear, impacting tribological performance. Chemical composition influences the surface energy and wettability of a material, affecting its ability to retain lubricants and form boundary or hydrodynamic lubrication films. Materials with lower surface energy may have better wetting and lubrication properties. The chemical composition can alter the intrinsic friction characteristics of a material due to variations in interatomic bonding, crystal structure and surface chemistry. For example, materials with self-lubricating properties (e.g., graphite or certain polymers) exhibit lower friction coefficients [47, 48]. Chemical composition influences the thermal stability and resistance to thermal degradation of materials under frictional heating. High-temperature applications require materials that can maintain their tribological properties without significant degradation. Chemical composition affects the wear mechanisms and resistance of materials. For instance, alloying elements can improve wear resistance by forming hard phases or solid solutions that resist abrasion and adhesive wear. The chemical composition determines the compatibility with lubricants and additives used in tribological applications.

6 Applications

6.1 Automotive Industry

Tribology is critical in engines, transmissions and braking systems to reduce friction and wear, enhancing efficiency and durability. Improved surface coatings and lubricants help reduce friction between moving parts in engines, transmissions and other mechanical systems. This leads to better fuel efficiency and longer component lifespan. Surface treatments such as plasma coatings and nanocomposites increase the durability of engine components, reducing maintenance costs and improving overall vehicle reliability. Green tribology principles are applied through the use of biodegradable lubricants and coatings, minimising the environmental footprint of automotive operations. By reducing friction losses, these advancements contribute to improved energy efficiency in vehicles, helping meet stringent emissions regulations and reducing fuel consumption. Surface engineering plays a crucial role in enhancing the efficiency and longevity of components in electric drivetrains, contributing to the performance and range of Electric Vehicles (EVs). Wan et al. explored how CrN and CrN/GLC coatings perform in terms of friction and wear when applied to automotive engine components [49]. CrN/GLC coatings exhibit lower friction coefficients compared to pure CrN coatings. The graphitic-like carbon layer in CrN/GLC coatings provides a lubricating effect, reducing friction between sliding surfaces. This reduction in friction can lead to improved fuel efficiency and reduced energy consumption in automotive engines. Figure 10 shows that the distinct wear rates of coatings differed from the various lubricated conditions. The lubricating properties of the GLC layer help in reducing wear by minimising direct contact between the sliding surfaces, which is critical for engine components that are subjected to high friction and wear conditions. Renondeau et al. examined the performance of DLC coatings under lubricated conditions, analysing how these coatings influence friction, wear, and overall component longevity in automotive applications [50].

6.2 Manufacturing

Tribological properties influence machining, forming processes and the lifespan of tools and dies. Engineers consider tribological properties when selecting materials for

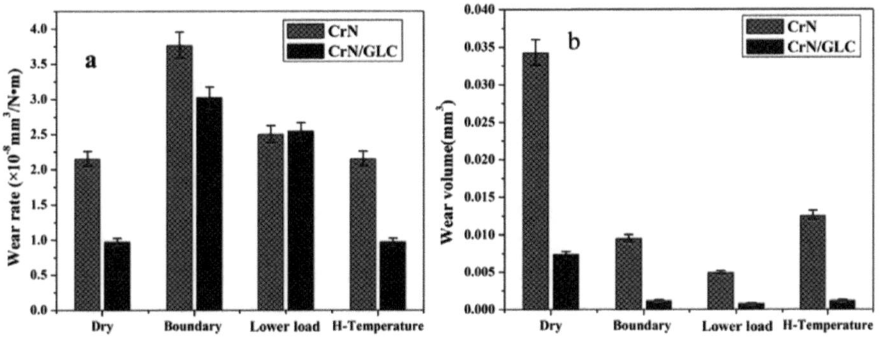

Fig. 10 **a** Wear rate of CrN and CrN/GLC coatings and **b** wear loss of the corresponding steel counterparts in different lubricated conditions [49]

components. For example, choosing materials with low friction coefficients and high wear resistance can improve the longevity and efficiency of machinery [49]. Techniques such as surface coatings, heat treatments and surface texturing are employed to modify surface properties to enhance wear resistance, reduce friction and improve lubricant retention. Proper lubrication is essential in manufacturing to reduce friction and wear between moving parts. Lubricants are selected based on their viscosity, additives and compatibility with materials to ensure optimal performance and longevity. Tribological considerations are critical in the design and selection of bearings and seals to minimise friction losses, prevent leakage and extend component life. Manufacturers conduct wear testing to evaluate materials and coatings under simulated operating conditions, helping to identify optimal materials and surface treatments. Techniques like surface texturing, polishing, and the application of low-friction coatings aim to reduce frictional losses in machinery, improve energy efficiency, and reduce operating costs. Understanding tribological properties allows manufacturers to implement effective maintenance strategies such as scheduled lubrication and inspection regimes to prevent premature wear and failures. Advanced computational tools simulate and model tribological interactions, aiding in the design optimisation of components and systems to achieve desired performance and reliability.

6.3 Biomedical Implants

Joint prostheses and dental implants require excellent wear resistance and biocompatibility materials to ensure longevity and performance. The choice of implant materials is crucial to ensure compatibility with biological tissues and to withstand the mechanical stresses encountered in the body. Tribological properties such as wear resistance, friction coefficient and biocompatibility influence material selection. Surface treatments and coatings are applied to implants to improve their tribological performance. Techniques like plasma spraying, ion implantation and DLC coatings can enhance wear resistance and reduce friction, promoting longevity and reducing the risk of adverse biological reactions. In some implants, particularly joint replacements like hip and knee prostheses, lubrication systems are incorporated to mimic natural joints. These systems reduce friction between articulating surfaces, improving mobility and reducing wear. Implants must resist wear over their lifespan to maintain functionality and avoid the release of wear debris, which can trigger inflammatory responses. Materials and coatings with high wear resistance are essential for long-term implant success. Tribological properties contribute to the overall biomechanical compatibility of implants. They affect how well implants integrate with surrounding tissues, function under physiological loads and avoid detrimental friction-induced damage to tissues. Kaur et al. provided an overview of the tribological performance of titanium alloys with various surface modifications, focusing on their applications in the biomedical field [51].

Various coatings, such as hydroxyapatite (HA), titanium nitride (TiN), and diamond-like carbon (DLC), are applied to titanium alloys to enhance their tribological properties. These coatings improve wear resistance and reduce friction, making them suitable for use in joint replacements and dental implants. Coatings like hydroxyapatite promote better integration with bone tissues, improving the success rate of implants and prosthetic devices. Titanium alloys with surface modifications show improved performance in joint replacements, including hip and knee implants. Surface-modified titanium alloys are also effective in dental implants, where improved bioactivity and reduced wear enhance the integration of implants with the jawbone and overall durability.

6.4 Aerospace

Aircraft components such as bearings and turbines need to withstand extreme conditions while maintaining low friction and wear. Tribological properties are critical in the design and manufacturing of engine components such as bearings, seals, gears and shafts. High-performance lubricants and coatings are used to minimise friction and wear, thereby improving efficiency and extending component life. Tribology plays a vital role in the operation and maintenance of landing gear systems. Lubrication and surface treatments are employed to reduce friction and wear between moving parts, ensuring smooth operation and safety during landings and takeoffs. Hydraulic and pneumatic actuators, as well as control surfaces in aircraft, rely on tribological principles to provide precise and reliable operation. Lubricants and materials with low friction coefficients are used to minimise energy losses and wear. Tribology is important for ensuring passenger comfort and safety in aircraft interiors. For example, in seat mechanisms and sliding components, tribological coatings and lubricants are applied to reduce noise, vibration and wear. In spacecraft, where reliability is critical due to the extreme environments of space, tribology is essential for mechanisms such as solar array deployment systems, antenna pointing mechanisms and robotic arms.

Lubrication and materials with high vacuum compatibility and resistance to temperature extremes are utilised. Aerospace applications often involve harsh environmental conditions, including high temperatures, pressures and exposure to contaminants. Among various coating technologies, WC–Co (Tungsten Carbide-Cobalt) plasma spray coatings are widely utilised for their excellent wear resistance and durability. Koutsomichalis et al. compared WC–Co plasma spray coatings with other types of tribological coatings used in aerospace and found that WC–Co coatings generally provide superior wear resistance and thermal stability compared to many alternative coatings for high-stress applications [52]. Swain et al. investigated the excellent wear resistance of the plasma-sprayed NiTi coatings [53]. The coatings effectively reduce wear rates and extend the service life of aerospace components by providing a durable surface layer that withstands abrasive conditions. The ability to withstand high wear and provide low friction can contribute to more efficient and reliable aerospace systems.

7 Recent Advances

7.1 Nanotribology

Study of tribological phenomena at the nanoscale involving advanced materials like graphene and carbon nanotubes for superior lubrication. Techniques such as AFM and nanoindentation have enabled precise measurement of mechanical properties at nanoscale dimensions. This includes mapping surface topography, measuring friction forces with atomic resolution and characterising material wear behaviour at microscopic scales. Advances in molecular dynamics simulations and theoretical models have provided more profound insights into the fundamental mechanisms governing friction and wear at the atomic and molecular levels. This includes understanding how surface asperities interact, the role of adhesion forces and the influence of surface chemistry on tribological behaviour. Nanomaterials such as graphene, carbon nanotubes and various nanostructured coatings have shown promise in enhancing tribological performance. These materials offer superior mechanical properties, high surface area-to-volume ratios and unique lubricating capabilities that can reduce friction and wear in diverse applications. Nanotechnology has enabled precise control over surface properties through techniques like self-assembly, molecular layer deposition and ion beam modification. These methods can tailor surface roughness, chemistry and topography to optimise tribological performance under specific operating conditions. Drawing inspiration from natural lubrication mechanisms found in biological systems (e.g., joints, muscles), researchers are developing biomimetic lubricants and coatings. These bio-inspired materials aim to replicate the superior lubrication efficiency and wear resistance observed in biological interfaces.

7.2 *Surface Engineering*

Techniques like coating and surface texturing enhance tribological properties. For example, DLC coatings provide high hardness and low friction. Techniques such as plasma treatments, laser surface modification and nanostructuring have been employed to achieve these improvements. Additionally, the development of biomimetic surfaces and smart coatings that respond to environmental stimuli represent cutting-edge innovations in this field.

7.3 *Green Tribology*

Focuses on environmentally friendly lubricants and materials to reduce ecological impact. Bio-lubricants derived from renewable resources are a key area of research. Key developments include the use of environmentally friendly lubricants such as biodegradable oils and greases derived from renewable sources like vegetable oils. Additionally, research has focused on surface texturing and coatings to enhance lubrication efficiency and reduce frictional losses. Novel approaches also include the development of bio-inspired lubricants that mimic natural lubrication mechanisms found in biological systems. Overall, these advancements aim to improve energy efficiency, reduce emissions and extend the lifespan of mechanical components sustainably.

8 Conclusion

Understanding and improving tribological properties is essential for developing efficient, durable, and sustainable mechanical systems. Advances in material science, surface engineering and lubrication technology continue to push the boundaries of what is possible in tribology, offering solutions to longstanding industrial challenges. Coating materials vary significantly in their ability to reduce friction between surfaces. Some coatings excel in minimising frictional forces, thereby improving efficiency and reducing energy consumption in mechanical systems. The wear resistance of coating materials is crucial for prolonging the lifespan of components and reducing maintenance costs. Coatings that exhibit high wear resistance are preferred for applications where components are subjected to abrasive or erosive environments. Certain coatings possess inherent lubricating properties or can be designed to enhance lubrication between interacting surfaces. This is particularly beneficial in reducing friction and wear, thereby improving the overall efficiency and reliability of machinery. Considerations such as resistance to corrosion, oxidation and chemical exposure are essential for determining the suitability of coating materials in specific operating environments. The selection of coating materials should be based on a comprehensive evaluation of their tribological properties relative to the specific requirements of the application. A well-chosen coating can significantly enhance the performance, durability and efficiency of mechanical systems by mitigating friction, improving wear resistance and enhancing lubrication.

References

1. Stachowiak GW, Batchelor AW (2013) Engineering tribology. Butterwoth Heinemann, London
2. Blau PJ (1992) Friction, lubrication, and wear technology. ASM Handb 18:490–513
3. Barber JR (2013) Multiscale surfaces and Amontons' law of friction. Tribol Lett 49:539–543. https://doi.org/10.1007/s11249-012-0094-6
4. Popov VL (2017) Coulomb's law of friction. In: Contact mechanics and friction. Springer, Berlin, Heidelberg. https://doi.org/10.1007/978-3-662-53081-8_10
5. Kikuuwe R, Takesue N, Sano A, Mochiyama H, Fujimoto H (2005) Fixed-step friction simulation: from classical Coulomb model to modern continuous models. In: 2005 IEEE/RSJ international conference on intelligent robots and systems. IEEE, pp 1009–1016
6. Toughe Z, Huixing Z (2000) Industrialisation of MEVVA source ion implantation. Surf Coat Technol 128–129:1–8. https://doi.org/10.1016/S0257-8972(00)00635-6
7. Devaraju A (2016) A review on important factors affecting dry sliding friction. J Surface Sci Technol 32(3–4):71–76. https://doi.org/10.18311/jsst/2016/7751
8. Devaraju A, Elayaperumal A, Alphonsa J, Kailas SV, Venugopal S (2012) Microstructure and dry sliding wear resistance evaluation of plasma nitrided austenitic stainless steel type AISI 316LN against different sliders. Surf Coat Technol 207:406–412. https://doi.org/10.1016/j.surfcoat.2012.07.031
9. You S, Tang J, Wang Q (2024) A new 3D plastoelastohydrodynamic lubrication model for rough surfaces. Friction 12(6):1176–1193. https://doi.org/10.1007/s40544-023-0793-z
10. Kopeliovich D (2013) Lubrication regimes, SubsTech (Substances & Technologies). https://www.substech.com/dokuwiki/doku.php?id=lubrication_regimes
11. Muratore C, Hu JJ, Voevodin AA (2009) Tribological coatings for lubrication over multiple thermal cycles. Surf Coat Technol 203(8):957–962. https://doi.org/10.1016/j.surfcoat.2008.08.073
12. Sui X, Liu J, Zhang S, Yang J, Hao J (2018) Microstructure, mechanical and tribological characterisation of CrN/DLC/Cr-DLC multilayer coating with improved adhesive wear resistance. Appl Surf Sci 439:24–32. https://doi.org/10.1016/j.apsusc.2017.12.266
13. Sargent GA, Saigal D (1986) Erosion of low-carbon steel by coal particles. A S L E Trans 29(2):256–266. https://doi.org/10.1080/05698198608981685
14. Bartenev GM, Lavrent'ev VV (1981) Friction and wear of polymers, vol 6. Elsevier
15. Foley AG, Chisholm CJ, McLees VA (1988) Wear of ceramic-protected agricultural subsoilers. Tribol Int 21:97–103. https://doi.org/10.1016/0301-679X(88)90081-3
16. Krell A, Blank P (1988) On abrasive wear of zirconia-toughened alumina. Wear 124(3):327–330. https://doi.org/10.1016/0043-1648(88)90221-9
17. Holtz D, Janssen R, Friedrich K, Claussen N (1989) Abrasive wear of ceramic-matrix composites. J Eur Ceram Soc 5:229–232
18. Krishna LR, Somaraju KRC, Sundararajan G (2003) The tribological performance of ultra-hard ceramic composite coatings obtained through microarc oxidation. Surf Coat Technol 163:484–490. https://doi.org/10.1016/S0257-8972(02)00646-1
19. Wang SH, Chen JY, Xue L (2006) A study of the abrasive wear behaviour of laser-clad tool steel coatings. 200(11):3446−3458. https://doi.org/10.1016/j.surfcoat.2004.10.125
20. Krella A (2020) Resistance of PVD coatings to erosive and wear processes: a review. Coatings 10(10):921. https://doi.org/10.3390/coatings10100921

21. Singh VP, Sil A, Jayaganthan R (2011) A study on sliding and erosive wear behaviour of atmospheric plasma sprayed conventional and nanostructured alumina coatings. Mater Surf Coat Technol Des 32(2):584–591. https://doi.org/10.1016/j.matdes.2010.08.019
22. Ji X, Luo C, Sun Y, Zhao J (2019) Corrosive wear of multilayer Fe-based coatings laser cladded from amorphous powders. Wear 438–439:203113. https://doi.org/10.1016/j.wear.2019.203113
23. Lee CK (2012) Corrosive wear and mechanical properties of Ni/Al2O3 micro-and nanoparticulates-reinforced coatings on Ti-6Al-4V Alloy. Tribol Trans 55(5):640–651. https://doi.org/10.1080/10402004.2012.689611
24. Yonekura D, Fujita J, Miki K (2015) Fatigue and wear properties of Ti–6Al–4V alloy with Cr/CrN multilayer coating. Surf Coat Technol 275:232–238. https://doi.org/10.1016/j.surfcoat.2015.05.014
25. Xie Y, Hawthorne HM (1999) Wear mechanism of plasma-sprayed alumina coating in sliding contacts with harder asperities. Wear 225:90–103. https://doi.org/10.1016/S0043-1648(98)00351-2
26. Drobný P, Mercier D, Koula V, Škrobáková SI, Čaplovič Ľ, Sahul M (2021) Evaluation of adhesion properties of hard coatings by means of indentation and acoustic emission. Coatings 11(8):919. https://doi.org/10.3390/coatings11080919
27. Cai F, Zhang J, Wang J, Zheng J, Wang Q, Zhang S (2021) Improved adhesion and erosion wear performance of CrSiN/Cr multi-layer coatings on Ti alloy by inserting ductile Cr layers. Tribol Int 153:106657. https://doi.org/10.1016/j.triboint.2020.106657
28. Markov DP (2022) Adhesion at friction and wear. Friction 10:1859–1878. https://doi.org/10.1007/s40544-021-0564-7
29. Schall JD, Milne ZB, Carpick RW, Harriso JA (2021) Molecular dynamics examination of sliding history-dependent adhesion in Si–Si nanocontacts: Connecting friction, wear, bond formation, and interfacial adhesion. Tribol Lett 69:1–19. https://doi.org/10.1007/s11249-021-01431-z
30. Packham DE (2003) Surface energy, surface topography and adhesion. Int J Adhes Adhes 23(6):437–448. https://doi.org/10.1016/S0143-7496(03)00068-X
31. Zhang N, Zhan X, Li Z, Liu Y, Ma J, Gao J (2024) Effect of squamous textured SiC on the dry-sliding tribological properties of graphite asperities. Wear 544:205297. https://doi.org/10.1016/j.wear.2024.205297
32. Chen Z, Etsion I (2019) The elastic-plastic contact behavior of rough surfaces with hard coatings. Tribol Int 134:435–442. https://doi.org/10.1016/j.triboint.2019.02.025
33. Barletta M, Pezzola S, Vesco S, Tagliaferri V, Trovalusci F (2014) Experimental evaluation of plowing and scratch hardness of aqueous two-component polyurethane (2K-PUR) coatings on glass and polycarbonate. Prog Org Coat 77(3):636–645. https://doi.org/10.1016/j.porgcoat.2013.11.025
34. Zhang Y, Kang Z, Bessho T (2017) Two-component spin-coated Ag/CNT composite films based on a silver heterogeneous nucleation mechanism adhesion-enhanced by mechanical interlocking and chemical grafting. Nanotechnol 28(10):105607. https://doi.org/10.1088/1361-6528/aa595f
35. Stegmüller MJ, Grant RJ, Schindele P (2019) Quantification of the interfacial roughness when coating stainless steel onto aluminium by friction surfacing. Surf Coat Technol 375:22–33. https://doi.org/10.1016/j.surfcoat.2019.06.060
36. Aggelis DG, Soulioti DV, Barkoula NM, Paipetis AS, Matikas TE (2012) Influence of fiber chemical coating on the acoustic emission behavior of steel fiber reinforced concrete. Cem Concr Compos 34(1):62–67. https://doi.org/10.1016/j.cemconcomp.2011.07.003
37. Oblak E, Kalin M (2015) Relationship between the nanoscale topographical and mechanical properties of tribochemical films on DLC coatings and their macroscopic friction behavior. Tribol Lett 59:1–16. https://doi.org/10.1007/s11249-015-0575-5
38. Chvedov D, Jones R (2004) Frictional behavior of rolled surfaces coated with polymer films. Surf Coat Technol 188:544–549. https://doi.org/10.1016/j.surfcoat.2004.07.013
39. Aihua L, Jianxin D, Haibing C, Yangyang C, Jun Z (2012) Friction and wear properties of TiN, TiAlN, AlTiN and CrAlN PVD nitride coatings. Int J Refract Hard Met 31:82–88. https://doi.org/10.1016/j.ijrmhm.2011.09.010

40. Madhuri A, Jena S, Gupta M (2023) Swain BP (2023) Tweaking the mechanical stability of $Al_{1-x}W_xN$ ternary nitride alloys for hard coating applications, and study on their structural, photoluminescence and surface chemical analysis. J Alloys Compd 968:171854. https://doi.org/10.1016/j.jallcom.2023.171854
41. Das S, Guha S, Das PP, Ghadai RK (2020) Analysis of morphological, microstructural, electrochemical and nano mechanical characteristics of TiCN coatings prepared under N_2 gas flow rate by chemical vapour deposition (CVD) process at higher temperature. Ceram Int 46(8):10292–10298. https://doi.org/10.1016/j.ceramint.2020.01.023
42. Ghadai RK, Das S, Kumar D, Mondal SC, Swain BP (2018) Correlation between structural and mechanical properties of silicon doped DLC thin films. Diam Relat Mater 82:25–32. https://doi.org/10.1016/j.diamond.2017.12.012
43. Das S, Guha S, Ghadai R, Kumar D, Swain BP (2017) Structural and mechanical properties of CVD deposited titanium aluminium nitride (TiAlN) thin films. Appl Phys A 123:1–5. https://doi.org/10.1007/s00339-017-1032-0
44. Yamamoto K, Matsukado K (2006) Effect of hydrogenated DLC coating hardness on the tribological properties under water lubrication. Tribol Int 39(12):1609–1614. https://doi.org/10.1016/j.triboint.2006.01.004
45. Tian SF, Jiang LT, Guo Q, Wu GH (2014) Effect of surface roughness on tribological properties of TiB_2/Al composites. Mater Des 53:129–136. https://doi.org/10.1016/j.matdes.2013.06.038
46. Góral A, Lityńska-Dobrzyńska L, Kot M (2017) Effect of surface roughness and structure features on tribological properties of electrodeposited nanocrystalline Ni and Ni/Al_2O_3 coatings. J Mater Eng Perform 26:2118–2128. https://doi.org/10.1007/s11665-017-2662-2
47. Ren F, Wang S, Yu M, Duan H, Su M, Ren M, Sun J (2020) Influence of graphite/graphene on the tribological behaviors of self-lubricating fabric composite. Materials 13(1):232. https://doi.org/10.3390/ma13010232
48. Sorrentino A (2018) Tribology of self-lubricating polymer nanocomposites. In: Menezes P, Rohatgi P, Omrani E (eds) Self-lubricating composites. Springer, Berlin, Heidelberg. https://doi.org/10.1007/978-3-662-56528-5_4
49. Wan S, Pu J, Li D, Zhang B, Tieu AK (2017) Tribological performance of CrN and CrN/GLC coated components for automotive engine applications. J Alloys Comp 695:433–442. https://doi.org/10.1016/j.jallcom.2016.11.118
50. Renondeau H, Papke BL, Pozebanchukz M, Parthasarathy PP (2009) Tribological properties of diamond-like carbon coatings in lubricated automotive applications. Proc Inst Mech Eng, Part J: J Eng Tribol 223(3):405–412. https://doi.org/10.1243/13506501JET548
51. Kaur S, Ghadirinejad K, Oskouei RH (2019) An overview on the tribological performance of titanium alloys with surface modifications for biomedical applications. Lubricants 7(8):65. https://doi.org/10.3390/lubricants7080065
52. Koutsomichalis A, Vaxevanidis N, Petropoulos G, Xatzaki E, Mourlas A, Antoniou S (2009) Tribological coatings for aerospace applications and the case of WC-Co plasma spray coatings. Tribol Ind 31(1&2):37
53. Swain B, Mantry S, Mohapatra SS, Mishra SC, Behera A (2022) Investigation of tribological behavior of plasma sprayed NiTi coating for aerospace application. J Therm Spray Tech 31:2342–2369. https://doi.org/10.1007/s11666-022-01452-7

Chapter 10
Corrosion Properties of Coating Materials

Sanketa Jena and Bibhu Prasad Swain

Abstract In recognition of their superior capabilities, nanocoatings are increasingly widely used in multiple industries as coating technology advances. Surface hardness, adhesive strength, long-term and/or high-temperature corrosion resistance, the improvement of tribological real estate, and many other benefits are provided by coatings. Moreover, coatings can be placed in thinner and smoother thicknesses, allowing for greater customisation for machinery, increased efficiency, reduced fuel consumption, reduced environmental impact, and cheaper costs for upkeep and operation. Effective use of coatings lessens the impact of harsh conditions. A coating is a material consisting of layers with dimensions not larger than a few nanometers in size or containing components at the microscopic level. Good adhesion and excellent physical coverage of the coated surface are made possible by the minute diameters of the nanomaterials and the high density of their underlying boundaries. However, these exquisite qualities could provide active locations for corrosion assault. The corrosion resistance of ceramic, metallic, non-metallic, and nanocomposite coatings over metallic substrate surfaces is reviewed in this chapter. It lists the elements influencing these substrates' corrosion and the circumstances in which these coatings were necessary to offer the required resistance.

Keywords Coating materials · Corrosion resistance · Metallic coatings · Ceramic coatings · Non-metallic coatings

1 Introduction

Corrosion is a natural process in which metal gradually deteriorates due to chemical or electrochemical environmental reactions. It typically happens due to exposure to moisture, oxygen, and other substances in the atmosphere or surrounding materials. This process often results in the formation of rust or other undesirable compounds

S. Jena (✉) · B. P. Swain
Department of Physics, National Institute of Technology Manipur, Langol, Imphal West, Manipur, India
e-mail: sanketajena1@gmail.com

B. P. Swain (ed.), *Advances in Mechanical Coating*, Materials Horizons: From Nature to Nanomaterials, https://doi.org/10.1007/978-981-96-7484-8_10

on the surface of the metal, weakening it over time. It's a significant concern in many industries, including construction, transportation, and manufacturing, because it may result in expensive repairs or replacements, decreased efficacy, and weakened structures. Corrosion is the surface disintegration of metals/alloys within a specific environment. It typically pertains to the unwanted chemical interaction between these substances and the liquids, gases, salts, and other substances that make up their surroundings. Because of a variety of reasons, including their chemical makeup and the nature of electrochemical processes, certain metals have more robust durability against corrosion compared to others. One way to characterise a material's susceptibility to corrosion is its capacity to tolerate harsh environments. This primarily impacts how long operating parts will last while in use. The International Union of Pure and Applied Chemistry (IUPAC) defines corrosion as "an irreversible interfacial interaction among a substance (metal, ceramic, and polymer) with its surroundings that concludes in the breakdown of the substance or collapse into the substance of an aspect of the atmosphere."

Nevertheless, there are numerous explanations for corrosion. Corrosion frequently, but not always, negatively impacts how the material under consideration is used. The term corrosion does not refer to only physical or mechanical processes like blistering or vaporisation, scratches, or mechanical fractures. Consequently, ISO 8044–1986 provides an additional description, stating, among other things: "Physicochemical interaction among a metal and its surroundings that culminates in modifications to the metal's properties, and which may frequently consequence in limitation of the metal's functioning, the atmosphere's function, or the technical framework of which these form a part." Most metals and alloys tend to corrode or react with oxygen and water found in their surroundings to reach their least energetic state—also known as their most stable state—because nature permits materials to remain in this condition. To sum up, nearly all metals tend to transform into their stable oxide state. Corrosion may seriously harm various structures, including buildings, bridges, cars, household appliances, drinking water systems, gas and petroleum pipelines, and the framework of industrial facilities.

When creating materials with outstanding properties, the way metals engage with their surroundings is crucial. For instance, iron and steel exposed to an industrial environment might develop a film of iron oxide called "rust." These rust coatings function as an effective barrier against corrosion, shielding the steel exterior in the early phases of environmental exposure. Nonetheless, the constituents of rust sometimes exhibit a certain degree of attraction for water or electrolytes. As a result, as time passes and imperfections/pores begin to appear, the barrier of defence degrades. Starting at the places of fault, the corrosion mechanism spreads.

In summary, the rust layer may be mechanically thick and firmly affixed to the metal surface initially, but it is inadequate to create an efficient and sturdy coating that shields the metal edge from oxygen and water. Based on the individual's perspective, films are essential for preventing corrosion, particularly for products and products exposed to corrosive conditions. Protecting the coated object for as much time as feasible is the primary objective of users of corrosion protection coatings. As a result, the coating serves as more than simply a layer of protection between the item and its surroundings; it also has to work to shield any tiny, locally accessible substrate portions from harm and to prevent them from spreading.

2 Major Drawbacks of Corrosion

Corrosion damages structural elements and can even lead to major collapses that cost much money and take lives. In parallel, they have worked to limit corrosion through the development of better alloys, modifications to the alloys' microscopic structure, the application of coatings robust to corrosion, environmental changes, and other means. Their main goal was to increase the substance's longevity as much as possible. Some of the significant drawbacks of corrosion are as follows:

- Structural Integrity: Corrosion weakens the structural integrity of metal objects, which can lead to catastrophic failures in critical infrastructure such as bridges, pipelines, and buildings.
- Safety Concerns: Corrosion compromises the safety of structures and equipment, increasing the risk of accidents, injuries, and even fatalities. For example, weakened metal structures may collapse unexpectedly.
- Financial Costs: Corrosion results in substantial financial costs due to the need for repairs, replacements, and maintenance of affected structures and equipment. These costs can be particularly burdensome for industries reliant on metal assets.
- Reduced Efficiency: Corrosion can impair the performance and efficiency of machinery and equipment, leading to decreased productivity, increased energy consumption, and operational disruptions.
- Environmental Impact: Corrosion products, such as rust and other metal oxides, can contaminate soil, water bodies, and air, posing environmental hazards and contributing to pollution.
- Aesthetic Degradation: Corrosion detracts from the appearance of metal surfaces, affecting the aesthetics of structures, vehicles, and other objects. This can diminish the perceived value and attractiveness of these items.
- Lifespan Reduction: Corrosion accelerates the degradation of metal components, reducing their lifespan and necessitating premature replacement, which further increases costs and resource consumption.

Overall, corrosion represents a pervasive and costly problem that affects various aspects of human activity, ranging from infrastructure and industry to environmental and economic considerations. Efforts to mitigate corrosion through preventive measures and advanced materials science are ongoing to minimise its negative impacts.

3 Mechanism Behind Corrosion Properties

Comprehending and controlling corrosion heavily relies on the field of electrochemistry and thermodynamics. According to these responses, corrosion may be divided into three categories: chemical, physical, and electrochemical. Figure 1a and b represents the real corrosion image in thin coated films and aluminium corrosion pitting on PVD coatings. The fundamental principles of chemical kinetics of heterogeneity processes apply exclusively to corrosion by chemicals. It describes instances of corrosion, such as the rusting of metals in non-electrolytes or dry gases, that do not produce an electric current. The metallic corrosion of solid metals in contact with liquid metal is a typical example of the physical mechanisms of corrosion. Most of the time, solid metal is attacked by liquid metal seeping into its grain boundaries, but occasionally, the solid metal breaks and creates an alloy with the molten metal. Anodic and cathodic responses are two categories into which electrochemical reactions fall. Anodic response, on the other hand, entails anodic disintegration. In addition, the cathodic reactions—metal reduction, metal deposition, oxygen consumption, or hydrogen evolution—depend on the surrounding conditions. Oxidation processes during corrosion responses may happen locally, in specific locations, or evenly over the substrate.

The majority of metallic films that are often utilised are inherently unstable. In its most basic form, corrosion is a mechanism by which a metal returns to its thermodynamic condition. For most, this entails the sulfide or oxide production. Most corrosion processes are electrochemical with two or more electrode reactions: anodic partial reaction for oxidation and cathodic partial reaction for reduction. Figure 2 shows the schematic of the corrosion process of Fe coating, where both oxidation and reduction are included. Electrochemical thermodynamics and electrochemical kinetics must be studied to comprehend corrosion reactions. For instance, anodic and cathodic partial processes usually involve iron corrosion in an acidic medium. The overall corrosion reaction of iron in acidic solution can be broken down into these partial reactions:

Anodic partial reaction (Oxidation)

$$Fe(s) \rightarrow Fe^{2+}(aq.) + 2e^{-} \tag{1}$$

This represents the oxidation of iron solid (Fe(s)) to iron ions (Fe^{2+}(aq)), releasing two electrons in the process. This step occurs at the anode, where the metal loses electrons and enters the solution as ions.

And cathodic partial reaction (Reduction)

$$2H^{+}(aq.) + 2e^{-} \rightarrow H_2(g) \tag{2}$$

This represents the reduction of hydrogen ions (H^+(aq)) from the acidic solution to form hydrogen gas (H_2(g)), consuming two electrons in thc proccss. This step occurs at the cathode, where reduction reactions take place.

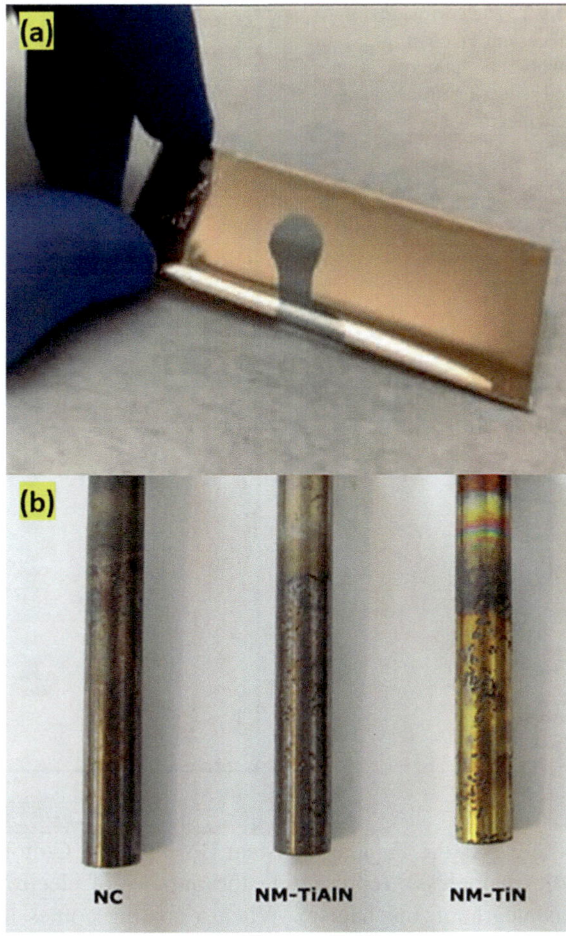

Fig. 1 **a** Corrosion in thin film coatings, **b** Aluminum corrosion pitting on PVD coatings [1]

The overall corrosion reaction of iron in the acidic medium can be represented by combining the anodic and cathodic reactions.

$$Fe(s) + 2H^{+}(aq.) \rightarrow Fe^{2+}(aq.) + H_2(g) \tag{3}$$

This equation shows the simultaneous occurrence of the anodic and cathodic reactions, leading to iron corrosion in an acidic solution.

The kinetics of both partial interactions at the electrode's surface determine the corrosion pace. Faraday's law states that the partial anodic current density for metal dissolution iaM and the metal dissolution rate at any potential VM have a linear correlation.

$$V_M = \frac{i_{aM}}{nF} \tag{4}$$

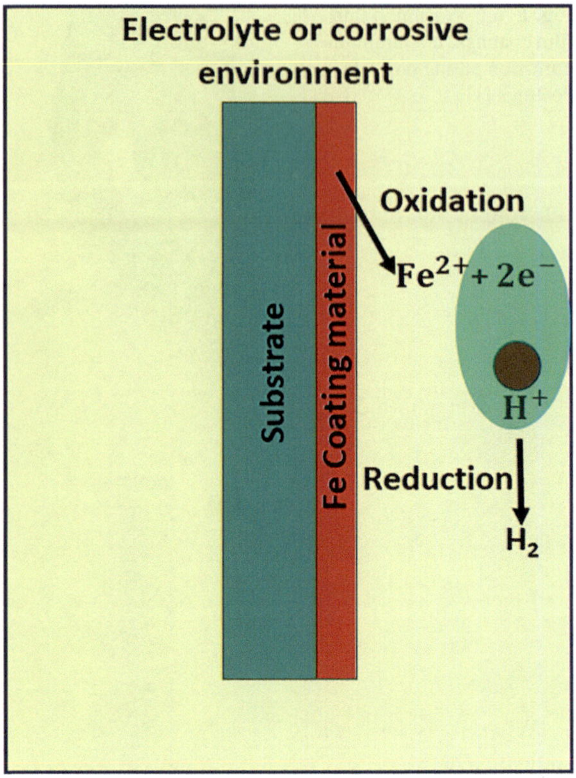

Fig. 2 The corrosion process of Fe coating involves both the oxidation and reduction processes in acid electrolytes

where F is the Faraday constant (F = 96,485 C/mol) and n is the charge number (dimensionless), representing the number of electrons transferred throughout the disintegration mechanism. When a coating comes into contact with an oxidising electrolytic environment without any additional polarisation, it naturally develops a particular potential known as the corrosion potential or E_{corr}. The degree of corrosion current density (i_{corr}) is equal to the partial anodic current density at the corrosion potential. Thus, Eq. (4) gets

$$V_{corr} = \frac{i_{corr}}{nF} \tag{5}$$

The optimum potentials of the anodic and cathodic partial processes are where the corrosion potential is located. Electrochemical thermodynamics predicts the partial processes' equilibrium potential. Any chemical reaction's total stoichiometry may be written as

$$0 = \varepsilon v_i \beta_i \tag{6}$$

where the reactants and their byproducts are denoted by β. The harmonious coefficient v_i of the products is positive, whereas that of the reactants is negative. The reaction's free enthalpy (ΔG) is

$$\Delta G = \varepsilon v_i \mu_i \tag{7}$$

wherein the participating series' chemical potential is represented by μ_i. The relevant equilibrium potential E_{rev}, obtained from Eq. (6) in an electrochemical cell, is presented by

$$\Delta G = -nFE_{rev} \tag{8}$$

Under standard conditions (all activities equal to one)

$$\Delta G^0 = -nFE^0 \tag{9}$$

where ΔG^0 represents the standard free enthalpy and E^0 represents the standard potential of the reaction.

Electrode reactions are commonly written in the form

$$\varepsilon v_o \beta_o + ne = \varepsilon v_r \beta_r \tag{10}$$

When β_o is on the left side of the equality sign alongside the free electrons, it indicates the stoichiometric parameter of the oxidised species, v_o. Conversely, when βr comes on the right side of the equality signal, it means the stoichiometric factor of the reducing species, V_r. Regarding positive stoichiometric coefficients Vo and Vr, the outcome described by Eq. (10) is a partial reduction. Where E^0 is the reaction's standard potential and ΔG^0 is its standard free enthalpy.

The default potential of the partial reduction process of Eq. (10) concerning the conventional hydrogen electrode may be defined by adjusting the conventional chemical potentials of the solvated proton and the molecular hydrogen to zero, $\mu^0_{H^+} = 0$, $\mu^0_{H_2} = 0$. The electrode interaction standard potential is in line with the total response.

$$\varepsilon v_o \beta_o + \frac{n}{2} H_{2(PH_2=1bar)} = \varepsilon v_r \beta_r + nH^+_{(aH^+=1)} \tag{11}$$

According to the formula, there has to be an opposite free energy shift for a specific response to occur.

$$\Delta G = -nFE \tag{12}$$

The cell potential has to be favourable for this to happen. The variation between two half-cell reactions, i.e., the one at the cathode, deducted from the one at the anode, is known as the cell potential.

$$E = \text{Cathode half cell} - \text{anode half cell} \tag{13}$$

Reaction is possible if the cell is positive. The reaction's generating power increases with the magnitude of this potential imbalance. Further variables will determine the occurrence and rate of corrosion. Ohm's law (I = E/R) governs the current flow and completed circuit requirements for corrosion to occur. The cell potential computed here represents the highest values for the scenario of two distinct responses. The cell potential would stay as determined if the resistance were boundless, but corrosion would not occur. The potential of each half-cell could surpass one another, and the degree of corrosion would appear limitless if the resistance of the circuit were zero. The potentials of the two half-cells started moving somewhat toward one another at an intermediate resistance in the circuit. Polarisation is the name given to this shift in potential. The resistivity of the medium, surface coatings, and the metal itself are some variables that affect the resistance in the circuit. Once the corrosion current has been established, the corrosion density may be computed by calculating the surface area. Faraday's principles can be used to calculate a corrosion rate expressed in terms of metal loss per unit of time. The degree of polarisation can be used to forecast the kind and severity of corrosion and its rates. Corrosion reduces when polarisation rises. Comprehending how atmospheric alterations impact polarisation might help manage corrosion.

4 Factors Affecting Corrosion

Corrosion is affected by many factors, including temperature, relative velocity between the respective coating and the medium, surface, grain orientation, micro stress, and time. Polarisation has three fundamental causes: activation, concentration, and potential decrease. Concentration polarisation is the outcome of either a species depleting that is essential to the corrosion process or an overabundance of a species that impedes the corrosion process. The corrosion reaction includes a rate-controlling phase that leads to activation polarisation. The voltage shift caused by environmental factors and the current circuit between the anode and cathode sites is known as the potential drop. The corrosion process depends on coating composition and structure size. Arukalam et al. studied the influence of the size of ZnO nanoparticles (30, 50, and 90 nm) on barrier coating and corrosion resistance of poly(dimethylsiloxane)-coated Q235 steel in an aqueous solution of 3.5 wt% NaCl/ Pseudomonas aeruginosa environment [2]. Selvi et al. studied the coating composition dependence on corrosion resistance of pulse and DC electrodeposited Co–P coatings (P wt. % varying from 7, 9, and 12 wt. %) at ambient temperature in a non-deaerated 3.5% NaCl solution [3]. For both pulse and DC deposited specimens, the potentiodynamic polarisation results for Co with 9 weight percent P showed a reduced corrosion current density and an increased polarisation resistance (Fig. 3a and b). DC deposited coatings' Nyquist and Bode diagrams, which were produced based on EIS observations, revealed one time constant, but the response of pulsed deposited coatings demonstrated a dual time constant. More significant amounts of charge transfer resistance (Rct) and fewer double-layer capacitance (Qdl) units have been observed for pulsed deposited coatings.

The pulse-deposited coatings had a lower corrosion rate than DC-deposited coatings [3]. Additional studies were done on the impact of grain size at various nanoscale dimensions. A more refined nanograin dimension was not always associated with the most effective corrosion resistance in the investigations, so a smaller or increased nanograin dimension of the nanomaterial itself does not always have a comparable impact when protecting against corrosion. Grain improvement by twin borders and grain borders can lower the corrosion pace of magnesium AZ31 alloy; however, grain borders outperform twin borders in enhancing corrosion resilience [4]. Even though the durability to corrosion deteriorates shortly after involvement for a certain period, grain refineries along twin borders may enhance corrosion rebellion promptly. An annealing process can enhance durability against corrosion by causing solute discrimination at twin boundaries and reducing crystal defects [4]. Mosavat et al. examined the resilience to corrosion of electrodeposited nanocrystalline Zn–Ni alloy films with a standard dimension of grains ranging from 13 to 55 nm [5]. Figure 3c illustrates polarisation resistance vs. Ni content weight percentage. Grain position and other microstructural variations, such as subsequent phases, will influence corrosion length. When aggressive conditions assault the edges of a material's grains, the speed at which its characteristics degrade is determined mainly by the grains' dimensions.

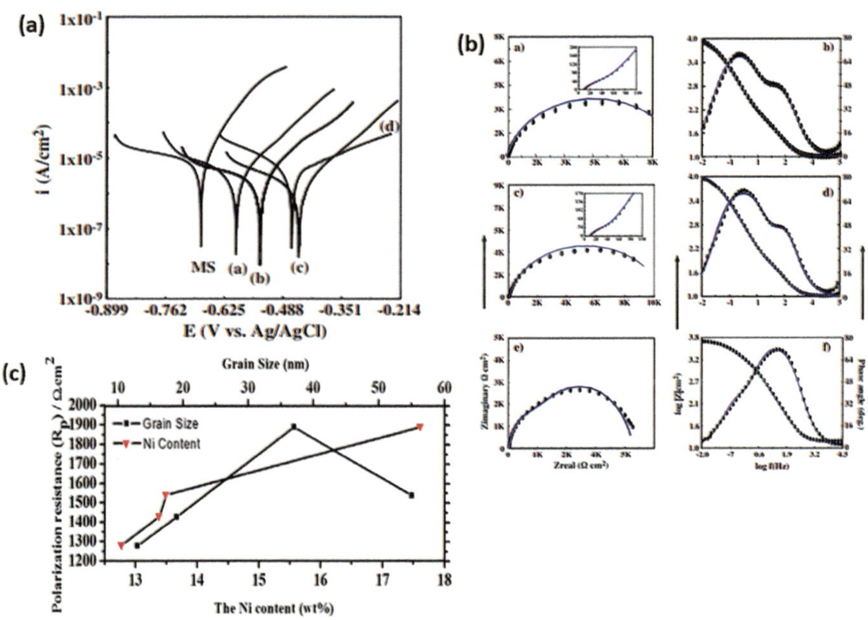

Fig. 3 **a** Tafel plot, **b** EIS plots Co–P films on MS wafers for pulse deposition of Co–P (7, 9, 12 wt.% P) [3]. **c** Effect of grain size and Ni content on polarisation resistance [5]

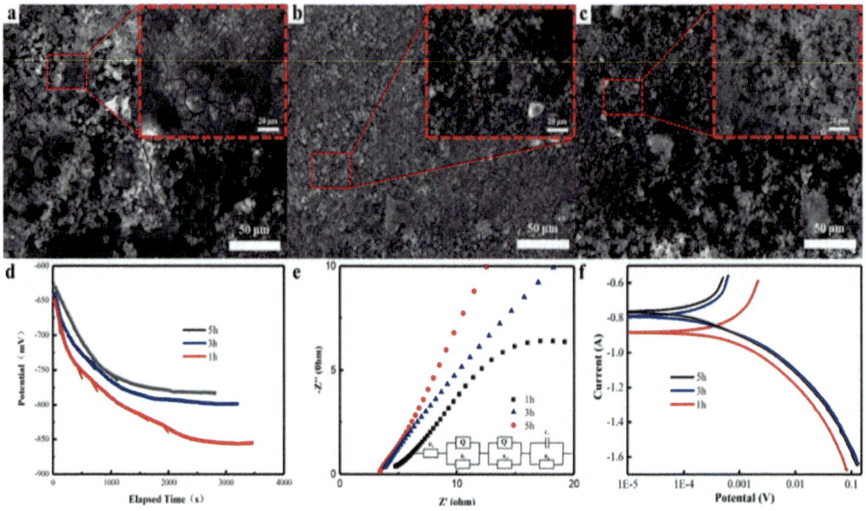

Fig. 4 **a**–**c** SEM of the specimens after electrochemical corrosion; **d** Open circuit potential of specimens **e** Nyquist plots; **f** Polarisation curves of specimens with varying annealing time (1, 3, 5 h) [7]

Because corrosion is an electrochemical process, its progression rises with temperature [6]. Time has an impact on the corrosion degree. The expression for the degree of corrosion is a function of time. Wang et al. investigated the impact of Yttrium concentration and an 800ºC thermal treatment method on the microstructure, insertions, and resilience to corrosion of shipbuilding steel, in addition to the electrochemical variables on corrosion performance [7]. He stated that after a 3-h heating procedure, the specimen treated with a 1-h heating process (at 800 ºC) showed fewer slope characteristics and more significant fractures. Figure 4a-c represents the SEM images of the samples after the corrosion process. After a 5-h heating process, the specimen developed a consistent, thick film of corrosion protection on the outermost layer, which can significantly reduce the pace at which ions diffuse and effectively shield the matrix from corrosion. The open-circuit voltage graph progressively advances (Fig. 4d). The specimen that is 5 h long has a higher voltage and exhibits the most significant susceptibility to corrosion (Fig. 4e and f). Temperature rises cause a boost in the diffusion coefficient of oxygen in the reaction, thereby limiting current density, which prevents the formation of a dense corrosion layer and speeds up corrosion. In certain circumstances, intensity polarisation will cause the pace of corrosion to rise as the corrodent's speed increases across the metals' interface. Because the higher velocity moves the cathodic polarisation contour to meet the anodic polarisation in the active corrosive zone eventually, it may nevertheless end in reduced concentration levels with inert materials.

The coating technique additionally influences the coating's corrosion characteristics. Concentrating cells may develop more readily in rough exteriors or narrow voids. Since films might serve as commencement locations, substrate purity is essential. Biological development alters the underlying surface chemistry and acts as an investment, which can accelerate corrosion. Microscopic fluctuations in the metal exterior might affect the corrosion mechanism. Microstress can affect the progression of normal corrosion, but is necessary for stress fractures or endurance. Although most corrosion rates are modest and nearly undetectable daily, others are swift and aggressive.

The durability of a substance and associated corrosion outcomes due to the potential and pH of aqueous solutions are visually represented by potential/pH illustrations, also known as Pourbaix graphs [8]. The potential is displayed on the vertical plane, and the pH is displayed on the horizontal. Since Pourbaix diagrams make identifying which species is dominant at equilibrium for a particular voltage at pH simple, they are commonly utilised in corrosion research. Wu and colleagues investigated the effects of a deaerated 0.1 M KF solution's pH (pH = 5, 7.5, 9) on the architecture, chemical makeup, and resilience of fluoride coatings formed on AZ31 Mg alloy at −1.4 V versus SCE [9]. Films grew narrower as the pH approached 9, which led to a decline in the film's defensive qualities. Concerning its chemical makeup and arrangement, the fluoride coating formed at pH 7.5 provided the optimum protection for the AZ31 Mg alloy. The Bode diagrams of the EIS of the AZ31 Mg alloy with various fluoride coatings (collected at varied pH) upon immersing in a 0.1 M NaCl solution are shown in Fig. 5a and b. Potentiodynamic polarisation research by Chianpairot et al. showed that, in an acidic atmosphere, the degree of nc Ni–W corrosion typically falls alongside [10]. Figure 3c and d shows the potentiodynamic profiles of mc Ni and several nc Ni–W compositions in a 3.5-weight percent NaCl solution at pH 10 and 3.

5 Calculation of Corrosion Parameters

In electrochemical kinetics, the Tafel model establishes a relationship between the overpotential and the progression of an electrochemical process. After being inferred empirically, the Tafel formula was eventually proved to be founded on an underlying theory. In honor of Swiss scientist Julius Tafel, the equation bears his name. For a straightforward, unimolecular redox process, it explains how the potential differential between the electrode's surface and the bulk electrolytes determines the electrical current flowing across the electrode's surface.

$$Ox + ne^{-} \leftrightarrows \text{Red} \tag{14}$$

The Tafel model is utilised on every electrode independently in cases when an electrochemical response splits into multiple half-reactions on different electrodes (Fig. 6). The Tafel expression on a single electrode is expressed as

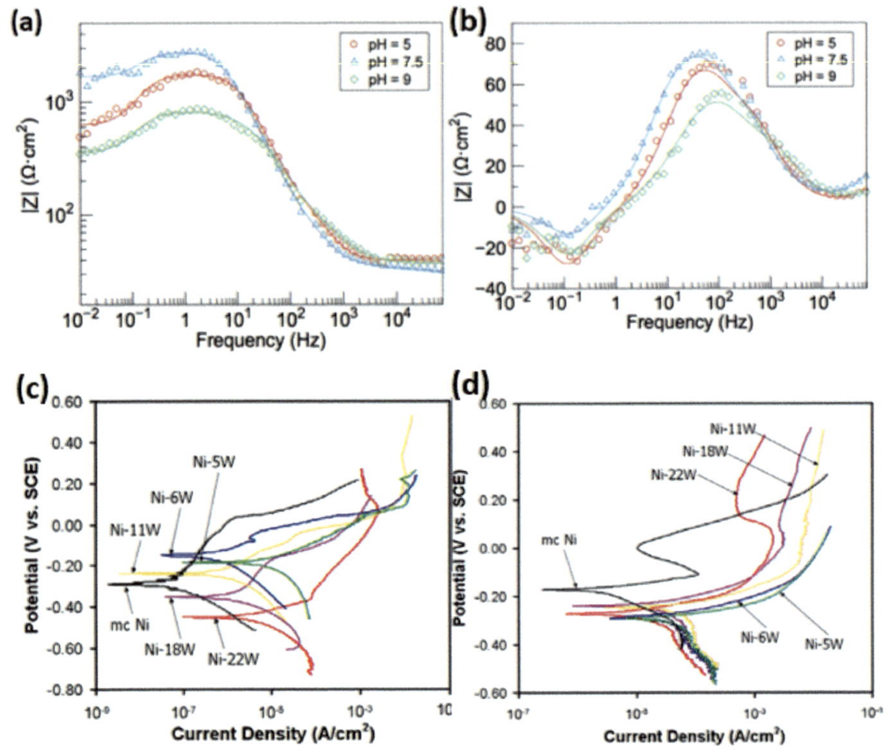

Fig. 5 **a** and **b** pH-dependent bode plots in 0.1 M NaCl solution [9], Potentiodynamic curves of mc Ni and various nc Ni–W alloys in 3.5 wt.% NaCl solution **c** at pH 10 **d** at pH 3 [10]

$$\eta = \pm A log\left(\frac{i}{i_{corr}}\right) \tag{15}$$

A (+) sign underneath the factor indicates an anodic response and a cathodic reaction by a negative (−) sign. The overpotential is represented by η, the Tafel inclination by *A*, the current density in A/m^2, and the exchange current densities in A/m^2 by i_0. When $|\eta|>0.1\ V$, the Tafel formula approximates the Butler-Volmer solution.

5.1 Corrosion Potential

The corrosion potential, often denoted as E_{corr}, is a measure used in electrochemistry to predict a metal's propensity to corrode in a particular setting. It displays the optimum potential of a metal-corroding electrode concerning a reference electrode, often a saturated calomel electrode (SCE) or a standard hydrogen electrode (SHE).

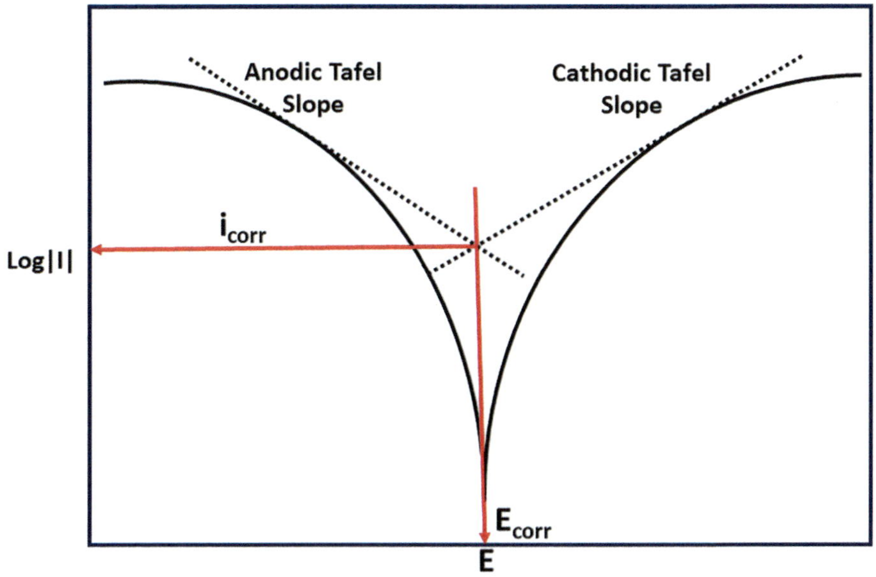

Fig. 6 Tafel plot for corrosion parameter analysis

The likelihood of metal corroding increases with a decrease in corrosion potential. E_{corr} is typically determined through electrochemical methods such as potentiodynamic polarisation or impedance spectroscopy. It's an essential parameter in corrosion science and engineering as it helps understand and predict metals' corrosion behavior in various environments.

5.2 Corrosion Current

Corrosion current, often denoted as i_{corr} is a term used in electrochemistry and corrosion science to describe the rate at which a material corrodes under specific conditions. It represents the flow of electrons associated with the corrosion reaction occurring at the metal surface. In electrochemical terms, corrosion current is typically determined utilising methods like potentiodynamic polarisation or electrochemical impedance spectroscopy. It's an essential parameter in corrosion studies as it provides quantitative information about the corrosion rate of metal under given situations. Corrosion current density (i_{corr}) is often expressed in units of amperes per unit area (such as mA/cm^2) and can be used to assess the severity of corrosion, evaluate the effectiveness of corrosion control measures, or compare the corrosion resistance of different materials.

The corrosion currents are required to compute the corrosion rates. The Tafel slope technique may determine the corrosion currents after understanding the reaction mechanisms behind the corrosion response. The Butler-Volmer equation provides the connection between the current (i) and the overpotential (η) under charge transfer (kinetic, or activation) control;

$$i = i_{corr}(e^{2.303\frac{\eta}{ba}} - e^{-2.303\frac{\eta}{bc}}) \quad (16)$$

where *icorr* is the corrosion current, the rate of charge transfer at the corrosion potential *Ecorr*; 2.303 is the conversion factor between *loge* and *log*10; η (*V*) is the overpotential, expressed as the separation between applied potential *E* and the corrosion potential E_{corr} and b_a (*V*) and b_c (*V*) are the Tafel slopes of the anodic and cathodic responses, accordingly. The corrosion potential E_{corr} is the open circuit potential of a corroding sample. The corrosion current i_{corr} and the Tafel slopes b_a and b_c can be computed from the experimental result.

The Tafel model for the anodic process optimises the Butler-Volmer solution for substantial anodic overpotentials (η/*ba*″ 1) and is represented as Eq. (17).

$$\eta = b_a log\left(\frac{i}{i_{corr}}\right) \quad (17)$$

Similarly, the Tafel solution for the cathodic response is provided for high cathodic overpotentials (η/b_c″−1);

$$\eta = -b_c log\left(\frac{i}{i_{corr}}\right) \quad (18)$$

The Tafel predicts a straight line above two equations for changing the current's logarithm with potential. Consequently, Tafel plots, which are semi-logarithmic graphs, are frequently used to display currents. Tafel slope evaluation is the name given to this kind of study.

5.3 Corrosion Rate

The pace at which a metal degrades due to corrosion is known as its corrosion rate. The kinetics of both cathodic (reduction) and anodic (oxidation) processes determine it. Faraday's law states that the corrosion current, C, and the metal dissolving rate, R, or corrosion rate (m*ml*/*year*), have a linear connection:

$$R_M = 3.17E^{-9}\frac{M}{\eta F \rho A} i_{corr} \quad (19)$$

where $3.17E^{-9}$ is the conversion factor from $cm\ s^{-1}$ to $mm\ year^{-1}$, M ($g\ mol^{-1}$) is the atomic weight of the samples, *n* is the total amount of electrons traded in the process, ρ ($g\ cm^{-3}$) is the density of the samples, *F* (96,485 $C\ mol^{-1}$) is the Faraday constant, and A (cm^2) is the sample's region. Another term for the ratio M/n is comparable weights.

5.4 Polarisation Resistance

Polarisation resistance is used in electrochemistry and corrosion science to describe the resistance to electron flow at an electrode surface under polarisation conditions. It measures how much a metal resists corrosion when subjected to an electrical potential or current. Higher polarisation resistance indicates better corrosion resistance, which means the metal is less prone to corrosion under the given conditions.

Mathematically, polarisation resistance (R_p) can be calculated from the Tafel slopes obtained from potentiodynamic polarisation curves or electrochemical impedance spectroscopy measurements. It's often used in corrosion studies to evaluate the effectiveness of corrosion inhibitors, coatings, or other protective measures applied to metal surfaces.

$$R_P = \frac{1}{i_{corr}}\left[\frac{b_a b_c}{2.303(b_a + b_c)}\right] \tag{20}$$

Here, ba and bc are the anodic and cathodic Tafel slopes.

6 Types of Corrosion

There are nine types of corrosion that any coating material can experience. Those are listed below.

1. Uniform corrosion
2. Intergranular corrosion
3. Galvanic corrosion
4. Crevice corrosion
5. Pitting corrosion
6. Erosion corrosion
7. Stress corrosion cracking
8. Biological corrosion
9. Selective leaching.

Specific substances or alloys may also experience additional kinds of corrosion. Corrosion may often be prevented or controlled by using suitable building materials, employing appropriate layout and setup methods, adhering to certain in-plant protocols, or amalgamating factors.

6.1 Uniform Corrosion

A coating substance prevents corrosion by creating a passive layer on the outermost layer. A naturally occurring coating forms when the metal is subjected to air for a while. A further method for forming it swifter is chemical processing. One kind of corrosion in which the substance corrodes uniformly along its outermost layer is called uniform corrosion. It is characterised by a relatively constant rate of corrosion over the entire exposed area of the material. In uniform corrosion, the metal surface gradually deteriorates, resulting in a consistent loss of material thickness [11].

This type of corrosion happens when the substance's surroundings remain relatively consistent, such as when the pH, temperature, and concentration of corrosive materials (such as acids or salts) are uniform across the surface. Uniform corrosion often takes the form of generalised rusting or tarnishing of metals.

Factors contributing to uniform corrosion include exposure to moisture, oxygen, acidic or alkaline substances, and various environmental pollutants. It is commonly observed in outdoor structures, pipelines, automotive bodies, and other metal components exposed to atmospheric conditions.

While uniform corrosion may not lead to sudden failure, it can still weaken the structural integrity of materials over time, leading to issues such as thinning of metal surfaces and reduction in load-bearing capacity. Preventive measures against uniform corrosion typically involve the usage of inhibitors and barrier coatings, alloying elements, or cathodic protection systems to mitigate its effects.

6.2 Intergranular Corrosion

Intergranular corrosion occurs along the grain boundaries of a metal rather than on the grain surfaces themselves. Grain boundaries are regions where the crystal structure of a metal changes from one orientation to another, and these areas are often more chemically reactive than the grain interiors. Intergranular corrosion can be particularly problematic because it can significantly weaken the material without much visible surface damage.

Intergranular corrosion typically occurs due to the segregation of impurities or alloying elements at the grain boundaries during the metal's processing or exposure to corrosive environments. These impurities can form localised galvanic cells, where the grain boundaries act as cathodic sites and the grain interiors act as anodic sites, promoting corrosion along the boundaries.

Several factors can contribute to intergranular corrosion, including:

- Sensitisation: Sensitisation occurs when certain alloying elements, such as chromium in stainless steels, migrate to the grain boundaries, depleting the surrounding regions of these elements and making them more susceptible to corrosion.

- Welding: The heat-affected zones around welds can become depleted in alloying elements and more prone to intergranular corrosion, particularly in certain types of stainless steels and other alloys.
- Environmental factors: Corrosive environments, such as those containing chlorides or acids, can exacerbate intergranular corrosion by accelerating the degradation of grain boundaries.

Intergranular corrosion can significantly deteriorate structural integrity and metal breakdown, especially in critical applications such as aerospace, automotive, and chemical processing industries. Preventive measures against intergranular corrosion include using corrosion-resistant alloys, proper heat treatment procedures, and selecting suitable welding techniques and filler materials to minimise sensitisation effects. Additionally, post-weld heat treatment and the application of corrosion inhibitors can lessen the likelihood of intergranular corrosion in susceptible materials.

6.3 Galvanic Corrosion

Whenever two distinct metals come into electrical interaction with one another and are subjected to an electrolyte—such as humidity or a destructive solution—an electrochemical procedure called galvanic corrosion, often referred to as bimetallic corrosion or different metal corrosion, takes place. One of the metals serves as an anode in this procedure and sheds electrons through corrosion, whereas the additional metal serves as the cathode and is mostly unharmed.

The mechanism of galvanic corrosion is the flow of electrons from the anodic metal (the one more likely to undergo oxidation) to the cathodic metal (the one more likely to undergo reduction) through the conductive medium (the electrolyte). This flow of electrons causes the metal ions to dissolve from the anodic metal into the electrolyte, resulting in corrosion.

Parameters that affect the severity of galvanic corrosion involve the electrochemical potential difference between the two metals (known as the Galvanic Series), the surface area ratio of the two metals, the conductivity of the electrolyte, and environmental conditions such as temperature and pH.

Galvanic corrosion results in localised destruction and degradation of the metal surfaces, which can compromise the structural integrity of components, reduce their lifespan, and result in costly repairs or replacements. It is commonly observed in various industries, including marine, automotive, aerospace, and construction.

Preventive measures against galvanic corrosion include:

- Avoiding direct contact between dissimilar metals in corrosive environments.
- Using compatible materials or applying protective coatings to minimise the potential for galvanic coupling.
- Implementing cathodic protection methods, such as sacrificial anode systems or impressed current systems, to control the corrosion potential of the metals.

- Designing structures with proper insulation or barriers to prevent the flow of electrolytes between dissimilar metals.
- To identify and treat galvanic corrosion effectively and prevent serious harm, periodic checks and upkeep are necessary.

6.4 Crevice Corrosion

Crevice corrosion is concentrated in small cracks, gaps, or stagnant areas on metal surfaces. These crevices can be formed by joints, seals, gaskets, threads, or any other irregularities where two surfaces come into close contact. Crevice corrosion is particularly problematic because it can occur even in environments with high overall stability against metal corrosion.

The mechanism of crevice corrosion involves the depletion of oxygen or other corrosion inhibitors within the crevice due to restricted diffusion or stagnant conditions. This depletion creates an environment where the metal becomes more susceptible to corrosion. Additionally, chloride ions or other aggressive species can further accelerate the corrosion process within the crevice.

Crevice corrosion typically manifests as localised attack along the metal surfaces within the crevice, often resulting in swirls or holes. The corrosion products may accumulate within the crevice, further exacerbating the corrosion process and making it difficult to detect until significant damage has occurred.

Factors that contribute to crevice corrosion include:

- Presence of crevices: Any situation where two metal surfaces come into close contact or where there are irregularities in the surface geometry can create crevices susceptible to corrosion.
- Environmental conditions: Crevice corrosion is often more pronounced in environments with high chloride concentrations, such as marine environments or industrial settings where saltwater or other corrosive solutions are present.
- Stagnant conditions: Restricted fluid flow or stagnant areas within the crevice can exacerbate the depletion of corrosion inhibitors and promote corrosion.

Preventive measures against crevice corrosion include:

- Designing components with smooth surfaces and minimising crevices or sharp edges where possible.
- Using materials with high corrosion resistance or applying protective coatings to surfaces vulnerable to crevice corrosion.
- Regular inspection and maintenance to detect and address potential crevice corrosion sites before significant damage occurs.
- Implementing proper sealing techniques or using inert fillers minimises crevice formation in joints or connections.
- Ensuring proper fluid flow and ventilation to prevent stagnant conditions within crevices.

6.5 Pitting Corrosion

A targeted kind of corrosion known as pitting corrosion is indicated by the appearance of tiny fissures or trenches on a material's exterior. The metal might be extensively punctured by these fractures, often leading to significant material loss and structural weakening, even though the corrosion rate may be relatively low.

The phenomenon of pitting corrosion involves the initiation of corrosion at localised sites on the material's exterior, typically due to small imperfections, defects, or impurities. Once initiated, the corrosion process continues to progress preferentially at these sites, creating pits that can penetrate through the covering of shielding oxidation on the outermost layer of material.

Several factors contribute to the initiation and propagation of pitting corrosion:

- Localised damage or imperfections: Surface scratches, nicks, weld defects, or other forms of mechanical damage can serve as initiation sites for pitting corrosion.
- Inclusions or impurities: Non-metallic inclusions, such as sulfides or oxides, trapped within the metal matrix can act as sites for localised corrosion initiation.
- Variations in composition or microstructure: Variations in composition or microstructure within the material, such as segregation of alloying elements or differences in grain boundaries, can create galvanic cells that promote pitting corrosion.
- Environmental factors: Pitting corrosion is often more pronounced in environments containing chloride ions, such as marine environments or industrial settings where saltwater or brines are present. Other aggressive ions or chemicals can also contribute to the beginning and spread of corrosion pits.

Pitting corrosion can be particularly insidious because it can occur even in the surroundings where the overall corrosion rate is low. The protective coatings or surface treatments may not entirely prevent pitting corrosion if they become compromised or damaged.

Preventive measures against pitting corrosion include:

- Regular inspection and maintenance to detect and address surface defects or damage that could serve as initiation sites for pitting corrosion.
- Using materials with high corrosion resistance or applying protective coatings to surfaces vulnerable to pitting corrosion.
- Controlling the environment to minimise exposure to corrosive agents, such as chlorides or other aggressive ions.
- Implementing proper design and fabrication practices to minimise the formation of defects or variations in composition that could promote pitting corrosion.
- Employing cathodic protection methods or inhibitors to mitigate the risk of pitting corrosion in susceptible materials.

6.6 Erosion Corrosion

Erosion corrosion is accelerated corrosion when the barrier oxide layer on a metal surface is continually removed or disrupted by mechanical forces, such as flowing fluids (liquid or gas) containing abrasive particles. This combination of corrosion and mechanical wear can lead to rapid material loss and degradation of the metal component.

The mechanism of erosion-corrosion involves three main components:

- Corrosion: A caustic atmosphere, such as an electrolyte solution or gas, initiates the corrosion phenomenon on the metal exterior. This typically results in a protective oxide layer, which helps to inhibit further corrosion under normal conditions.
- Fluid flow: The presence of flowing fluids, either liquid or gas, increases the metal surface's ability to lose its defensive oxide coating. The mechanical action of the fluid can physically wear away the oxide layer, exposing fresh metal to a corrosive environment.
- Abrasive particles: The fluid may contain abrasive particles, such as sand, sediment, or solid particles entrained in the flow. These particles can act as abrasives, further damaging the protective oxide layer and accelerating corrosion by promoting mechanical wear on the metal surface.
- Erosion corrosion is commonly observed in industrial environments with high fluid flow rates, such as piping systems, valves, pumps, and turbine components. It can also occur in marine environments, where the movement of seawater can erode metal surfaces, especially in areas where turbulence or impingement occurs.

  Preventive measures against erosion-corrosion include:

  - Selecting materials with high resistance to erosion and corrosion for components exposed to high-flow environments.
  - Implementing proper design and fabrication techniques minimises fluid turbulence and reduces fluid flow's impact on metal surfaces.
  - Protective coatings or linings enhance the durability against corrosion and abrasion of the material's exteriors.
  - Regular inspection and maintenance to detect and address erosion-corrosion damage before it leads to significant material loss or failure.
  - Adjusting fluid flow rates or modifying operating conditions to reduce the severity of erosion-corrosion in susceptible components.

6.7 Stress Corrosion Cracking

Tensile tension and corrosive conditions cause a corrosion breakdown known as stress corrosion cracking (SCC). Unlike most forms of corrosion, SCC involves the brittle fracture of a material, often without significant visible corrosion on the surface. This makes it particularly insidious and challenging to detect.

The mechanism of stress corrosion cracking typically involves three main factors:

- Tensile stress: SCC requires the presence of tensile stress, which can be applied externally (e.g., mechanical loading, residual stress from manufacturing processes) or arise internally due to factors such as thermal expansion or metallurgical changes. The tensile stress reduces the material's resistance to cracking, making it more susceptible to fracture.

- Corrosive environment: SCC occurs in a specific corrosive atmosphere interacting with the material to promote crack initiation and propagation. The corrosive environment can differ based on the substance and application, but often includes temperature, pH, chemical composition, and specific ions (e.g., chlorides, sulfides).
- Material susceptibility: Certain materials are more susceptible to SCC than others. Factors influencing susceptibility include the material's composition, microstructure, and mechanical properties. For example, some alloys used in the aerospace or nuclear industries are particularly prone to SCC under specific conditions.

SCC can occur in various materials, including metals, polymers, and ceramics, although it is most commonly associated with metals such as stainless steel, aluminium alloys, and certain high-strength steel.

Preventive measures against stress corrosion cracking include:

- Material selection: Choosing materials with high resistance to SCC for applications with a significant risk of cracking. This may involve selecting alloys designed to resist SCC or using alternative materials less susceptible to cracking.
- Stress reduction: Minimising or controlling the tensile stress level applied to the material through proper design, fabrication, and operational practices. This may include stress relief treatments, design modifications to reduce stress concentrations, or implementing load management strategies.
- Environmental control: Managing the corrosive environment to minimise the risk of SCC by controlling factors such as temperature, pH, chemical composition, and harmful ions. This may involve selecting corrosion-resistant coatings or inhibitors, controlling water chemistry in cooling systems, or implementing proper water treatment practices.
- Monitoring and inspection: Regularly inspecting components for signs of SCC, such as cracking, corrosion, or changes in material properties. Early detection allows for timely intervention and preventive maintenance to mitigate the risk of catastrophic failure.

6.8 Biological Corrosion

Biological corrosion, also known as microbiologically influenced corrosion (MIC), occurs due to the presence and activity of microorganisms on metal surfaces. These microorganisms, including bacteria, fungi, algae, and archaea, can colonise metal surfaces in various environments, including soil, water, and industrial systems.

The mechanism of biological corrosion involves several processes:

- Biofilm formation: Microorganisms affix to a metallic body and produce a biofilm, a slimy layer of microbial cells, extracellular polymeric substances (EPS), and other organic and inorganic compounds. The biofilm provides a protective environment for microorganisms, shielding them from external factors and facilitating their metabolic activities.

- Metabolic activities: Microorganisms within the biofilm metabolise nutrients and produce byproducts, such as organic acids, sulfides, and other corrosive substances. These metabolic byproducts can create localised chemical environments that promote metal surface corrosion.
- Electron transfer: Some microorganisms can transfer electrons between different chemical species, including metal ions and oxygen, through metabolic processes such as microbial respiration or metal reduction. This electron transfer can lead to electrochemical reactions on the metal surface, accelerating corrosion.

Biological corrosion can occur in a wide variety of industrial and natural environments, including:

- Marine environments: Seawater contains diverse microbial communities that can colonise metal structures such as ships, offshore platforms, and marine pipelines, leading to biofouling and corrosion.
- Soil environments: Microorganisms in soil can attach to buried metal structures, such as underground pipelines and storage tanks, causing corrosion and degradation.
- Industrial systems: Microorganisms can thrive in industrial environments such as cooling water systems, pulp and paper mills, and oil and gas production facilities, where they can cause corrosion of metal equipment and infrastructure.

Preventive measures against biological corrosion include:

- Biocide treatment: Using biocides or antimicrobial agents to control the growth of microorganisms and prevent biofilm formation on metal surfaces.
- Cathodic protection: Implementing cathodic protection systems to mitigate the electrochemical reactions that drive corrosion, thereby reducing the susceptibility of metal surfaces to microbial attack.
- Surface coatings: Applying protective coatings or linings to metal surfaces to create a barrier against microbial attachment and corrosion.
- Environmental control: Managing ecological factors such as temperature, pH, and nutrient availability to minimise the growth of microorganisms and their corrosive activities.
- Regular monitoring and maintenance: Conduct regular inspections and maintenance of metal structures to detect and address microbial colonisation and corrosion before significant damage occurs.

6.9 Selective Leaching

A kind of corrosion procedure called preferential dumping is often called the dealloying process or separating corrosion, in which one component of an alloy is preferentially removed from the material, leaving behind a porous and weakened structure.

This type of corrosion occurs when the alloy contains two or more metals with significantly different electrochemical potentials, and it typically happens in aggressive environments such as acidic or alkaline solutions.

The mechanism of selective leaching entails the breakdown of a single alloying material from the material while leaving the other elements relatively unaffected. This preferential dissolution occurs due to the formation of galvanic cells within the alloy, where another material functions as the cathode and is shielded from corrosion, whereas another material serves as the anode (sacrificial material).

Selective leaching can lead to the formation of porous structures or networks of tunnels within the material, weakening its mechanical properties and compromising its structural integrity. The extent and rate of selective leaching depend on various factors, including the composition of the alloy, the type of corroding surroundings, and the circumstances of pressure and temperature.

Common examples of selective leaching include:

- Dezincification: Dezincification is a type of selective leaching that occurs in brass alloys containing zinc and copper. In aggressive environments such as acidic or low-pH solutions, zinc preferentially dissolves from the alloy, leaving behind a porous and weakened copper-rich structure.
- Dealloying of aluminium alloys: Certain aluminium alloys containing alloying materials such as copper, silicon, or magnesium are susceptible to selective leaching in alkaline environments. This can result in removing the less noble alloying elements, leaving a porous aluminium matrix behind.

Preventive measures against selective leaching include:

- Alloy selection: Choosing alloys with compatible electrochemical properties and similar corrosion resistance to minimise the risk of selective leaching.
- Environmental control: Avoiding exposure to aggressive environments or controlling surrounding parameters such as pH, temperature, and chemical composition to minimise the likelihood of selective leaching.
- Surface treatments: Applying protective coatings or surface treatments to the material creates a barrier against corrosive attack and inhibits selective leaching.

Regular inspection and maintenance: Conduct regular inspections and maintenance of alloy components to detect signs of selective leaching and address any corrosion damage before it leads to structural failure.

7 Types of Nanocoatings for Corrosion Resistance

Nanocoating materials are generally categorised into metallic, ceramic, and nanocomposite coatings.

7.1 Metallic and Their Composite Coatings

A metallic coating refers to a layer of metal applied to a substrate for various purposes, such as corrosion protection, conductivity enhancement, or decorative purposes. Metallic coatings can be used through different processes, each with advantages and applications. Metallic coatings involve one or more than one of the pure metals such as Cd, Ni, W, Zn, P, Co, Fe, Cu, etc. [12, 13]. Pure metal or metal that has been alloyed to improve certain qualities can be used for nanotechnologies. It is possible to create metallic coatings by multiple techniques such as sputtering, CVD, electrodeposition, and multi-arc ion plating. Introducing metals in nanocoating improves their physical [14], chemical [15], mechanical [16], and thermal [17] properties. The welding capability and workmanship issues that occur while the layer thickness of the zinc barrier must be raised to offer the required safeguards are resolved by a lighter zinc coating [15]. Because pure zinc's resilience to corrosion is inadequate under extreme oxidation and elevated temperatures, it is amalgamated with metallic components to increase its durability against corrosion [15]. It has been demonstrated that Zn–Ni alloy films are harder than zinc and cadmium alone.

Additionally, due to cadmium's inadequate sticky resilience to wear, zinc-nickel alloys are the front-runners to replace it for aeronautical purposes [16]. Another example of efficient nanocoating is nanocrystalline cobalt and its alloys. They have been recognised as a cost-effective substitute for diligently, poisonous chromium that is resistant to wear and corrosion. At levels below the nanoscale, alloyed metal has better corrosion resistance than pure metal. When contrasted with particular pure metals, the $(Cu_{0.6}Hf_{0.25}Ti_{0.15})90Nb10$ BMG composite, for instance, exhibits exceptional durability against corrosion [18]. The development of Hf-, Ti-, and Nb-enriched highly repellent surface coatings underexposure in media comprising acid and chloride ions is responsible for their outstanding resistance [18]. When nickel and copper are alloyed, nickel is more resistant to corrosion in water-based environments and experiences less blistering in saltwater [19].

Furthermore, phosphorous-nickel alloying facilitates the protective coating's passivation in corrosive situations more quickly than nickel [20]. Rajaraman et al. used polarisation tests and EIS to assess the corrosion characteristics of HfN film on titanium screws [21]. According to his study, the group capped with HfN exhibited the most potent repellent to corrosion because of its greater resistance, with an average E_{corr} of −0.452 V and an i_{corr} of 0.0354 $\mu A/cm^2$. Uncoated Ti screws showed $E_{corr} = -0.088$ V, and $i_{corr} = 0.1527$ $\mu A/cm^2$, respectively [21]. The EIS and polarisation plots of HfN coating over titanium screws and uncoated Ti screws are represented in Fig. 7. The corrosion parameters of some metals and their composite films are described in Table 1.

Metallic coating made with Nickel film can enhance the oxygen resistance by a factor of two and corrosion resistance by a factor of 3 of hot work tool steels (HWT) in red sea water [31]. The Tafel plot of the HWT steels with and without Nickel coating is shown in Fig. 8. More investigation was done on the impact of grain size at various nanoscale levels. Figure 9 shows the optical and SEM micrographs of the

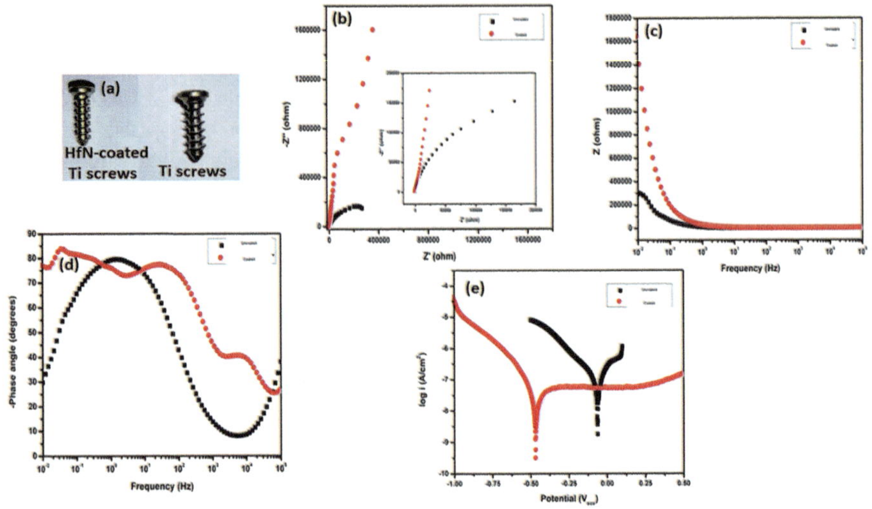

Fig. 7 **a** Real image of HfN-coated titanium screws and commercially pure titanium screws, **b** Nyquist plot, **c** Bode impedance, **d** Bode phase, **e** Tafel plot of HfN-coated titanium screws and commercially pure titanium screws [21]

coating of substrate, coating, and epoxy and lamellar structure and oxides of coating, respectively.

7.2 Ceramic Coatings

Ceramic coatings are protective coatings that include inorganic materials, typically oxides, carbides, nitrides, or silicides, which form a thin layer over a substrate surface. These coatings are known for their exceptional hardness, high-temperature resistance, corrosion resistance, and thermal insulation properties. Ceramic coatings are extensively employed across several sectors, including automotive, aerospace, electronics, and manufacturing. Ceramic nanoparticles scattered throughout an adhesive or substrate substance make up ceramic coatings. The ceramic particles provide hardness and durability, while the binder helps adhere the coating to the wafer. Ceramic coatings can be applied using several techniques, including thermal spraying (like plasma or flame spraying), CVD, PVD, dip coating, and sol–gel processes. Each method offers unique coating thickness, uniformity, and application efficiency advantages. Because of considerably increased toughness and durability, ceramic oxide films provide an improved surface than metallic or organic oxide treatments despite their reduced thickness. Ceramic coatings are currently used in many sectors because of their appealing thermal and electrical qualities and more excellent resistance to oxidation, corrosion, and wear than metals in elevated temperature settings. The corrosion resistance of films made of Cr-CrN, Ti-TiN, Zr-ZrN, and

Table 1 Corrosion parameters of some metals and their composite coatings

	Coatings	Substrates	Coating method	Electrolyte	E_{corr} (V)	I_{corr} (μA/cm^2)	Remarks	Refs.
1	HfN	Ti screws	Chemical synthesis	Simulated body fluid (SBF)	−0.452	0.0354	HfN-coated Ti screws showed increased durability against corrosion and impedance	[21]
2	Ta-coating	Ti6Al4V	Microplasma spray	4 M NaCl	–	0.0034	Ta-coating on Ti6Al4V showed better corrosion than Ti-coated Ti6Al4V and uncoated Ti6Al4V substrates	[22]
3	Ni–Ti coatings	316L stainless steel	Mechanical alloying method	3.5% NaCl	−0.364	9.5	The coatings exhibited higher corrosion potential compared with the 316L substrate (about −0.85)	[23]
4	Ni/CrC	52,100 steel	Cold Spraying	DI water + 0.5 mol/L H_2SO_4 + 0.5 mol/L HNO_3	−0.046	3.9	The Ni/CrC on 52,100 steels showed lower I_{corr} and higher E_{corr} than the uncoated sample (0.568 V and 18.37). The coated sample shows less corrosion rate (18.95 mpy) as compared to the uncoated sample (88.48 mpy)	[24]

(continued)

Table 1 (continued)

	Coatings	Substrates	Coating method	Electrolyte	E_{corr} (V)	I_{corr} ($\mu A/cm^2$)	Remarks	Refs.
5	TiNii–xCu	Tii–Ni	Laser surface alloying	0.9 wt% NaCl solution	−0.4296	0.0949	Because of its most vulnerable inactive coating and the most significant proportion of $TiNi_2Cu$ that dissolves preferentially, TiNi–26.0 Cu exhibits an elevated rate of corrosion	[25]
6	Ni–Coi–TiN/CeO_2	Ni–Co matrix	Jet electrodeposition	3.5 wt% NaCl solution	−0.428	10.07	When contrasted with the Ni–Co framework, the anti-corrosion effectiveness of Ni-Co–TiN/CeO_2 coatings is outstanding	[26]
7	WC/Ni	Q345R steel	Oscillating laser-cladding	3.5 wt.% NaCl solution	−0.472	4.769	When 20% WC is added, polarisation resistance is 16% lower than that when 10% WC is added	[27]

(continued)

Table 1 (continued)

	Coatings	Substrates	Coating method	Electrolyte	E_{corr} (V)	I_{corr} ($\mu A/cm^2$)	Remarks	Refs.
8	Fe-based coating	Q235 steel	High-speed laser cladding	3.5% NaCl solution	−0.3370	1.52	The cladded Fe-based films exhibit a greater barrier to corrosion with a smaller I_{corr} and a substantially larger E_{corr} as contrasted with Q235 steel	[28]
9	Cr–Fe–Ni–Co–Mo–Si–B amorphous coating	316L steel	Atmospheric plasma spraying	3.5 wt% NaCl solution	−0.506	5.609	The amorphous coating exhibited the lowest I_{corr} and the maximum E_{corr} when contrasted with subjected heat films. Because of the comparatively large amount of Cr_2O_3, the passivity film on the amorphous coating supplied the most effective shielding	[29]
10	WCi–Cr_3C_2 composite coating	316L stainless steel	High-velocity oxy-fuel (HVOF) spraying	36% HCL solution	−0.2591	21.933	The ability to withstand corrosion rose steadily as the Cr_3C_2 hard phase grew, the E_{corr} increased, and the I_{corr} dropped	[30]

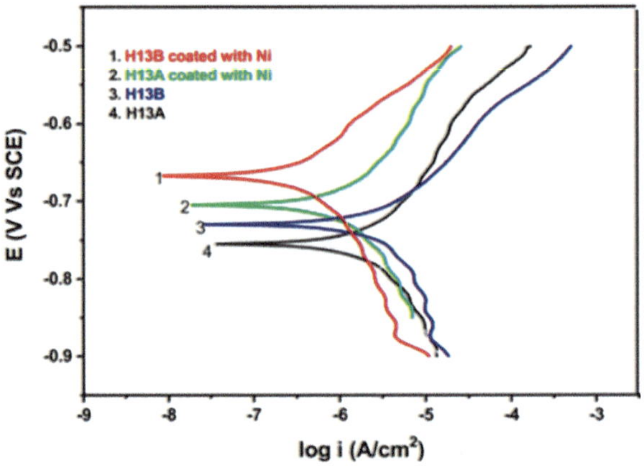

Fig. 8 Tafel plot of the HWT steels with and without Nickel coating [31]

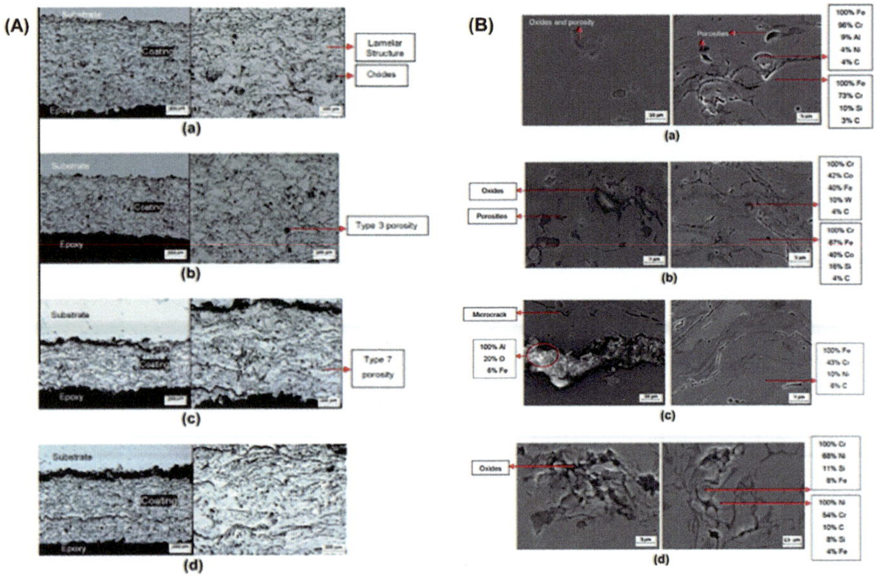

Fig. 9 Optical and SEM micrograph of coating: **a** substrate, coating, and epoxy and **b** lamellar structure and oxides of coating [32]

Mo-MoN on Ti-6Al-4 V alloy was investigated by Tao et al. using the PVD (Filtered vacuum cathode deposition (FCVAD) + controlled accelerated arc deposition (CAA-PVD)) technique [33]. The Zr-ZrN coating, according to him, possesses the highest corrosion resistance in the 3.0% NaCl solution, with polarisation resistance of 210.9 kOhm, corrosion penetration of 1.09 μm/year, and I_{corr} of 0.123 $\mu A/cm^2$, correspondingly. Employing the selective laser melting (SLM) and sol–gel processes,

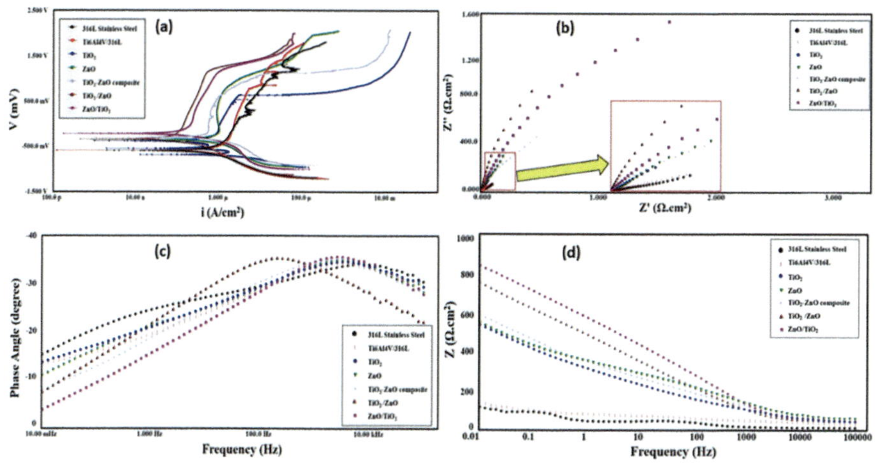

Fig. 10 **a** Tafel, **b–d** EIS plot of ZnO/TiO2 treated and untreated Ti6Al4V/316L substrates [34]

Yetim et al. developed single and duplex ZnO/TiO_2 ceramic coatings on 316L stainless steel and Ti6Al4V platform [34]. The potentiodynamic polarisation curves and EIS plots of ZnO/TiO_2 treated and untreated Ti6Al4V/316L substrates are represented in Fig. 10a–d. The corroded microstructure of ZnO/TiO_2 treated and untreated Ti6Al4V/316L substrates is expressed in Fig. 11 of all the coated designs, and the ZnO/TiO_2 multilayer coating had superior tribological and corrosion resistance. Following the sol–gel coating, the specimens' corrosion performance was improved by the ceramic barrier coatings made of the Ti6Al4V/316L surface. Compared to unprotected substrates, surfaces coated with monolayers (TiO_2 and ZnO) and composites (TiO_2-ZnO) incurred fewer harms.

Furthermore, the specimens decorated with many layers (ZnO/TiO_2, TiO_2/ZnO) were primarily intact, and the inflexible ceramic layer remained intact during the corrosion test [34]. Yan et al. observed enhanced corrosion resistance at a current density of 15 A dm^{-2} after examining the impact of micro-arc oxidation conditions on the morphology and characteristics of Al_2O_3 films produced on sintering 2024 aluminium alloys [35]. Nayak et al. studied the effect of coating thickness of alumina (Al_2O_3)-yttria-stabilised zirconia (YSZ) atmospheric by plasma spray (APS) on a cast iron substrate for thermal barrier coating application [36]. He stated that TBC's resilience to thermal deterioration has been improved and molten salt cannot penetrate YSZ due to the dense Al_2O_3 outermost layer covering YSZ in the YSZ/Al_2O_3 composition [36]. In Feng et al., supersonic plasma sprayed Al_2O_3 films on Q235 lower carbon steel substrates using an alternate sealant coating method [37]. The EIS and immersed corrosive study support the conclusion that the alternate sealing method successfully stops connection gaps from forming.

Furthermore, compared to conventional sealant coatings, it has a more substantial inhibitory impact on the migration inside corrosive fluids. Improved permanent durability against corrosion is conferred by the alternate sealed covering, which has

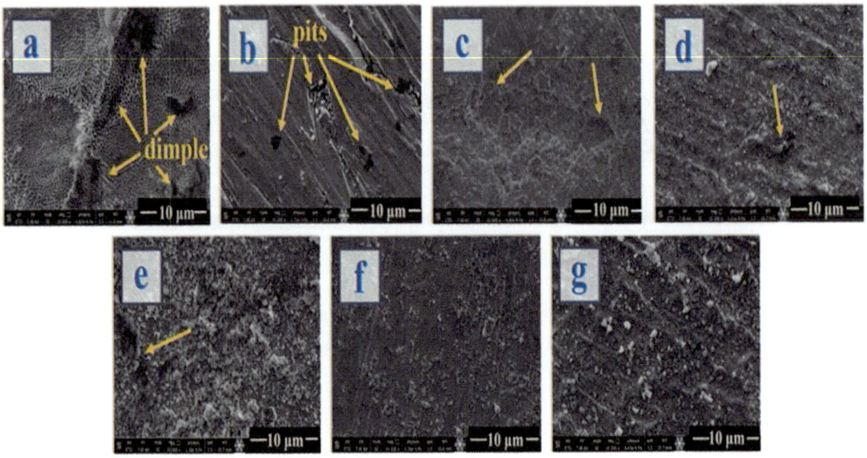

Fig. 11 The corroded surfaces SEM image of ZnO/TiO_2 treated and untreated Ti6Al4V/316L substrates [34]

a permeability drop of 70.5%, a rise in corrosive potential to 74 mV, and a reduction in corrosion current density of almost 78% in contrast to the typical sealed film. The corrosion parameters of some ceramics and their composite coatings are represented in Table 2.

7.3 Non-metallic Coating

A non-metallic coating refers to any material applied to a surface to provide protection, insulation, or aesthetic enhancement that does not contain metal as a primary component. These coatings can contain various substances, such as polymers, ceramics, or organic compounds. They are often used for thermal insulation, anti-corrosion, and cosmetic reasons in the automobile, aviation, building, and semiconductor sectors. Examples of non-metallic coatings include paints, varnishes, enamels, and polymer coatings like Teflon.

Non-metallic films are often employed for corrosion resistance in various applications. Some standard non-metallic films used for corrosion resistance include:

1. Paints and Primers: Many paints and primers contain corrosion-resistant additives or are explicitly formulated to protect metal surfaces from corrosion. Epoxy, polyurethane, and acrylic paints are commonly used for this purpose.
2. Polymer Coatings: Polymer coatings such as polyethylene, polypropylene, and fluoropolymers like Teflon offer excellent corrosion resistance and are often used in multiple sectors, such as chemical processing, oil and gas, and marine applications.

Table 2 Corrosion parameters of some ceramics and their composite coatings

	Coatings	Substrates	Coating method	Electrolyte	E_{corr} (V)	I_{corr} ($\mu A/cm^2$)	Remarks	Refs.
1	TiO_2 Nano Tube/Cu–Chitosan coating	Ti6Al4V sheet	Anodic oxidation (AO) + electro-cathodic deposition (ECD) + electrophoretic deposition (EPD)	Simulated body fluid (SBF)	−0.4141	11.3785	The TNT−Cu2/90/Ch sample showed an excellent corrosion-preventing efficacy of 79.81% and an inadequate corrosion progression of 0.0611 mm/year. due to protective chitosan coating	[38]
2	TiN composite coating	AISI 420	Selective laser melting	6 wt% ferric chloride solution	–	–	TiN/AISI 420 composite coating demonstrates improved mechanical and corrosion resistance	[39]
3	Cr−CrN	Ti−6Al−4 V alloy	Filtered vacuum cathode deposition (FCVAD) + controlled accelerated arc deposition (CAA-PVD)	3% aqueous NaCl solution	−0.026 (Cr−CrN), 0.095 (Mo−MoN), −0.078 (Ti−TiN), −0.017 (Zr−ZrN)	0.218 (Cr−CrN), 8.190 (Mo−MoN), 1.480 (T−TiN), −0.017 (Zr−ZrN)		[33]
4	Ni/316L alloy	Q235 steel	High-speed laser-cladding	3.5% NaCl solution	−0.305	0.89	Resilience to corrosion could be improved by annealing	[40]

(continued)

Table 2 (continued)

	Coatings	Substrates	Coating method	Electrolyte	E_{corr} (V)	I_{corr} ($\mu A/cm^2$)	Remarks	Refs.
5	ZnO/TiO_2	Ti6Al4V/316L	Selective laser melting (SLM) + sol gel process	Simulated body fluid (SBF)	−0.195	11.45	Comparing composite films to samples without treatment, a substantial decrease in I_{corr} was seen, along with improved durability against corrosion	[34]
6	Al_2O_3	Powder metallurgy (PM) 2024 aluminium alloy	Micro-arc oxidation (MAO)	3.5 wt.% NaCl solution	0.001	–	The MAO films with the maximum durability against corrosion were achieved at an I_{corr} of 15 A dm^{-2}	[35]
7	Al_2O_3 + SiO2	Q235 low-carbon steel	Sealing coating technology	5 wt.% NaCl solution	−0.490	1.8383	Increasing the thickness of the sealing layer improves the corrosion resistance	[37]
8	CrN	Plasma-enhanced magnetron sputtering (PEMS)	316L austenitic steel and silicon wafers	3.5 wt.% NaCl solution	−0.0048	0.0742	Enhanced corrosion resistance is observed	[41]

(continued)

Table 2 (continued)

	Coatings	Substrates	Coating method	Electrolyte	E_{corr} (V)	I_{corr} ($\mu A/cm^2$)	Remarks	Refs.
9	CrN-PtIr	Plasma-enhanced magnetron sputtering (PEMS)	316L austenitic steel and silicon wafers	3.5 wt.% NaCl solution	−0.2236	0.212	Compared to the CrN treatment, the CrN−PtIr film corrodes much more slowly	[41]
10	CrSiCN	Carbon steel C45, 6MnCr5 (1.7131), and X155CrVMo12	Reactive cathodic arc evaporation (CAE)	Distilled water and sand (10 g/L)	0.054	0.021	Because films with a more significant proportion of C demonstrated better corrosion resistance, the C/N ratio was crucial	[42]

3. Ceramic Coatings: Ceramic coatings provide durable protection in opposition to corrosion, abrasion, and chemical exposure. They are often applied to metal surfaces through thermal or plasma spraying.
4. Powder Coatings: Powder coatings are utilised after drying the powder and heating it to dry out to create a robust and defensive covering. They offer good corrosion resistance and are commonly used in automotive, appliance, and architectural applications.
5. Anodising: Anodising is an electrochemical process that creates a protective oxide layer on the surface of metals like aluminium. This oxide layer enhances corrosion resistance and provides a colouring or dyeing surface.
6. Organic Coatings: Organic coatings such as epoxy, polyurethane, and acrylic films offer barrier protection against corrosion and can be applied by various methods, including spraying, dipping, or brushing.

These non-metallic coatings offer a range of properties such as chemical resistance, UV resistance, flexibility, and adhesion, making them suitable for different corrosion protection requirements across various industries. In a 3.5-weight percent NaCl remedy, Savameri et al. examined the corrosion endurance evaluation of nano-silica and polytetrafluoroethylene (PTFE) films on steel substrates [43]. By plugging in a few of the coating's holes and flaws, nanoparticles made of SiO_2 increase the ability to withstand corrosion of PTFE protective coatings according to an evaluation of the impact of introducing the microscopic particles to the topcoat top. An analysis of coatings' durability against corrosion revealed that PTFE films had a higher durability against corrosion than nano-silica films. The PTFE coatings with glass bead microparticles in the primer film and nanoparticles of silicon dioxide in the topcoat layer, which was the most effective specimen generated in this investigation, decreased the extent of corrosion by about 80 times. Thicker coatings exhibit lower corrosion current densities and consequently higher corrosion resistance. Micro Al_2O_3 with nano SiO_2 exhibited a corrosion potential and current density of -0.577 V and 0.5750 $\mu A/cm^2$ respectively [43]. Figure 12a–d represents the corrosion test results and the SEM microstructure image of nano-silica and PTFE coatings on MS substrate is represented in Fig. 13. Abdolmaleki et al. Ni-W-PTFE nanocomposite films on copper by electrodeposition technique [44]. Corrosion studies showed that, in the initial stage, bath levels of PTFE (up to $8gL^{-1}$) enhanced the ability to withstand the corrosion of deposits. The homogeneous distribution of strengthened nanoparticles in sediments is primarily responsible for these improved qualities. A solution containing 8 g L^{-1} PTFE has the maximum durability against corrosion (13,293 $\Omega\,cm^2$) [44]. Li et al. created nano-zinc oxide on steel substrates by inserting 3-methacryloxypropyltrimethoxysilane (KH570), altered silicon dioxide nanoparticles through the stratification spacing. This formed nano-zinc oxide when combined with the carboxyl group in the acid stearic [45]. To create an extremely hydrophobic covering, PTFE was bonded together, which showed an impedance value of 6.87 × 109 Ω in sodium chloride solution. Following being submerged in seawater for seven days, the extremely hydrophobic coating's E_{corr} and i_{corr} values are the following:

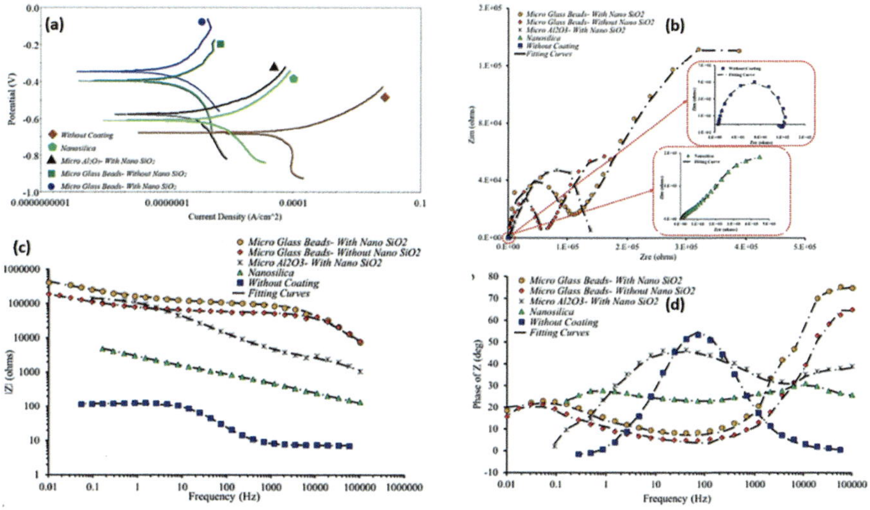

Fig. 12 **a** Tafel plot, **b–d** EIS plot of nano-silica and PTFE on MS substrate [43]

−0.227 V and 7.68 × 10−11 A cm^{-2}. Furthermore, the super-hydrophilic film on the metal surface exhibits a 99.99% corrosion prevention effectiveness [45].

8 Other Advanced Materials

There are some future promising materials that provide advanced corrosion resistance. Based on their constituent parts and atomic configurations, two-dimensional materials can be broadly divided into four groups, namely the graphene family, Xenes, chalcogenides, and 2D oxides. 2D materials such as graphene and MXenes contribute to improved corrosion resistance primarily due to their unique structural and chemical properties [11]. Graphene and MXenes have a layered structure with a high aspect ratio (thickness in nanometers, lateral dimension in micrometers) [46, 47]. This structure can form a physical barrier that restricts the diffusion of corrosive species (e.g., oxygen, water, ions) towards the underlying material substrate. This barrier effect can significantly slow down corrosion by reducing exposure to corrosive agents [48]. Graphene and MXenes can alter the electrochemical behavior of the material they protect. They can act as inhibitors or passivating agents, modifying the local electrochemical environment at the material surface [49]. For instance, they can stabilise the passive oxide layer on metals, making it less prone to dissolution in corrosive media. These 2D materials can be functionalised or chemically modified to enhance their interaction with the substrate or provide specific surface properties that resist corrosion and provide better mechanical properties [50, 51]. Functional groups or dopants on graphene or MXene sheets can tailor their affinity to certain

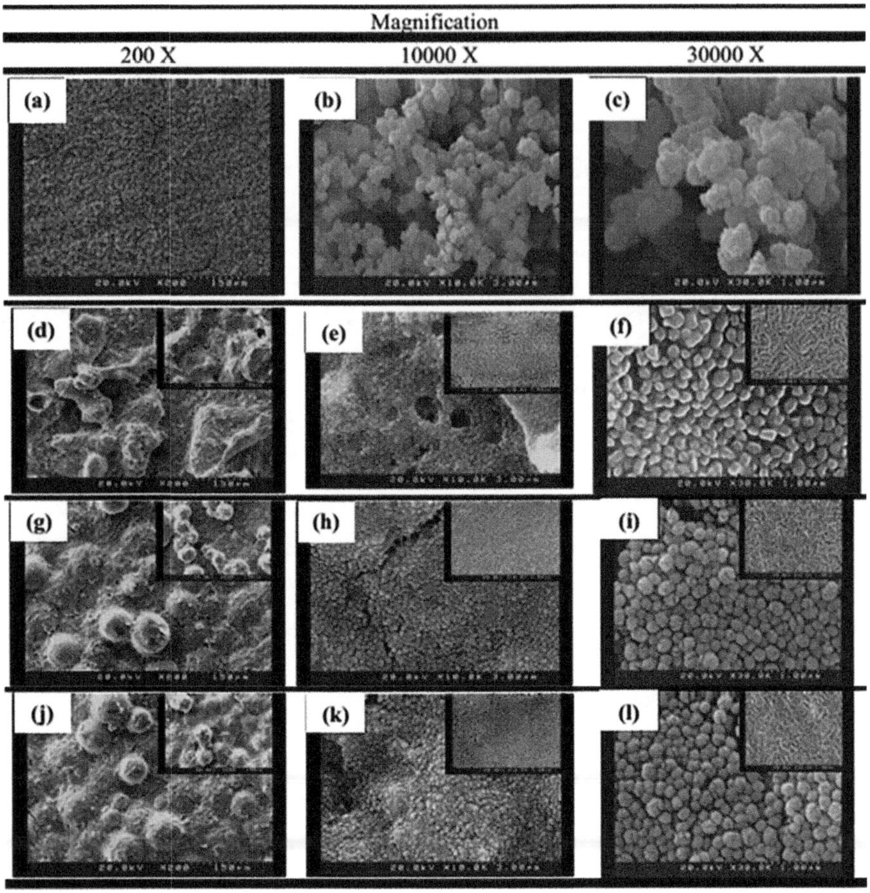

Fig. 13 SEM pictures of PTFE and nanosilica coatings on MS substrate. **a–c** nanolayer of silica 46. **d–l** Morphology pictures of the primer film (big image) and the prime and topcoat films (small image) are included in the PTFE films. **d–f** Nano SiO_2 combined with microscopic Al_2O_3 **g–i** Micro glass beads: **j–l** micro glass beads—with nano SiO_2 and without nano SiO_2 [43]

corrosive species or enhance their adherence to the substrate. When used as coatings or composites, graphene and MXenes can reinforce the mechanical properties of the material substrate, making it more resistant to physical damage that could lead to accelerated corrosion. 2D graphene and MXenes can be tailored to be compatible with various materials, including metals, polymers, and ceramics, allowing for versatile applications in different environments where corrosion resistance is critical [52]. At the nanoscale, the properties of 2D materials can be finely tuned to optimise corrosion resistance without significantly altering the bulk properties of the material they protect.

9 Future Challenges

A material's corrosion resistance determines its resilience and longevity; thus, it's critical to consider it while evaluating the material's effectiveness. The basic kinetics and principles of the corrosion phenomenon are explained by corrosion investigation. Ultrafine components found in nanocoating may have an impact on the final surface's lattice arrangement, grain size, porosity, arrangement of intermetallic fragments, surface condition, etc. The extremely thin and dense grain limits of these components make it difficult to create new corrosion hypotheses for how they relate to the exterior. For instance, smoothing the exterior improves the ultrathin authenticity, uniformity, and endurance efficiency of films, which reduces the likelihood of pit formation at such surfaces. However, the extra smoothness that comes from having nanoparticles saturating the surface could decrease the coating's compliance and lead to some of the coating's portions detaching. Multilayer coatings can provide superior corrosion resistance to single-layer coatings [53]. Moreover, a surface roughness reduction might increase the possibility of preferred intergranular degradation, allowing an additional permeable and defective film to form at the threefold junction's boundary with the grain. It can be tough to determine which of these two theories best describes the two opposing ways in how the nanomaterials engage with the surface. Furthermore, oxide formation is impacted by the existence of these nanomaterials on the substrate's surface; as a result, the alteration of the outermost state's trajectory is variable and will be challenging to identify. Induced is the change in surface condition from passivation to pit initiation and ultimately to film collapse. The current state is established based on the original condition of the untreated surface and the final coated surface's cleanliness, which will impact the overall interface corrosion. Owing to the uneven dispersion of the nanoparticles across the coated substrate's surface, ions may build up and produce weak spots of elevated potential, which in turn trigger the formation of pits. Conversely, a build-up of coating could physically separate the electrolyte ions from the substrate surface. It is difficult to estimate the changeover mechanism and more corrosion study are needed.

10 Conclusion

When it comes to improving the corrosion performance of surfaces, nanocoatings hold much more promise than micromaterial coatings. Since nanocrystalline structures have finer grain sizes, better space-filling and higher coating surface integrity than microstructures, they are better for enhancing corrosion. The substrate's surface becomes more resilient and stronger after applying nanocoating, which also improves the substrate's adhesive qualities. Nonetheless, the coating's composition and thickness must be planned to avoid lessening its resistance to corrosive and eroding forces. Nanocoatings achieve improved corrosion resistance by various methods, some of which may have unfavorable impacts. The nanocoatings' small diameters on the material's surface provide a consistent physical barrier.

Additionally, due to their grain borders' substantial density, nanoparticles have enhanced adhesive qualities, which boost the substrate's durability against corrosion. However, the grain boundary fraction and irregular surface caused by the tiny particle agglomeration may increase the possibility of anodic site formation, increasing the surface's vulnerability to corrosion assault. To obtain the desired corrosion protection, it is crucial to consider all of the atmospheric aspects connected to films and substrates. It should be highlighted that no single factor can influence the corrosion resistance in a single dimension, nor can any of them alone influence the nanocoating's corrosion behavior. However, each one of them influences how well the nanocoating resists corrosion. A filler consisting of metallic or ceramic nanoparticles is mixed into the host matrix to create nanocomposite films. This improves the matrix's physical characteristics and makes it suitable for nanocoating. Nanocomposite material films offer safeguards against corrosion by forming an efficient obstacle and obstructing charge transfer processes, such as oxygen sensitivity and ion movement.

Furthermore, the coating of nanocomposite materials enhances various other characteristics that contribute to improving the material's corrosion performance. These characteristics include cohesive and adhesive qualities, hydrophobicity, accumulation, and dissemination and dispersal capabilities. The same elements listed above have an impact on nanocomposite coating corrosion. In addition to these, the kind, volume, and technique of nanocomposite synthesis, as well as the state of the coated substrate's heating process, all have an impact on the ability to withstand corrosion of mixed coatings. SiO_2, SiC, and SiO_2 show favorable corrosion responses when introduced to the Ni–P matrices in a solution of sodium chloride for electroless Ni coating; on the other hand, Al2O3 shows enhanced resistance to corrosion when mixed in a Ni matrix for electrodeposition application. Conductivity has grown in importance when it comes to polymer composite material coverings.

References

1. Naghashnejad FFM, Nushari MN (2019) Evaluation of adhesion and erosion/corrosion resistance of nano-composite and nano-multilayer thin films in molten aluminum alloy. https://doi.org/10.1007/s42452-019-1299-1
2. Arukalam IO, Uzochukwu IN, Tüzün B et al (2024) Influence of ZnO nanoparticle size on barrier performance and corrosion protection of poly(dimethylsiloxane)-coated Q235 steel in chloride environment: bode and computational simulation investigations. Chem Afr 7:853–863. https://doi.org/10.1007/s42250-023-00791-4
3. Selvi VE, Seenivasan H, Rajam KS (2012) Electrochemical corrosion behavior of pulse and DC electrodeposited Co–P coatings. Surf Coat Technol 206:2199–2206. https://doi.org/10.1016/j.surfcoat.2011.09.063
4. Cheng Y, Zhao L, Xin Y et al (2023) Corrosion behavior of Mg-3Al-1Zn alloy with different grain refinement strategies. J Mater Sci. https://doi.org/10.1007/s10853-023-09315-4

5. Mosavat SH, Shariat MH, Bahrololoom ME (2012) Study of corrosion performance of electrodeposited nanocrystalline Zn–Ni alloy coatings. Corros Sci 59:81–87. https://doi.org/10.1016/j.corsci.2012.02.012
6. Al-obaidi NS, Ali FK, Abd AN, Mahmoud ZH (2024) Preparation of nanocomposites for corrosion treatment. https://doi.org/10.4314/bcse.v38i2.17
7. Wang B, Jin Q, Du X et al (2024) Corrosion mechanism and microstructure evolution of yttrium-doped marine steel. J Mater Res Technol 28:2752–2766. https://doi.org/10.1016/j.jmrt.2023.12.189
8. Ward J, Bowden D, Prestat E et al (2018) Corrosion performance of Ti3SiC2, Ti3AlC2, Ti2AlC and Cr2AlC MAX phases in simulated primary water conditions. Corros Sci 139:444–453. https://doi.org/10.1016/j.corsci.2018.04.034
9. Wu L, Liu C, Wei J et al (2023) Influence of pH on the formation, composition and protectiveness of fluoride conversion film deposited on AZ31 magnesium alloy in KF solution. Acta Metall Sin (Engl Lett) 36:1397–1408. https://doi.org/10.1007/s40195-023-01571-2
10. Chianpairot A, Lothongkum G, Schuh CA, Boonyongmaneerat Y (2011) Corrosion of nanocrystalline Ni–W alloys in alkaline and acidic 3.5wt.% NaCl solutions. Corros Sci 53:1066–1071. https://doi.org/10.1016/j.corsci.2010.12.001
11. Jena S, Laha S, Swain BP (2024) Ti2NTx MXene/rGO nanocomposites: optical tunning, functional bonding and electrochemical analysis for supercapacitor electrode application. Appl Surf Sci 666:160403. https://doi.org/10.1016/j.apsusc.2024.160403
12. Jena S, Madhuri A, Gupta M, Swain BP (2023) Investigation of interfacial and Interdiffusion study of Ti2N MXene phase from TiN/Ti multilayers. Thin Solid Films 780:139969. https://doi.org/10.1016/j.tsf.2023.139969
13. Jena S, Madhuri A, Dutta S, Swain BP (2024) Investigation of structural, phonon modes and electronic environment of GaP:Si nanostructure thin films. Physica B Condens Matter 677:415664. https://doi.org/10.1016/j.physb.2024.415664
14. Afshari V, Dehghanian C (2009) Effects of grain size on the electrochemical corrosion behaviour of electrodeposited nanocrystalline Fe coatings in alkaline solution. Corros Sci 51:1844–1849. https://doi.org/10.1016/j.corsci.2009.05.015
15. Youssef KhMS, Koch CC, Fedkiw PS (2004) Improved corrosion behavior of nanocrystalline zinc produced by pulse-current electrodeposition. Corros Sci 46:51–64. https://doi.org/10.1016/S0010-938X(03)00142-2
16. Sriraman KR, Strauss HW, Brahimi S et al (2012) Tribological behavior of electrodeposited Zn, Zn–Ni, Cd and Cd–Ti coatings on low carbon steel substrates. Tribol Int 56:107–120. https://doi.org/10.1016/j.triboint.2012.06.008
17. Klement U, Da Silva M (2007) Thermal stability of electrodeposited nanocrystalline Ni-and Co-based materials. J Iron Steel Res, Int 14:173–178. https://doi.org/10.1016/S1006-706X(08)60074-6
18. Qin CL, Zhang W, Asami K et al (2006) A novel Cu-based BMG composite with high corrosion resistance and excellent mechanical properties. Acta Mater 54:3713–3719. https://doi.org/10.1016/j.actamat.2006.04.005
19. Ghosh SK, Dey GK, Dusane RO, Grover AK (2006) Improved pitting corrosion behaviour of electrodeposited nanocrystalline Ni–Cu alloys in 3.0wt.% NaCl solution. J Alloys Compd 426:235–243. https://doi.org/10.1016/j.jallcom.2005.12.094
20. Singh DDN, Ghosh R (2006) Electroless nickel–phosphorus coatings to protect steel reinforcement bars from chloride induced corrosion. Surf Coat Technol 201:90–101. https://doi.org/10.1016/j.surfcoat.2005.10.045
21. Rajaraman V, Ariga P, Ramalingam K, Sekaran S (2024) Evaluation of corrosive properties of hafnium nitride coating over titanium screws: an in vitro study. 16. https://doi.org/10.7759/cureus.55456
22. Alontseva D, (Yantsen) YS, Voinarovych S et al (2024) Biocompatibility and corrosion of microplasma-sprayed titanium and tantalum coatings versus titanium alloy. Coatings
23. Chen C, Cai T, Lin H, Lu H (2024) Effects of process parameters on microstructure and properties of Ni–Ti coatings synthesised by mechanical alloying method. Results Surf Interfaces 14:100198. https://doi.org/10.1016/j.rsurfi.2024.100198

24. Lupu FC, Munteanu C, Müftü S et al (2024) Evaluation of the wear properties and corrosion resistance of 52100 steel coated with Ni/CrC by cold spraying. Coatings
25. Qiao Q, Cristino VAM, Tam LM, Kwok CT (2023) Corrosion properties of Ti-Ni-Cu coatings fabricated by laser surface alloying. Corros Sci 222:111426. https://doi.org/10.1016/j.corsci.2023.111426
26. Chen Y, Li H, Fang C et al (2023) Preparation and characterisation of Ni-Co-TiN/CeO2 composite coating by jet electrodeposition for improved mechanical and anti-corrosion properties. Surf Topogr 11
27. Coating OLN, Li X, Zhang S et al (2023) Effect of WC content on the wear and corrosion properties of oscillating laser-cladding-produced nickel-based coating
28. Jian Y, Liu Y, Qi H et al (2024) Effects of scanning speed on the microstructure, hardness and corrosion properties of high-speed laser cladding Fe-based stainless coatings. J Market Res 29:3380–3392. https://doi.org/10.1016/j.jmrt.2024.02.087
29. Liang Y, Yang M, Kong D et al (2024) Effect of annealing on corrosion resistance and mechanical properties of Cr–Fe–Ni–Co–Mo–Si–B amorphous coating. J Mater Res Technol 29:4622–4635. https://doi.org/10.1016/j.jmrt.2024.02.104
30. Yang M, Song P, Kong D et al (2024) Wear and corrosion properties of HVOF sprayed WC-Cr3C2 composite coating for application in polysilicon cyclone separator. J Mater Res Technol 29:78–89. https://doi.org/10.1016/j.jmrt.2024.01.114
31. Hadidi H, Saminathan R, Zouli N (2024) Enhancement of diffusive oxidation resistance and topographical corrosion resistance of hot work tool steels by nickel multifunctional coatings. J Mater Res Technol 28:1227–1232. https://doi.org/10.1016/j.jmrt.2023.12.052
32. Sá Brito VRS, Bastos IN, Costa HRM (2012) Corrosion resistance and characterisation of metallic coatings deposited by thermal spray on carbon steel. Mater Des 41:282–288. https://doi.org/10.1016/j.matdes.2012.05.008
33. Cr-crn R, Tao H, Zhylinski V et al (2023) Comparison of the mechanical properties and corrosion resistance of the Cr-CrN, Ti-TiN, Zr-ZrN, and Mo-MoN coatings
34. Arslan KO, Aksakal R, Cakmak B Synthesis and characterisation of single and duplex ZnO/TiO_2 ceramic films on additively manufactured bimetallic material of 316L stainless steel and Ti6Al4V Synthesis and characterisation of single and duplex ZnO/TiO_2 ceramic fi lms on additively m
35. Yan H (2024) Effects of micro-arc oxidation process parameters on micro-structure and properties of Al2O3 coatings prepared on sintered 2024 aluminum alloy. J Mater Eng Perform 33:1862–1873. https://doi.org/10.1007/s11665-023-08093-z
36. Nayak H, Kollur S, Suresh KP (2024) Development of equal proportional YSZ+Al2O3 thermal barrier coating and effect of coating thickness on the corrosion behaviour on cast iron substrate. High Temp Corros Mater 101:291–308. https://doi.org/10.1007/s11085-024-10222-5
37. Feng Y, Liu M, Jia L et al (2024) Study on the corrosion resistance of supersonic plasma spraying Al2O3 thin layer and SiO2 sealer alternately deposited coating. Coatings 14. https://doi.org/10.3390/coatings14010078
38. Raju KAK, Biswas A (2024) Structural and corrosion analysis of copper/chitosan coating on nanotubular TiO2/Ti6Al4V implant surface. Surf Coat Technol 482:130673. https://doi.org/10.1016/j.surfcoat.2024.130673
39. Tran D, Lin C-K, Tung P-C et al (2024) Effects of TiN content and heat treatment on microstructural changes, mechanical strength, and corrosion resistance in selective laser melting of TiN/AISI 420 composites. Mater Sci Eng: A 900:146438. https://doi.org/10.1016/j.msea.2024.146438
40. Dong H, Guo P-F, Han Y et al (2023) Enhanced corrosion resistance of high speed laser-cladded Ni/316L alloy coating by heat treatment. J Mater Res Technol 24:952–962. https://doi.org/10.1016/j.jmrt.2023.03.070
41. Yang D, Yan F, Zhang W, Xie Z (2024) Microstructure, wear resistance and corrosion resistance of CrN coating with platinum-iridium co-doping. Coatings 14. https://doi.org/10.3390/coatings14020238

42. Pana I, Parau AC, Dinu M et al (2024) Structural, mechanical, wear and anticorrosive properties of CrSiCN coatings used for industrial woodworking applications. Heliyon 10:e29496. https://doi.org/10.1016/j.heliyon.2024.e29496
43. Haji-Savameri M, Irannejad A, Norouzi-Apourvari S et al (2022) Evaluation of corrosion performance of superhydrophobic PTFE and nanosilica coatings. Sci Rep 12:1–14. https://doi.org/10.1038/s41598-022-20729-z
44. Abdolmaleki M, Ghamari K, Ahmadiyeh S et al (2024) Electrodeposition of Ni-W/PTFE nanocomposite coating and investigation of its corrosion resistance and microhardness. Metall Mater Trans A Phys Metall Mater Sci 55:352–359. https://doi.org/10.1007/s11661-023-07253-z
45. Li J, Liu Z, Wang Z (2023) Study of nano-ZnO improvement of the mechanical properties and corrosion resistance of modified-SiO2/PTFE superhydrophobic nanocomposite coatings by one-step spraying. New J Chem 47:6246–6257. https://doi.org/10.1039/D2NJ06376J
46. Saini A, Madhuri A, Sahoo SK et al (2024) Investigation of microstructural, chemical bonding and optical properties of Fe-Cu/rGO nanocomposites. J Alloy Metall Syst 5:100053. https://doi.org/10.1016/j.jalmes.2023.100053
47. Sahoo SK, Madhuri A, Saini A et al (2023) Correlation of microstructural and chemical bonding of FeNi-rGO nanocomposites. J Alloy Metall Syst 4:100035. https://doi.org/10.1016/j.jalmes.2023.100035
48. Sinha S, Singh WI, Nongthombam S et al (2022) Optical properties, electrochemical analysis and corrosion resistance studies of polyaniline/reduced graphene Oxide/ZrO2 for supercapacitor application. J Phys Chem Solids 161:110478. https://doi.org/10.1016/j.jpcs.2021.110478
49. Chanu SN, Sinha S, Devi PS et al (2022) Optical, electrochemical and corrosion resistance properties of iron oxide/reduced graphene oxide/polyvinylpyrrolidone nanocomposite as supercapacitor electrode material. Bull Mater Sci 45:1–11. https://doi.org/10.1007/s12034-022-02704-6
50. Jena S, Madhuri A, Gupta M, Swain BP (2024) Comparative assessment and experimental analysis of atomic coordination and mechanical properties of sputter deposited Ti-rich TiN, TiC, and TiCN thin films. J Alloys Compd 1002:175217. https://doi.org/10.1016/j.jallcom.2024.175217
51. Madhuri A, Jena S, Gupta M, Swain BP (2023) Tweaking the mechanical stability of Al1−xWxN ternary nitride alloys for hard coating applications, and study on their structural, photoluminescence and surface chemical analysis. J Alloys Compd 968:171854. https://doi.org/10.1016/j.jallcom.2023.171854
52. Chanu SN, Swain BP (2024) Synthesis and characterisation of electrospun reduced graphene oxide/nickel oxide/poly(caprolactone) nanofibers for enhanced corrosion resistance and energy storage performance. Next Nanotechnol 5:100049. https://doi.org/10.1016/j.nxnano.2024.100049
53. Jena S, Gupta M, Swain BP (2024) Effect of interlayer thickness on electrochemical properties and corrosion resistance of sputtered TiNxnm/Tixnm multilayer thin films. Mater Sci Semicond Process 182:108729. https://doi.org/10.1016/j.mssp.2024.108729

Chapter 11
Surface Passivation and Surface Treatment of Coating Materials

Sanketa Jena and Bibhu Prasad Swain

Abstract The book chapter deals with the significance of surface passivation and surface treatment methods in enhancing coatings' performance, durability, and adhesion. Surface passivation involves forming a protective layer on material surfaces to deter corrosion and environmental deterioration. Chemical, electrochemical, organic, and inorganic techniques can be employed, each offering distinct advantages tailored to specific applications. Surface treatment encompasses mechanical, chemical, and physical procedures to prepare substrates for coating application. Mechanical treatments like abrasive blasting and grinding enhance surface roughness for improved mechanical interlocking. Chemical treatments such as acid etching and alkaline cleaning boost surface energy and cleanliness, while physical methods like plasma treatment, ion implantation, and laser surface treatment alter surface properties to enhance coating adhesion and performance. The chapter investigates how these processes heighten coating adhesion, prevent corrosion, and strengthen wear resistance. It also explores characterisation and evaluation methods like surface energy measurement, chemical analysis, topographical analysis, and adhesion testing to gauge treatment effectiveness. Real-world examples from the aerospace, automotive, and electronics sectors demonstrate the practical applications and advantages of surface passivation and treatment techniques. The chapter concludes by examining current obstacles and prospects, emphasising the necessity for eco-friendly and sustainable practices and technological advancements to meet the evolving demands of diverse industries. The comprehensive overview underscores the critical role of surface passivation and treatment in ensuring optimal performance and longevity of coating materials.

Keywords Surface treatment · Surface passivation · Applications · Future challenges

S. Jena (✉) · B. P. Swain
Department of Physics, National Institute of Technology Manipur, Langol, Imphal West, Manipur, India
e-mail: sanketajena.res@gmail.com

B. P. Swain (ed.), *Advances in Mechanical Coating*, Materials Horizons: From Nature to Nanomaterials, https://doi.org/10.1007/978-981-96-7484-8_11

1 Introduction

Surface passivation and treatment are pivotal in enhancing coating materials' performance, durability, and adhesion across various industries. These techniques involve creating protective layers and applying mechanical, chemical, and physical procedures to prepare surfaces for coatings. Surface passivation aims to prevent corrosion and environmental degradation by forming a shield on material surfaces, while surface treatment methods such as abrasive blasting, acid etching, and plasma treatment are utilised to optimise adhesion and performance [1]. In this introduction, we delve into the mechanisms, methods, and applications of surface passivation and treatment, highlighting their importance in improving coating properties and ensuring longevity in industrial applications. Surface passivation and surface treatment are critical processes in preparing substrates for coating, ensuring enhanced adhesion, durability, and resistance to environmental factors.

Surface passivation and treatment are essential processes that significantly enhance the performance, durability, and adhesion of coatings in various industries. By creating protective layers on material surfaces, surface passivation effectively prevents corrosion and environmental degradation, extending the coated materials' lifespan. On the other hand, surface treatment techniques like abrasive blasting, acid etching, and plasma treatment are crucial for preparing surfaces before coating application. These treatments improve surface roughness, enhance cleanliness and surface energy, and modify surface properties to promote better adhesion and coating performance. Altogether, surface passivation and treatment are critical for increasing coating adhesion, preventing corrosion, and enhancing wear resistance, ensuring that coatings maintain their optimal properties and functionality over time.

2 History of Passivation

Researchers accomplishing electrochemical investigations identified passivation, yet they failed to realise how useful the technique was for everyday applications instantly. Chemist James Keir noted in 1790 that iron does not rust in an aggressive nitric acid treatment. He saw that the iron rusted instantly in the identical solutions when the water was altered, producing a bubbling solution containing dark brown water. Swiss scientist Christian Friedrich Schönbein expanded on the experiment's specifications by 1836. As Keir pointed out, he showed how a sliver of iron dissolved in weaker nitric acid produced hydrogen. However, the iron could tolerate the diluted acid if first immersed in an acidic solution. For an extended period, the harmful substances of water appeared to be kept away. Nitric acid's effects on the surroundings and public health grew increasingly noticeable in the 1900s when the passivation of stainless steel with the chemical gained popularity. A German manufacturing business, Adolf Coors' study, found that citric acid was a good substitute. Citric acid, a safer and greener substitute for nitric acid, gained popularity among businesses in the 1990s. It

Fig. 1 **a** Surface passivated equipment, **b** Passivation process service, **c** Effect of passivation

was British electrochemist Michael Faraday who originally described the passivation phenomenon. He suggested to Schönbein that the passive state might be brought on by an oxide skin produced by the powerful acid. Figure 1 represents the real images of surface passivated equipment, passivation process service and effect of passivation process.

3 Mechanisms of Surface Passivation

Chemical passivation involves producing a stable shielding layer on the material's exterior through chemical reactions. Surface passivation refers to making a material's surface less reactive, typically by applying a thin layer of another material that acts as a barrier. This is commonly used in semiconductors and metals to prevent corrosion or unwanted chemical reactions on metals like stainless steel and aluminium surfaces. The formation of oxide layers is a fundamental method of surface passivation used to protect metals from corrosion and environmental degradation. Oxide layers are typically created through chemical reactions between the metal surface and oxygen from the air or an oxidising agent. These oxide layers serve as protective barriers that prevent further oxidation and corrosion of the underlying metal. The passivation layer can be applied through various methods such as chemical treatments, deposition techniques like plasma-enhanced chemical vapor deposition (PECVD), or thermal oxidation. In semiconductor manufacturing, for instance, passivation layers are crucial to reduce surface recombination velocities and thereby enhance solar cell efficiency or improve integrated circuits' performance and reliability. This section explores the mechanisms, methods, and benefits of oxide layer formation for surface passivation.

Medium and low-carbon steel is the primary raw material for stainless steel. To change the final qualities, they are alloyed with several different metals. For instance, chromium and nickel provide hardness and resistance to corrosion. Additional metals improve tensile and shear strength, malleability, toughness, ductility, and other qualities. One essential procedure for improving the corrosion resilience of stainless-steel components is passivation. Stainless steel passivation is utilised in aerospace, pharmaceuticals, food processing, and marine applications where corrosion resistance and surface integrity are critical. Properly passivated stainless-steel components can maintain their performance and appearance over extended periods in harsh environments. The primary goal of passivating stainless steel is to remove free iron and other contaminants from the surface that could initiate corrosion. Despite its name, stainless steel can corrode under certain conditions if not properly treated. Passivation helps restore and enhance its inherent corrosion resistance by forming a protective oxide layer on the surface [2].

The passivation process typically involves several steps:

1. Cleaning: The surface is thoroughly cleaned to remove dirt, grease, oil, scale, and other contaminants. This step is crucial to ensure effective passivation.
2. Acid Treatment: After cleaning, the stainless steel is treated with an acid solution, often a nitric acid-based solution. This acid removes any embedded iron or iron oxide particles from the surface, which could otherwise lead to rust spots or corrosion.
3. Rinsing: The acid-treated surface is properly washed with clean water to eliminate residual acid and contaminants.
4. Passivation Solution: The stainless steel is then immersed in a passivation solution, typically a nitric or citric acid-based solution. This solution facilitates the formation of a passive oxide layer on the surface, enhancing the steel's corrosion resistance.
5. Rinsing and Drying: After the desired passivation time, the components are rinsed again with water to remove any remaining passivation solution, followed by drying to prevent water stains.

Passivation significantly improves the corrosion resistance of stainless steel by promoting the formation of a stable oxide layer. It ensures that the surface is clean and free from contaminants that could degrade performance over time. Passivation can improve the appearance of stainless steel by removing discoloration and restoring its original finish. Since passivation frequently uses acidic solutions, it is imperative to take security precautions, such as ventilated and wearing personal protection equipment (PPE). Depending on the application and location, there may be regulatory requirements or environmental considerations concerning the disposal of passivation chemicals.

3.1 Citric Acid and Nitric Acid Passivation

Citric acid and nitric acid are both commonly used in passivation, a method used to enhance the corrosion resistance of stainless steel and other metals. Citric and nitric acid passivation involves immersing the metal in an acid solution to remove free iron and other surface contaminants, promoting the formation of a passive oxide layer. Citric acid passivation is preferred over nitric acid in many applications due to environmental and safety concerns associated with nitric acid. It involves immersing the metal in a citric acid and water solution, typically at elevated temperatures (often around 50–70 °C). Citric acid passivation effectively removes free iron from the surface of stainless steel, which helps prevent rust and corrosion. Additionally, it offers a greener substitute for conventional nitric acid passivation.

Nitric acid passivation has been a traditional method used for stainless steel passivation. It involves immersing the metal in a nitric acid solution at room temperature, typically around 20–45% concentration. Nitric acid passivation removes contaminants from the surface and enhances the formation of a protective oxide layer on the metal's surface. This layer improves corrosion resistance.

- **Citric Versus Nitric Acid Passivation**

Citric acid is generally considered safer and more environmentally friendly than nitric acid, which is hazardous and requires careful handling and disposal. Both methods effectively passivate stainless steel, but citric acid has gained popularity due to its effectiveness and reduced environmental impact. Citric acid passivation may comply more with environmental regulations and workplace safety guidelines than nitric acid. While both citric acid and nitric acid are used for passivation, they are often preferred today due to their effectiveness, safety, and environmental considerations. Figure 2 represents the citric and nitric acid passivation treatment.

3.2 Alkaline Passivation

This approach involves the cleaning of metal parts in an alkaline passivation solution. This solution is typically alkaline, often containing sodium hydroxide (NaOH) or potassium hydroxide (KOH) as the primary chemical agents. It removes any remaining iron contaminants from the surface of the metal. It promotes the production of a uniform and stable passive oxide layer on the metal surface. During immersion, the alkaline solution acts on the metal surface, removing free iron atoms and other contaminants through dissolution. Simultaneously, the surface undergoes oxidation, forming a passive oxide layer (typically chromium oxide for stainless steel) that provides corrosion resistance. After the desired passivation time, the metal parts are thoroughly rinsed to remove any residual alkaline solution and contaminants. Proper rinsing is critical to prevent alkaline residue from interfering with the performance

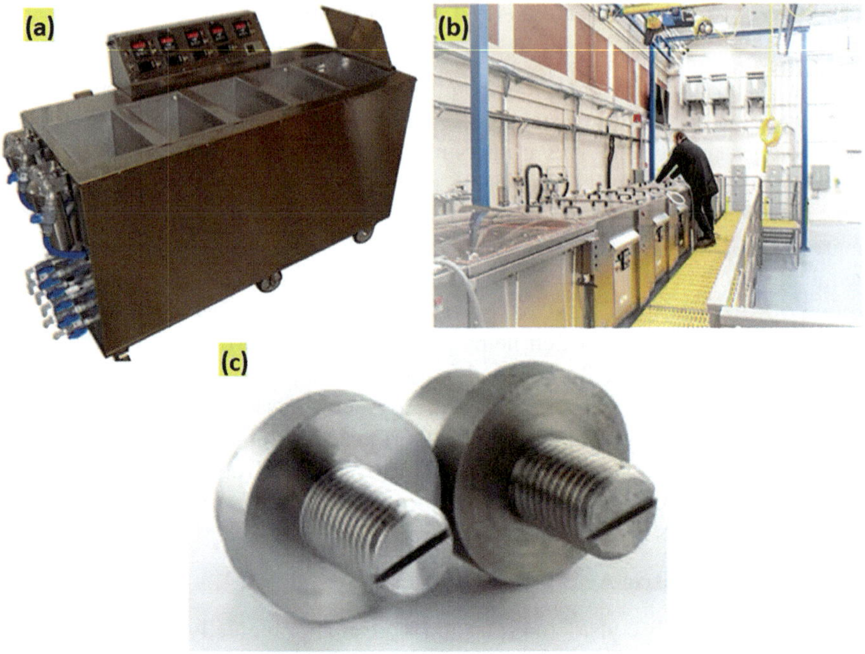

Fig. 2 Citric and nitric acid passivation process

of the passive oxide layer. The parts are then dried to avoid water spots and ensure a clean surface.

Depending on the application requirements, the effectiveness of the alkaline passivation can be verified through various tests, such as salt spray testing, humidity testing, or electrochemical tests, to confirm the corrosion resistance of the treated metal.

Advantages of Alkaline Passivation

- Effective Removal of Contaminants: Alkaline solutions clean and remove surface contaminants, preparing the metal to produce a passive oxide layer.
- Enhanced Corrosion Resistance: The passive oxide layer formed during alkaline passivation significantly improves the corrosion resistance properties of stainless steel and other metals.
- Suitability for Complex Geometries: Alkaline passivation can be applied to complex shapes and geometries, ensuring uniform coverage and protection.
- Cost-Effective: Compared to other passivation methods, alkaline passivation is generally more cost-effective and straightforward.

Alkaline passivation is widely used in industries where stainless-steel components require enhanced corrosion resistance, such as automotive, aerospace, pharmaceutical, and food processing equipment. It provides a reliable method to extend the lifespan and durability of metal components exposed to harsh environments or corrosive substances.

3.3 Oxide-Layer Formation

Surface passivation in semiconductor devices often involves the formation of an oxide layer. This oxide layer serves several crucial purposes in enhancing the device performance. The number of surface states can be significantly reduced by forming an oxide layer (typically silicon dioxide, SiO_2) on the semiconductor surface. Surface states can trap charge carriers, increasing recombination rates and decreasing device performance. The oxide layer acts as a barrier, minimising the impact of these surface states. A high-quality oxide layer can improve the minority carrier lifetime in the semiconductor material. This is particularly important in photovoltaic cells and other semiconductor devices, where the efficiency depends on the ability of carriers to move through the material without recombining.

Silicon dioxide (SiO_2) is an excellent electrical insulator. By forming an oxide layer, it is possible to isolate different regions of the semiconductor device electrically, such as isolating the gate electrode from the semiconductor channel in MOSFETs (Metal–Oxide–Semiconductor Field-Effect Transistors) [3]. Oxide layers can also protect the underlying semiconductor material from chemical reactions or environmental contamination. This is important for maintaining the integrity and performance of the semiconductor device over time.

The basic chemical reaction for oxide layer formation involves the metal reacting with oxygen to form a metal oxide. This process is typically formed through thermal oxidation, where metal reacts with oxygen at elevated temperatures. For instance,

$$2Al + 3O_2 \rightarrow Al_2O_3$$

$$4Fe + 3O_2 \rightarrow 2Fe_2O_3$$

$$Si + O_2 \rightarrow SiO_2$$

This process is carefully controlled in semiconductor manufacturing to achieve a uniform and high-quality oxide layer. Common methods include:

1. Dry Oxidation: Oxidation in an oxygen or dry oxygen atmosphere at high temperatures (typically 800–1200 °C).
2. Wet Oxidation: Oxidation in an atmosphere containing steam (H_2O) at lower temperatures (typically 600–1000 °C).

The preferred technique is determined by variables like the desired thickness of the oxide layer, the required electrical properties, and the specific semiconductor material used. Black et al. explored the ALD method's passivation studies of MoOx,

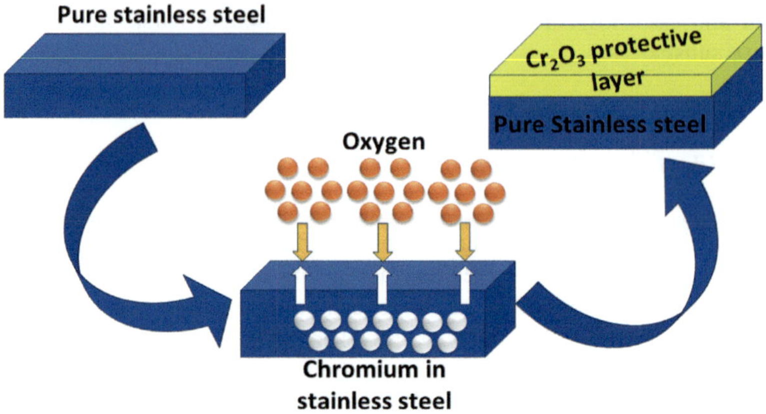

Fig. 3 Surface passivation of stainless steel by oxide layer formation

ZnO, TiOx, POx, and Nb2O5 coating to design, optimise, and passivate contacts in applying c-Si solar cells [4]. ALD-synthesised aluminium oxide (Al_2O_3) nanolayers were introduced in 2006 as a unique passivation method for p- and n-type crystalline Si (c-Si) substrates. Comparing the sputtering process to ALD or PECVD, it seems that less passivation is produced in Al_2O_3 [5]. In addition to bulk real estate, surface passivation critically depends on the characteristics of the Al_2O_3-Si contact. At this point of contact, a thin (~1–2 nm) SiO_x layer was discovered. Hoex et al. observed a (very tiny) rise in SiO_x thickness of ~ 0.3 nm during annealing at ~425 °C, near the TEM precision [6]. Figure 3 represents the schematic of the passivation process in stainless steel.

3.4 *Chemical Treatment (Cathodic and Anodic)*

In semiconductor manufacturing, chemical treatments such as cathodic and anodic processes are used in surface passivation to modify the properties of the semiconductor surface, particularly by forming oxide layers. Cathodic treatment involves applying a negative potential (cathode) to the semiconductor material submerged in an electrolytic solution. On the other hand, anodic treatment consists of using a positive potential (anode) to the semiconductor material in an electrolytic solution. Here's an overview of cathodic and anodic treatments in this context:

- Electrolytic Solution: The semiconductor material is immersed in an electrolyte that facilitates the dissolution of the semiconductor material.
- Application of Voltage: A negative potential is applied to the semiconductor material (cathode), causing electrons to flow from the solution to the semiconductor surface. A positive potential is applied to the semiconductor material (anode), causing electrons to flow from the semiconductor into the solution.

- Oxide Formation: Because of the cathodic reaction, oxide layers can form on the semiconductor surface. For instance, silicon (Si) can react with oxygen (O) in the electrolyte to form silicon dioxide (SiO_2). Anodic oxidation can occur when the semiconductor material (e.g., silicon) reacts with the oxygen (O_2) in the electrolyte to form an oxide layer. This process is often used to grow thick and uniform oxide layers.

Cathodic and anodic treatments in surface passivation involve applying controlled electrical potentials in electrolytic solutions to form oxide layers on semiconductor surfaces. These processes are fundamental in enhancing semiconductor devices' electrical performance and reliability across various applications, including MOSFETs, solar cells, and integrated circuits.

3.5 Plasma Implantation

Plasma implantation, also known as plasma immersion ion implantation (PIII), is a technique used in semiconductor processing that can be applied in surface passivation to modify the properties of the semiconductor surface. Figure 4 represents the schematic of the plasma implantation process for surface passivation. Here's how plasma implantation works and its role in surface passivation:

1. Plasma Generation: Plasma, a highly ionised gas containing positively and negatively charged ions, is generated in a vacuum chamber. This plasma can consist of various gases depending on the desired ion species to be implanted into the semiconductor surface.
2. Acceleration and Implantation: The semiconductor wafer is placed in the vacuum chamber, and a high-voltage pulse is applied to accelerate ions from the plasma towards the wafer surface. The ions penetrate the surface of the semiconductor material to a controlled depth.

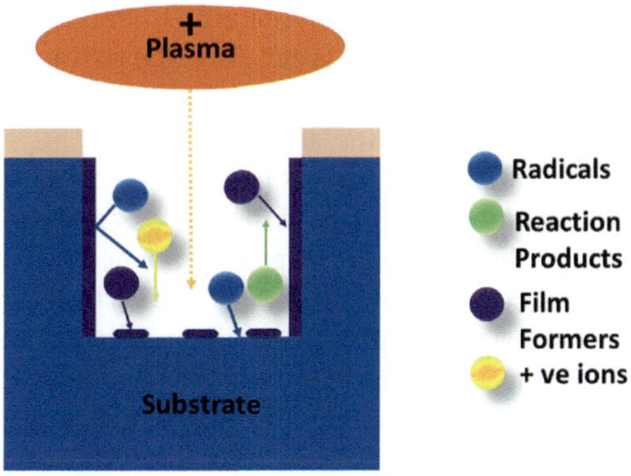

Fig. 4 Plasma implantation process for surface passivation

3. Ion Implantation: When ions from the plasma strike the semiconductor surface, they can embed themselves within the material. This process allows for precise doping or modification of the semiconductor surface layer without high temperatures.

In the context of surface passivation, plasma implantation can achieve several objectives, such as surface Modification, doping control, and oxide layer modification, which are explained here. Plasma implantation can introduce specific ions into the semiconductor surface, altering its electrical properties. For instance, implanting nitrogen or hydrogen ions can passivate dangling bonds on the silicon surface, reducing surface recombination rates and improving carrier lifetimes. This process allows precise control over doping concentrations near the surface, crucial for optimising device performance in semiconductor devices such as MOSFETs and solar cells. Plasma treatments can also modify or grow oxide layers on the semiconductor surface. For example, a thin oxide layer can be formed by implanting oxygen ions, which enhances surface passivation and provides electrical insulation.

3.6 Forced Passivation

Forced passivation is a technique used in semiconductor processing to deliberately alter or improve the passivation layer on the semiconductor surface. The passivation layer typically consists of an oxide (e.g., silicon dioxide, SiO_2) that protects the underlying semiconductor material and improves device performance. The main objective of forced passivation is to enhance the quality and effectiveness of the passivation layer. This can involve optimising the oxide layer's thickness, composition, and electrical properties to reduce surface recombination, minimise leakage currents, and improve the overall reliability of semiconductor devices. Silicon dioxide (SiO_2) is a common material used in forced passivation due to its excellent electrical insulation properties and compatibility with silicon semiconductors [7, 8]. Other materials, such as silicon nitride (Si_3N_4) and silicon Carbide (SiC), may also be used depending on specific device requirements [9]. Forced passivation is used in MOSFETs, solar cells, integrated circuits etc. Forced passivation plays a crucial role in semiconductor manufacturing by optimising the quality and effectiveness of the passivation layer on the semiconductor surface.

By utilising techniques like thermal oxidation, PECVD, and ALD, manufacturers can tailor the properties of the oxide or nitride layers to meet specific device requirements, thereby improving performance, reliability, and overall device longevity. Enhanced passivation layers reduce surface recombination velocity, which improves charge carrier lifetime and efficiency in devices such as solar cells and MOSFETs. The protection against environmental factors (e.g., moisture, contaminants) and electrical isolation between different semiconductor device regions are essential for reliability. Techniques like ALD and PECVD offer precise control over layer thickness and uniformity, critical for achieving consistent device performance. The following methods are involved in the forced passivation process.

- Thermal Oxidation: Controlled oxidation at elevated temperatures to grow a high-quality oxide layer. This method ensures uniform coverage and the desired thickness of the passivation layer.
- Plasma Enhanced Chemical Vapor Deposition (PECVD): Deposition of oxide or nitride layers using plasma-generated reactive species, enabling exact oversight of the qualities and dimensions of the film.
- Atomic-Layer Deposition (ALD): Atomic layers of elements are deposited one after the other, providing exact oversight of the conformity and depth of the passivation layer.
- Annealing: Heat treatment processes help heal defects and improve the electrical properties of the passivation layer.

3.7 Atomic-Layer Deposition

Atomic-layer deposition (ALD) is an accurate and regulated thin-film fabrication method that can be utilised for passivation purposes, especially in the context of semiconductor devices and other advanced materials. ALD passivation finds applications in various industries, including semiconductor manufacturing (for device isolation and protection), solar cell fabrication (to enhance efficiency and stability), and advanced medical device and optics coatings.

Principle of ALD

ALD is based on sequential, self-limiting surface reactions. It involves alternating exposures of a substrate to two or more gaseous precursor chemicals. One atomic layer at a moment, these building blocks interact with the exterior of the substrate in a self-limiting way to generate a monolayer of substance.

3.8 Passivation with ALD

The primary goal of passivation using ALD is to create a thin, uniform, and conformal layer on the surface of the substrate. Common materials used for passivation via ALD include oxides (alumina, silica), nitrides (silicon nitride), and other dielectric materials. The substrate (such as a silicon wafer or metal surface) is placed inside an ALD chamber. The chamber is evacuated, and the substrate surface is prepared by cleaning and activation processes to ensure uniformity. Sequential pulses of precursor gases are used when entering the chamber, each pulse reacting with the substrate surface to form a single atomic layer of the desired material. After each pulse, any excess precursor and reaction by-products are purged from the chamber.

The process repeats until the desired thickness of the passivation layer is achieved.

Advantages of ALD Passivation

- Conformality: ALD provides excellent conformality, meaning the passivation layer conforms precisely to the contours and features of the substrate, including high-aspect-ratio structures.
- Thickness Control: ALD allows accurate control over the thickness of the passivation layer down to the atomic level.
- Uniformity: The layer deposited by ALD is highly uniform across the substrate surface, ensuring consistent passivation performance.
- Quality: ALD-produced layers typically have high purity and low defect densities, which are crucial for effective passivation.

3.9 Physical Vapor Deposition

Physical Vapor Deposition (PVD) can be utilised as a method for passivation, particularly in depositing thin films of materials that enhance corrosion resistance and other functional properties on surfaces. Materials commonly used for PVD passivation films include titanium nitride (TiN), titanium oxide (TiO_2), chromium nitride (CrN), and aluminium oxide (Al_2O_3), among others [10, 11].

Advantages of PVD for Passivation

- Uniform Coating: PVD techniques can produce highly uniform and conformal coatings over complex geometries, ensuring consistent protection across the entire surface.
- Controlled Properties: The physical vapor deposition process allows precise control over film thickness, composition, and structure, tailoring the properties of the passivation layer to specific requirements.
- Enhanced Performance: PVD coatings can improve corrosion resistance, wear resistance, hardness, and other functional properties of the substrate material.

Applications

PVD coatings are commonly used to enhance the durability and performance of cutting tools by improving wear resistance and reducing friction. PVD coatings can be applied to medical implants to enhance biocompatibility and corrosion resistance. PVD coatings are used on lenses and mirrors to enhance reflectivity and durability.

The substrate material must be compatible with the PVD process and the deposited material to ensure proper adhesion and functionality of the passivation layer. Physical Vapor Deposition (PVD) is a versatile method for passivation, offering precise control over film properties and enabling the deposition of thin, protective coatings that enhance the corrosion resistance and functional performance of various materials in diverse applications.

3.10 Chemical Vapor Deposition (CVD)

Chemical Vapor Deposition (CVD) is a versatile method used in various industries to passivate materials. CVD is a powerful method for passivation because of its capacity to form precise, uniform layers of multiple materials that can significantly enhance the corrosion resistance and other functional characteristics of materials throughout various uses [12–15]. Passivation, in this context, typically refers to making a material more resistant to corrosion by forming a protective oxide layer on its surface. Here's how CVD can be used for passivation:

CVD Passivation Process

A suitable material that can form a stable oxide layer on the substrate material by a CVD technique. Common materials include silicon dioxide (SiO_2), silicon nitride (Si_3N_4), titanium nitride (TiN), aluminium oxide (Al_2O_3), etc., depending on the specific application and substrate material. Clean the substrate thoroughly to remove everything that can prevent the CVD film from adhering. Depending on the material and application, this typically involves cleaning with solvents, acids, or other suitable methods.

CVD can provide highly uniform and conformal coatings, even on complex geometries. The properties of the passivation layer can be tailored by adjusting fabrication parameters, such as temperature, pressure, and precursor gas composition. CVD coatings can improve corrosion resistance, wear resistance, and other functional properties of the substrate material. CVD passivation of semiconductor devices to protect against environmental factors.

4 Thermodynamics and Kinetics

The thermodynamic properties of the metal and oxide, such as Gibbs free energy, influence the stability and formation rate of oxide layers. Passivation involves the creation of a stable oxide layer. According to thermodynamics, for passivation to occur spontaneously, the formation of this oxide layer must lower the Gibbs free energy of the system. This means the oxide layer should be thermodynamically stable under the given conditions (temperature, pressure, etc.). The stability of the oxide layer depends on factors such as the metal's reactivity, the composition of the surrounding environment (e.g., presence of oxygen, pH of the solution), and temperature. Thermodynamically, metals tend to form oxides that are stable under ambient conditions. For example, stainless steel passivates due to the stable chromium oxide (Cr_2O_3) layer that forms on its surface.

The kinetics of oxidation involve factors like temperature, partial pressure of oxygen, and surface condition of the metal. Kinetics deals with the rate at which passivation occurs. Even if passivation is thermodynamically favourable, it may not occur spontaneously if the activation energy barrier for oxide formation is too high.

This activation energy depends on factors like the availability of oxygen or other oxidising agents, temperature, and the surface condition of the metal. The metal's surface reactivity influences the kinetics of passivation. Metals with a higher affinity for oxygen or other oxidants will tend to passivate more readily. The presence of contaminants or imperfections on the metal surface can hinder passivation kinetics by blocking sites where oxide formation could occur.

Thermodynamics dictates whether passivation can occur under given conditions, considering the stability of the oxide layer. However, kinetics determines how quickly passivation occurs. Metal might be thermodynamically capable of passivating, but the kinetics could be slow if the conditions do not favour rapid oxide formation. In real-world applications, both thermodynamics and kinetics must be considered. Engineers often manipulate conditions such as temperature, pH, and surface preparation to optimise passivation kinetics while ensuring the thermodynamic stability of the oxide layer. Thermodynamics governs the feasibility of passivation by ensuring the stability of the oxide layer, while kinetics dictates the rate at which this protective layer forms. Both factors are essential in understanding and controlling the passivation process in metals to prevent corrosion effectively.

5 Types of Oxide Layers

Classifications of oxide formation processes include native oxides, thermally grown oxides, anodic oxides, chemical passivation, electrochemical passivation, application of organic layers, and inorganic coatings. Native oxides spontaneously form when metals are exposed to oxygen-rich environments, such as aluminium forming aluminium oxide (Al_2O_3) or titanium forming titanium dioxide (TiO_2). Thermally grown oxides are produced by heating metals in oxygen-containing atmospheres to accelerate oxidation, as seen with silicon wafers in semiconductor applications. Anodic oxides are made through anodising, where metals serve as the anode in an electrochemical cell, for example, anodising aluminium to create a thick, porous aluminium oxide layer. Chemical passivation involves forming protective oxide layers to prevent further oxidation, like passivating stainless steel with nitric or citric acid to generate a chromium oxide layer. Electrochemical passivation uses anodic or cathodic processes to develop protective films on metal surfaces, for example, anodising aluminium to form a durable oxide layer. Organic layers like paints and polymers act as environmental barriers, as seen when epoxy coatings are applied to steel structures for corrosion protection. Inorganic coatings include depositing thin films of materials such as silicon dioxide or titanium dioxide employing methods such as PVD or CVD [16].

6 Surface Treatment Techniques

Mechanical treatments like abrasive blasting, grinding, and polishing aid in preparing surfaces for coating application. Abrasive blasting involves materials like sand or glass beads to clean and roughen surfaces, promoting mechanical interlocking for coatings. Grinding and polishing are employed to achieve the desired surface roughness and texture for optimal coating adhesion. Chemical treatments encompass acid etching and alkaline cleaning techniques. Acid etching utilises acidic solutions to create micro-roughness on surfaces, activating them for improved adhesion. Alkaline cleaning, on the other hand, removes organic contaminants and enhances surface energy to facilitate better coating adherence. Chromate and phosphate coatings enhance corrosion resistance and improve paint adhesion, and they are commonly applied to materials like steel and aluminium for added protection. Physical treatments involve plasma treatment, ion implantation, and laser surface treatment. Plasma treatment alters surface chemistry and boosts surface energy using ionised gas, while ion implantation introduces ions into the surface layer to modify properties such as hardness and wear resistance. Laser surface treatment employs laser beams to clean, texture, or alloy surfaces, enhancing their properties to promote better adherence to coatings [17, 18].

Applying thin films or coatings to alter surface properties such as hardness, friction, corrosion resistance, or optical characteristics. Techniques include PVD, CVD, and electroplating. Introducing functional groups or chemical species onto the surface to impart specific functionalities, such as hydrophobic or biocompatible properties. Controlled heating or annealing processes are used to modify the morphology and exterior. Surface treatment methods are widely used across automotive, aerospace, electronics, and biomedical industries. Examples include using coatings to improve engine components' wear resistance, applying anti-reflection coatings on lenses, or modifying the surface of implants to promote osseointegration in medical devices.

7 Role of Surface Passivation and Treatment in Coating Adhesion

Enhancing adhesion involves using surface treatments to promote strong adhesive bonds between coatings and substrates. Preventing corrosion is achieved by applying passivation layers that serve as protective barriers against environmental corrosion triggers. Improving wear resistance entails employing surface treatments to boost the hardness and longevity of coatings, thereby extending their operational life. Controlling surface energy involves treatments that effectively adjust surface energy levels to optimise wetting and bonding properties.

Passivation involves forming a protective oxide layer on the metal surface, which improves its resistance to corrosion. This oxide layer can act as a stable base for subsequent coatings, preventing the metal from reacting with its environment and ensuring the long-term durability of the coating. Passivation often includes cleaning processes that remove contaminants, oils, and residues from the metal surface. A clean surface promotes adhesion by allowing the coating to adhere directly to the metal without interference from impurities. Passivation treatments can modify the surface properties of the metal, such as surface energy and roughness, to make it more receptive to coatings. For instance, specific passivation methods can increase surface wettability, allowing coatings to spread more evenly and adhere better.

Surface treatment for passivation involves the following steps:

1. Surface Roughening: Mechanical or chemical treatments can roughen the metal surface, providing more surface area and creating micro-anchors for the coating material. This mechanical interlocking enhances adhesion by physically gripping the coating.
2. Chemical Modification: Treatments like priming or conversion coatings chemically modify the metal surface, creating a reactive layer that promotes bonding with the coating material. For example, conversion coatings such as phosphate or chromate treatments can enhance adhesion by providing a chemically compatible interface.
3. Adhesion Promoters: Some surface treatments are specifically designed as adhesion promoters. These treatments modify the surface chemistry to facilitate bonding with the coating material, improving the overall adhesion strength.

7.1 Surface Energy and Wetting

Surface passivation and treatment directly influence the bond strength between the metal substrate and the films. A properly prepared surface ensures intimate contact and chemical compatibility between the substrate and the coating material, which is crucial for long-term adhesion. Passivation treatments improve the corrosion resistance of the metal substrate, which in turn protects the coating from premature degradation due to exposure to environmental factors. This enhances the durability and performance of the coated metal. Surface treatments also contribute to the uniformity and appearance of the coating. A smooth and clean substrate promotes even coating application and reduces defects such as blistering or peeling. Different coatings (e.g., organic and inorganic coatings) require specific surface preparations for optimal adhesion. Surface passivation and treatment methods are tailored to meet the requirements of the coating material, ensuring compatibility and maximising adhesion performance.

The easiest way to characterise the intricate interfaces in an adhesive and a metal adherend is as an interphase, where the essential parameters are expressed in nanometers. Sophisticated spectroscopic methods have also confirmed that covalent bonds impact the adhesive bond's hydrolytic durability when presented alongside coupling mediators [19]. Calculations generated for the scenario of a liquid drop trapped on a flat, uniform substrate in equilibrium with the liquid vapor are commonly used to characterise the reactions involving an adherend and an adhesive in thermodynamic

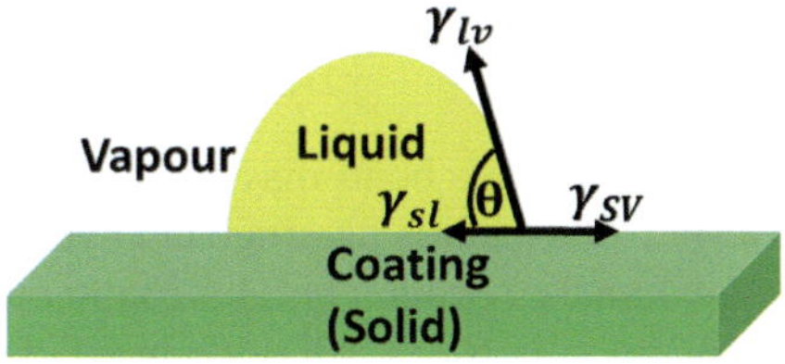

Fig. 5 Balance of surface tensions for a liquid drop on a coating surface

terminology. The Young formula can be used to represent the forces that are balanced amongst the liquid drop, solid substrate, and liquid–vapor:

$$\gamma_{SV} = \gamma_{sl} + \gamma_{lv} cos\theta$$

where θ is the optimal contact angle and γ is the appropriate surface energy at the three-phase point of contact (solid–vapor (sv), solid–liquid (sl), and liquid–vapor (lv). The minimisation of surface energy determines the critical value of the contact angle. The migration of a liquid with a lower surface energy, like water, minimises the outermost energy of a clean, high-energy surface like aluminium oxide. This results in minimal or zero contact angles on clean surfaces. The model representing the balance of surface tensions for a liquid drop on a coating surface is described in Fig. 5.

7.2 *Bondline Pressurisation and Adhesive Cure*

Customised rates of heating are used to fix the framework film adhesives thermally. The adherents are compressed mechanically or hydrostatically throughout their heating process. The adhesive's viscosity decreases with rising temperatures until it reaches its lowest point as the polymer bridging processes start. The adhesive flows through the low-viscosity phase of a pressurised sandwich consisting of two metal plates linked by a film adhesive, reducing plate spacing. Underneath the glue, a quadratic pressure distribution develops. Hydrostatic pressure-loaded thin adherend sheets may distort due to high pressure, resulting from the pressure in the adhesive at the sandwich's centre being greater than the force exerted on the substrate. When vacuum bag techniques are used for cures, the bond line thickness at the plate corners may be less than at the center. In addition to applying a hydrostatic limitation to bubble development in the adhesives, a high-pressure distribution frequently causes vacancies at the edges of repairs when a vacuum bag is used for bond pressurisation. Architectural film adhesives are usually engineered to minimise bubble formation for a positive pressure limitation of volatile gases.

7.3 Adhesive Bond Performance

A solid adhesive bond may not always translate into a robust or enduring binding. Despite other fluids like fuel and hydraulic fluid that might weaken a bond, water is the external environment. This chapter will concentrate on how vital adherend surface preparation is for an adhesive bond exposed to a humid environment regarding its climatic endurance. To guarantee, over time, resistance to the environment, surface treatment and restoration bonding are essential. Surface procedures for repairing bonding are the subject of studies, considering the bond interface's quantum nature and the connection between microscopic activity and macroscopic mechanical properties. The degree to which an adhesive bond resists environmental changes is primarily determined by its interfacial chemistry.

7.4 Characterisation and Evaluation of Surface Treatments

Surface energy measurement techniques involve utilising contact angle goniometry to evaluate surface energy and wettability characteristics. This assessment aids in understanding how materials interact with liquids on a surface, which is crucial for optimising wetting and bonding properties. Chemical analysis methods like XPS, AES, and FTIR scrutinise surface composition. These techniques unveil the chemical elements and functional groups present, offering insights into surface properties crucial for various applications. Topographical analysis uses tools like AFM and SEM to delve into the microstructure. These techniques provide detailed information on surface roughness, texture, and topographical features, assisting in understanding how these factors influence adhesion and other surface properties. Adhesion testing methodologies involving peel tests, pull-off tests, and scratch tests are pivotal for gauging the efficacy of surface treatments in enhancing adhesion. These tests determine the strength of the bond between coatings and substrates, helping assess the durability and performance of the treated surfaces.

8 Application of Surface Passivation and Surface Treatment

Surface passivation and surface treatment find wide applications across various industries where metals are used, particularly in sectors where corrosion resistance, adhesion of coatings, and surface durability are critical. Figure 6 represents the passivation service in aerospace, automotive, electronic, and medical industries. Here are some key applications of these processes:

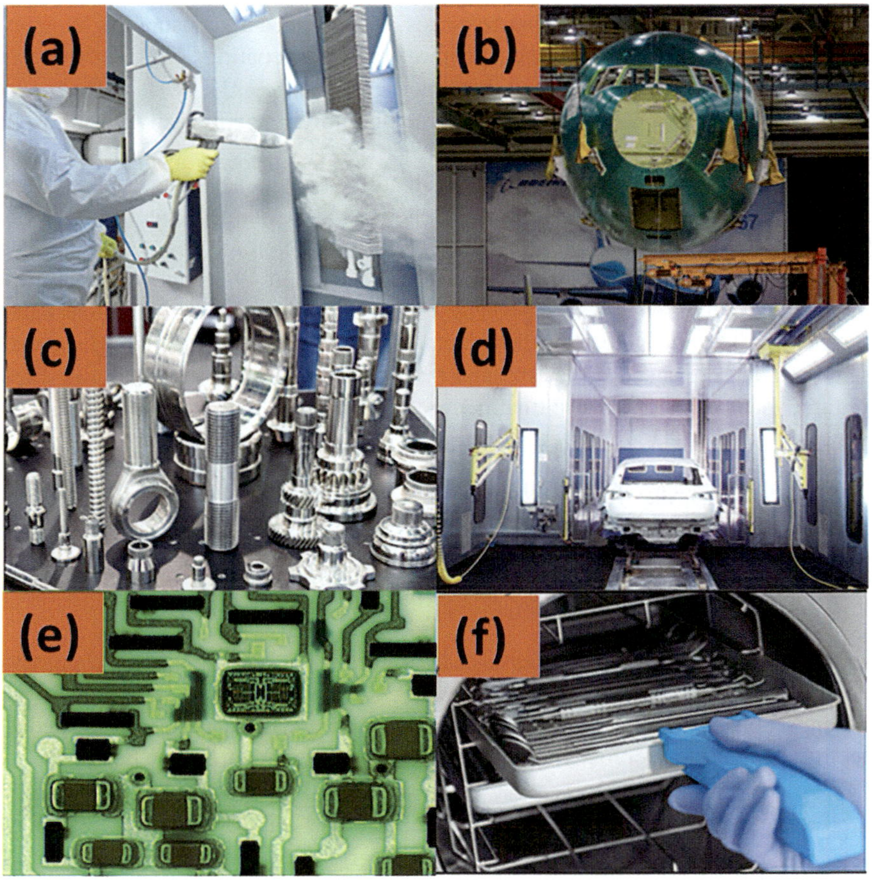

Fig. 6 Passivation service in **a** and **b** aerospace, **c** and **d** automotive industry, **e** electronic industry, and **f** Medical service

8.1 Aerospace Industry

The aerospace industry adheres to stringent regulatory standards (e.g., FAA in the US, EASA in Europe) regarding materials, manufacturing processes, and maintenance procedures to ensure the safety and reliability of aircraft. Passivation and surface treatments are essential to meet these standards, ensuring components can withstand the demanding operational environments encountered during flight. Aircraft components made from aluminium, titanium, and other metals require high resistance to corrosion due to exposure to atmospheric conditions and operational stresses [20]. Passivation preserves the mechanical integrity and strength of materials by preventing surface degradation. Ensures critical components such as fasteners, brackets, and structural elements maintain performance over extended periods. Surface passivation treatments such as anodising, chromate conversion coatings or passivation of stainless steels are applied to enhance corrosion resistance and ensure the longevity of parts. Aircraft components often require protective coatings for weather resistance, thermal protection, and aerodynamic performance [21]. Surface treatments ensure strong adhesion of these coatings, maintaining their integrity during flight and operational conditions.

Regarding use in aviation, Tiburcio et al. studied frequency-time domains characterising electrochemical noise of passivated AM350 stainless steel [22]. It was determined how AM 350 stainless steel would react to remedies comprising 5% NaCl and 1% H_2SO_4 after being passivated for 50 min at 49 °C and 70 °C in nitric acid and citric acid remedies, respectively. He reported a better corrosion resistance of the AM350 stainless steel passivated in nitric acid in citric acid [22]. Surface treatments provide enhanced protection and performance in the aerospace field. There are three main types of surface treatments carried out:

- **Plating**: Electroplating with metals like nickel, cadmium, or chrome to enhance corrosion resistance and hardness.
- **Shot Peening**: Increases longevity during fatigue and resilience to stress corrosion cracks by introducing compressive forces.
- **CVD and PVD Coatings**: Deposit thin films of ceramics or metals to provide thermal protection or improve wear resistance.

Surface treatments improve aerodynamic efficiency, reduce wear, and enhance resistance to high temperatures and erosion. Components such as antennas and housings benefit from coatings that will enhance electrical conductivity, reduce electromagnetic interference, or provide thermal management, termed avionics.

8.2 *Automotive Industry*

Vehicle bodies and components undergo surface treatments to guarantee proper adhesion of paint and coatings. Pretreatment processes such as phosphate or electroplating, plasma Spraying, thermal spraying, and conversion coating enhance adhesion and improve corrosion resistance, contributing to the overall durability and aesthetic appeal of automotive finishes. Engine components and chassis often require specific surface treatments to improve wear resistance, reduce friction, and enhance heat dissipation. These treatments ensure optimal performance and longevity of automotive systems under demanding operational conditions. In the automotive industry, passivation is commonly used for components such as fasteners and bolts, engine components, and exhaust Systems.

8.3 Electronics Industry

Passivation and surface treatment techniques are essential in the electronic industry to ensure electronic components and devices' reliability, performance, and longevity. They address issues such as corrosion, electrical performance, and mechanical durability, thereby contributing to electronic products' overall quality and functionality. Surface treatments are integral to PCB manufacturing, semiconductor fabrication, electronic component assembly (e.g., solderability), and ensuring reliability in harsh environments (e.g., aerospace and automotive electronics). The passivation and surface treatment in the electronic industry involve several procedures such as coatings, etching, plating, cleaning, and surface preparation. Applying thin films or layers of materials (e.g., polymers, metals, ceramics) to alter surface properties like conductivity, adhesion, or chemical resistance.

8.4 Medical Devices

Medical device manufacturers must adhere to strict regulatory standards (e.g., FDA regulations in the US, ISO standards internationally) regarding materials, processes, and biocompatibility testing. Passivation and surface treatments ensure devices meet these standards, contributing to their safety and efficacy in clinical settings.

Passivation in medical devices prevents metal component corrosion from bodily fluids or cleaning agents, which could compromise device integrity and safety. It ensures the material does not interact with the body in a harmful way, reducing the risk of adverse reactions or contamination. Smooth surfaces achieved through passivation make medical devices easier to clean and sterilise, which is crucial for maintaining aseptic conditions [23].

There are various methods by which this passivation can be done. Among them, two are majorly carried out, which are explained below;

- **Citric Acid Passivation**: Widely utilised because of its efficacy, safety, and environmentally amiable disposition compared to traditional methods like nitric acid.
- **Electropolishing**: Combines passivation with material removal to achieve a smooth, clean, and corrosion-resistant surface.

The passivation treatment is used to maintain surgical instruments like implants, prosthetics, and endoscopic devices. Scalpels, forceps, and other metal instruments benefit from passivation to preserve sharpness, cleanliness, and resistance to corrosion. Stainless steel and titanium implants require passivation to ensure biocompatibility and longevity within the body [24, 25]. Passivation helps maintain the cleanliness and functionality of intricate surfaces within endoscopes and similar devices.

Surface treatment helps to enhance biocompatibility, functional coatings, durability, and performance. Three main types of surface treatment techniques are used in the medical sector. Those are:

- Coatings: Thin films or layers applied to provide specific properties like hydrophilicity, anti-fouling, or biocompatibility
- Anodisation: Creates a durable oxide layer on metals like titanium to improve corrosion resistance and biocompatibility.
- Plasma Treatment: Modifies surface chemistry to enhance adhesion of coatings or improve surface cleanliness.

Medical devices' passive and surface treatment techniques are essential for maintaining biocompatibility, corrosion resistance, cleanliness, and performance. They ensure that devices can withstand the rigors of clinical use while minimising risks to patients and healthcare providers.

8.5 Oil and Gas Industry

Surface passivation and treatment play pivotal roles in the oil and gas industry by safeguarding equipment against corrosion, improving longevity, and enhancing operational reliability. These practices are essential for maintaining safety standards, complying with regulations, and optimising operational costs in this demanding sector. Passivation is a process that reduces corrosion rates on metal surfaces, particularly in the oil and gas industry. It involves using acids or chemical solutions to clean and treat stainless steel surfaces, ensuring their durability and safety. Regulations like NACE standards specify passivation procedures. Coatings like paints, epoxies, and polymers protect metal surfaces from corrosion in oil and gas operations. Surface roughness affects adhesion and fluid flow efficiency. Surface treatments can enhance properties like wear resistance and thermal stability. Thorough cleaning and pretreatment processes like degreasing and etching prepare surfaces for subsequent treatments. Surface treatments are essential for pipelines, tanks, valves, fittings, and marine structures to prevent corrosion, maintain flow efficiency, extend service life, and prevent structural degradation under high-pressure conditions.

9 Recent Advancements in Surface Passivation and Surface Treatment

Konicek et al. investigated how surface passivation affected the wear and friction characteristics of coatings of tetrahedral amorphous carbon and ultra-nanocrystalline diamond [26]. Without any liquid lubrication, strongly sp3-bonded, almost hydrogen-free carbon-based compounds can display exceptionally minimal friction and wear;

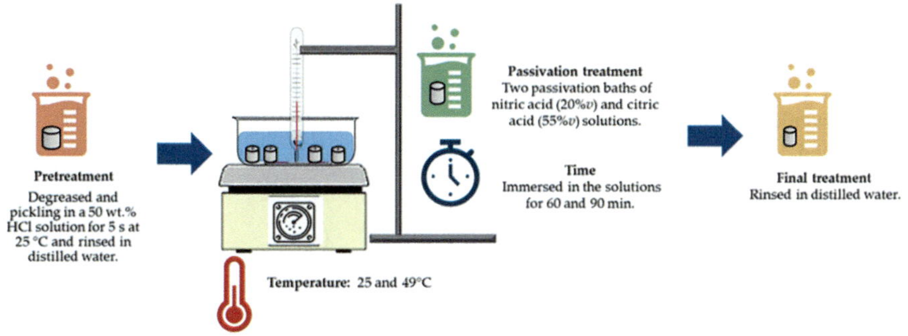

Fig. 7 Schematic of the passivation procedure of PHSS in citric and nitric acid baths

however, the vapor microenvironment limits these physical properties. The main process of solid lubrication is the passivation of dangling bonds by OH and H from the breakdown of vapor-phase H2O. With friction coefficients as small as 0.078 for ta-C and 0.008 for UNCD, this vapor-phase lubrication process is incredibly efficient; thousands of sliding passes are needed to create merely a few nanometers of wear [26].

Precipitation-hardening stainless steels are utilised in aerospace because of their resilience to corrosion and mechanics. The electrochemical behaviour of these steels was evaluated in NaCl and H_2SO_4 solutions by Tovar et al. for aerospace applications [27]. Figure 7 represents the schematic of the passivation treatment of PHSS in citric acid and nitric acid baths. The results showed two mechanisms: secondary passivation in H_2SO_4 and pseudo-passivation in NaCl. In certain circumstances, the anti-corrosion property of stainless steel corresponds to that of nitric acid, and citric acid functions as a passivating agent. The NaCl solution had current densities ranging from 10^{-4} to 10^{-5} mA/cm^2, whereas the H_2SO_4 solutions had current densities among 10^{-2} and 10^{-3} mA/cm^2 [27].

Since it is non-toxic and environmentally benign, the citric acid passivation method applied to PHSS may be an environmentally friendly choice compared to the nitric acid passivation method in practice [28]. The representation of passivation treatments in baths of citric and nitric acid for stainless steels 15-5PH, 17-4PH, CUSTOM 450, and AM 350 subjected to a 5 weight percent NaCl solution and a 1 weight percent H_2SO_4 solution, is shown in Fig. 8.

The materials passivated in citric acid had lower pitting potentials than PHSS passivated in nitric acid and submerged in NaCl. The potentiodynamic polarisation diagrams for coated PHSS in nitric acid are displayed in Fig. 9. The martensitic PHSS was characterised by optical microscopy (OM) and showed a microstructure with a martensitic (α0) phase. At the same time, the semi-austenitic PHSS had a morphology with austenite (γ) and delta (δ) ferrite stages, as shown in Fig. 10. The martensitic 15–5 PH (23.5), 17–4 PH (17.5), and CUSTOM 450 (20.9) shown in Fig. 9 demonstrated lower corrosion resistance than the AM 350 (semi-austenitic) (29.80), according to the PREN readings.

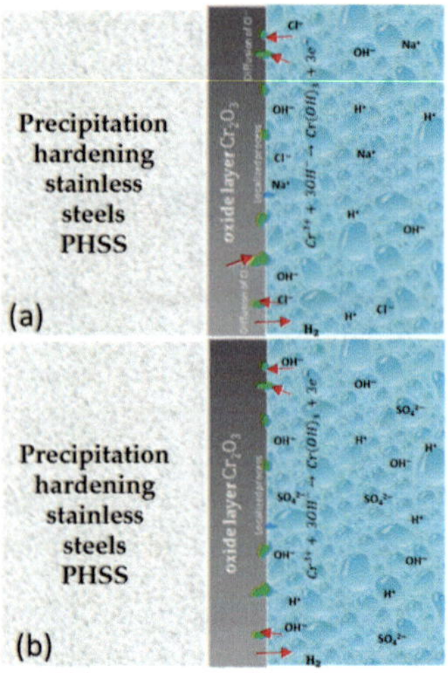

Fig. 8 Model of passivation treatment in citric and nitric acid baths for 15-5PH, 17–4PH, CUSTOM 450, and AM 350 stainless steels exposed to **a** 5 wt.% NaCl solution; **b** 1 wt.% H_2SO_4 solution [27]

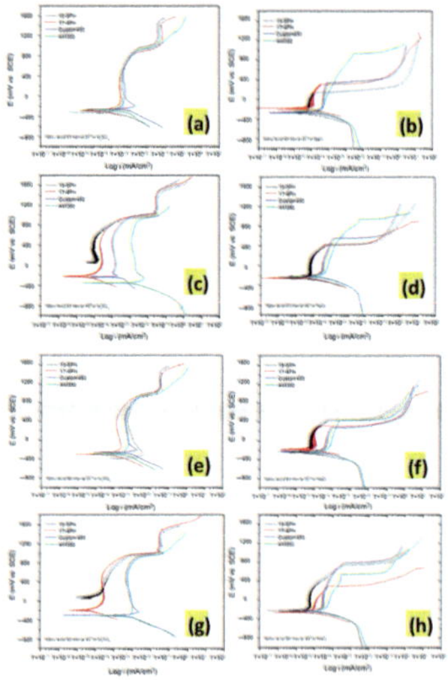

Fig. 9 Potentiodynamic polarisation diagrams for passivated PHSS in nitric acid, at 25 and 49 °C for 60 min: **a**, **c** H_2SO_4 and **b**, **d** NaCl and in nitric acid, for 90 min: **e**, **g** H_2SO_4 and **f**, **h** NaCl [27]

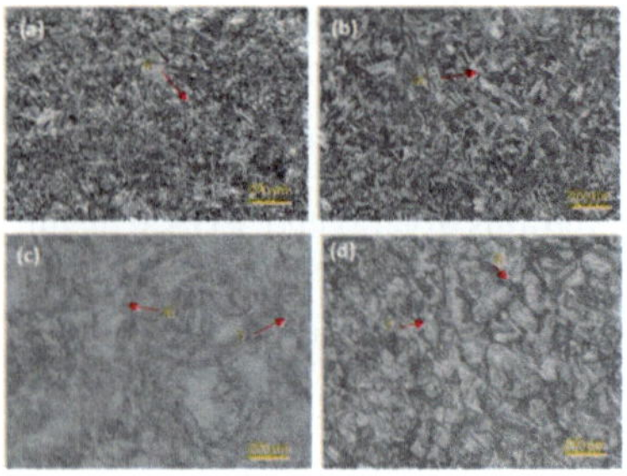

Fig. 10 OM morphology of PHSS. **a** 15–5PH, **b** 17–4PH, **c** Custom 450, and **d** AM 350 [27]

Passivation and coupled antireflection were additionally observed on PECVD silicon carbo-nitrides (SiC_xN_y) [29, 30]. One of the more often used dielectric films in semiconductors is thermally produced passivated silicon dioxide [31, 32]. TiO_x was superseded by hydrogenated silicon nitride (SiNx), although having superior optical qualities, because TiO_x proved ineffective for the adequate passivation of the front exterior of silicon solar cells, which was necessary for the development of silicon solar cells [33]. Later, Liao et al. demonstrated good n- and p-type silicon passivation employing TiO_x coatings; however, they synthesised the films via ALD [34]. Liu et al. demonstrated that passing sheets helps enhance the mechanisms of charge separation and transfer across semiconductor-liquid interfaces, thus improving the total conversion of solar power efficiency [35]. These thin films provide passivation properties that lower the recombination of charges at surface states, speed up reactions, and shield the semiconductor from chemical erosion. Si, ZnO, Cu_2O, and certain binary or ternary metal sulfides are among the semiconductor photographic electrodes unstable at high oxidative or reductive potentials needed for water oxidation and in highly alkaline pH solutions. In the beginning, photo corrosion was identified as a major issue in a PEC cell by Bard et al. and Gerischer et al. Photocorrosion is comparable with water oxidation (or reduction) when the anodic (or cathodic) corrosion potential lies within the band gap [36, 37]. These situations result in unstable photoelectrodes because the semiconductor degradation reaction's potential is usually lower and hence thermodynamically more favourable than the HER/OER interaction [38]. Certain metal oxides exhibit kinetic stability in strongly oxidative environments, whereas certain non-oxide semiconductors or low-valence metal oxides demonstrate kinetic stability in the face of photogenerated electron reductions. The use of sulfur as an electrical passivation reagent for III-V semiconductors has been investigated [39].

Consider GaAs. For instance, in a reactive S-containing solution, the native oxide of GaAs may be eliminated and chemical processes may result in the formation of the sulfides of Ga (Ga_xS_y) and (As_xS_y). S-bonds lower the density of surface states and provide shielding from additional oxidation. Most sulfurs employed to generate S terminate III-V surfaces were $(NH_4)_2S$ formulations [40]. It is currently noted that the passivation impact of alcohol solutions is better than that of aqueous sulfide concentrations. When Lebedev et al. studied a GaSb semiconductor passivated with $(NH_4)_2S$ dissolved in water as opposed to isopropanol, they discovered that Ga_2O_3 was discovered utilising an aqueous $(NH_4)_2S$ solution and that there was nearly no residual oxygen on the GaSb passivated with $(NH_4)_2S$ in isopropanol [41].

Additionally, they discovered that using isopropanol decreased the C contamination by almost 4 times, whereas using a water-based remedy raised it by 1.25 times. The interface's steady state density had been significantly reduced by $(NH_4)_2S$ passivation; nevertheless, air pollution and the passivation's poor chemical resistance in an atmospheric setting were clear disadvantages. Due to their considerable mobility, III-V compound semiconductors are considered viable alternatives to the Si channel in next-generation MOSFETs.

Ignited fascination with poly-Si/SiO_x interactions has occurred subsequently. The microelectronics sector initially developed this technique for high-speed bipolar junction transistors (BJTs) [42, 43]. Pariona et al. used a 2 kW Yb-fiber laser (IPG YLR-2000S) to investigate deterioration in adverse surroundings and perform laser surface remelting on an Al–1.5 weight percent Fe alloy [44]. He stated that the outermost coating was altered by the laser surface remelting technique, leading to increased durability against corrosion, a wide passivation assortment, and a smaller hysteresis loop region.

10 Challenges and Future Directions

Surface passivation primarily aims to reduce reactivity and enhance stability, especially in semiconductor and metal applications prone to corrosion or degradation. Surface treatment is a broader term encompassing various techniques to modify surface properties beyond passivation, including coatings, cleaning, and functionalisation. While passivation is crucial in electronics and metal industries, surface treatments have broad applications across different sectors for improving performance, durability, and functionality. While surface passivation focuses on reducing reactivity and stabilising surfaces, surface treatment covers various techniques to modify surface properties for specific purposes across diverse industries. Enhancing adhesion involves employing surface treatments to foster robust adhesive bonds between coatings and substrates. This not only bolsters adhesion but also aids in addressing environmental and safety concerns. For instance, strategies involve minimising hazardous chemicals like hexavalent chromium in passivation processes and advancing towards eco-friendly and sustainable surface treatment methods. Multilayer coatings can be beneficial as they offer superior corrosion resistance compared to single-layer coatings [45].

Passivation layers serve as protective barriers against environmental factors. Techniques also evolve to embrace advanced surface treatment technologies, such as nanotechnology innovations that lead to self-healing and smart coatings. Additionally, methods utilising advanced plasma and laser techniques allow for precise modifications of surfaces, enhancing the overall durability against corrosion and wear.

Improving wear resistance encompasses applying surface treatments to enhance the hardness and durability of coatings, thereby prolonging their lifespans. These efforts are increasingly integrated with coating processes to optimise performance and efficiency. For example, integrated systems are being developed to combine surface treatment and coating applications, streamlining processes and ensuring better control and automation for enhanced consistency and productivity.

11 Conclusion

The current state of the surface passivation technique for semiconductor coating technology was discussed in the preceding section. The passivation methods, including chemical passivation (nitric and citric acid passivation), physical vapor deposition, and chemical vapor deposition, are discussed in this chapter. Reducing surface roughness and stress requires precise plasma energy and film thickness optimising during plasma treatment and nitridation processes. In contrast, the ALD passivation technique was more user-friendly and effective, which may have made integrating ALD passivation into device production simpler. Enhancing reproducibility in the real world would be the main challenge of numerous of these passivation approaches, many of which were effective in laboratory settings. Later on, the passivation and surface treatment applications in various fields such as aerospace, medical, electronic, and automobile are thoroughly discussed. At last, the future challenges and strategies for improving advanced coatings and devices are reported here.

References

1. Gonzalez-Canche NG, Flores-Johnson EA, Cortes P, Carrillo JG (2018) Evaluation of surface treatments on 5052–H32 aluminum alloy for enhancing the interfacial adhesion of thermoplastic-based fiber metal laminates. Int J Adhes Adhes 82:90–99. https://doi.org/10.1016/j.ijadhadh.2018.01.003
2. Jena S, Laha S, Swain BP (2024) Ti2NTx MXene/rGO nanocomposites: Optical tunning, functional bonding and electrochemical analysis for supercapacitor electrode application. Appl Surf Sci 666:160403. https://doi.org/10.1016/j.apsusc.2024.160403
3. Agarwal A (2021) Strain effectiveness of gate-all-around silicon nanowire n-MOSFET transistors with physical analysis
4. Black LE, van de Loo BWH, Macco B et al (2018) Explorative studies of novel silicon surface passivation materials: considerations and lessons learned. Sol Energy Mater Sol Cells 188:182–189. https://doi.org/10.1016/j.solmat.2018.07.003
5. Dingemans G, Kessels WMM (2012) Status and prospects of Al2O3-based surface passivation schemes for silicon solar cells. J Vac Sci Technol, A: Vac, Surf Films 30. https://doi.org/10.1116/1.4728205
6. Bonilla RS, Hoex B, Hamer P, Wilshaw PR (2017) Dielectric surface passivation for silicon solar cells: a review. Phys Status Solidi (A) Appl Mater Sci 214:1–30. https://doi.org/10.1002/pssa.201700293
7. Rizal U, Swain BP (2020) Effect of positive substrate bias on microstructural, structural, and phonon characteristics of nanocrystalline silicon thin films. AIP Conf Proc 2273. https://doi.org/10.1063/5.0024228
8. Swain BS, Swain BP, Mahmood K et al (2015) Structural and optical properties of H2 diluted c-Si/a-SiOx core-shell silicon nanowire. Appl Phys A Mater Sci Process 118:269–274. https://doi.org/10.1007/s00339-014-8675-x
9. Wang X, Gao X, Zhang Z et al (2021) Advances in modifications and high-temperature applications of silicon carbide ceramic matrix composites in aerospace: a focused review. J Eur Ceram Soc 41:4671–4688. https://doi.org/10.1016/j.jeurceramsoc.2021.03.051
10. Jena S, Madhuri A, Gupta M, Swain BP (2023) Investigation of interfacial and interdiffusion study of Ti2N MXene phase from TiN/Ti multilayers. Thin Solid Films 780:139969. https://doi.org/10.1016/j.tsf.2023.139969

11. Jena S, Madhuri A, Gupta M, Swain BP (2024) Comparative assessment and experimental analysis of atomic coordination and mechanical properties of sputter deposited Ti-rich TiN, TiC, and TiCN thin films. J Alloys Compd 1002:175217. https://doi.org/10.1016/j.jallcom.2024.175217
12. Jena S, Madhuri A, Dutta S, Swain BP (2024) Investigation of structural, phonon modes and electronic environment of GaP:Si nanostructure thin films. Physica B Condens Matter 677:415664. https://doi.org/10.1016/j.physb.2024.415664
13. Ghadai RK, Das S, Kalita K et al (2021) Structural and mechanical analysis of APCVD deposited diamond-like carbon thin films. SILICON 13:4453–4462. https://doi.org/10.1007/s12633-020-00760-3
14. Ghadai RK, Kalita K, Mondal SC, Swain BP (2019) Genetically optimised diamond-like carbon thin film coatings. Mater Manuf Processes 34:1476–1487. https://doi.org/10.1080/10426914.2019.1594273
15. Das S, Guha S, Ghadai R, Swain BP (2021) A comparative analysis over different properties of TiN, TiAlN and TiAlSiN thin film coatings grown in nitrogen gas atmosphere. Mater Chem Phys 258:123866. https://doi.org/10.1016/j.matchemphys.2020.123866
16. Chen CH, Lee CY, Der GM et al (2022) The effect of oxalic acid as the pre-activator for the electropolishing of additive manufactured titanium-based materials and its characterisation. Polym (Basel) 14. https://doi.org/10.3390/polym14194198
17. Mohanty S, Basak S, Saran D et al (2024) Advanced surface engineering approaches for exotic applications. Korean Soc Precis Eng
18. Yerokhin AL, Nie X, Leyland A et al (1999) Plasma electrolysis for surface engineering. Surf Coat Technol 122:73–93. https://doi.org/10.1016/S0257-8972(99)00441-7
19. Abel M, Digby RP, Fletcher IW, Watts JF (2000) Evidence of specific interaction between γ-glycidoxypropyltrimethoxysilane and oxidised aluminium using high-mass resolution ToF-SIMS. Surf Interface Anal 29:115–125. https://doi.org/10.1002/(sici)1096-9918(200002)29:2%3c115::aid-sia702%3e3.3.co;2-y
20. Pop AB, Pop GI, Gusan V, Ţîţu AM (2023) Surface treatment in aerospace industry: a study on acid pickling and paint adhesion. Int J Mechatron Appl Mech 2023:58–69. https://doi.org/10.17683/ijomam/issue13.7
21. Dam JPB, van Tiringer U, Abrahami ST et al (2024) Surface engineering of aerospace aluminium alloys: understanding alloying effects on chemical pre-treatment and sol-gel coating adhesion. Surf Coat Technol 485:14. https://doi.org/10.1016/j.surfcoat.2024.130901
22. Tiburcio CG, Samaniego-Gámez O, Jáquez-Muñoz JM et al (2022) Frequency-time domain analysis of electrochemical noise of passivated AM350 stainless steel for aeronautical applications. Int J Electrochem Sci 17:220950. https://doi.org/10.20964/2022.09.49
23. Liu X, Chu PK, Ding C (2004) Surface modification of titanium, titanium alloys, and related materials for biomedical applications. Mater Sci Eng R Rep 47:49–121. https://doi.org/10.1016/j.mser.2004.11.001
24. Yan S, Song GL, Li Z et al (2018) A state-of-the-art review on passivation and biofouling of Ti and its alloys in marine environments. J Mater Sci Technol 34:421–435. https://doi.org/10.1016/j.jmst.2017.11.021
25. Xie B, Gao K (2023) Research progress of surface treatment technologies on titanium alloys: a mini review. Coatings 13. https://doi.org/10.3390/coatings13091486
26. Konicek AR, Grierson DS, Sumant AV et al (2012) Influence of surface passivation on the friction and wear behavior of ultrananocrystalline diamond and tetrahedral amorphous carbon thin films. Phys Rev B Condens Matter Mater Phys 85:1–13. https://doi.org/10.1103/PhysRevB.85.155448
27. Villegas-Tovar J, Gaona-Tiburcio C, Lara-Banda M et al (2023) Electrochemical corrosion behavior of passivated precipitation hardening stainless steels for aerospace applications. Metals (Basel) 13. https://doi.org/10.3390/met13050835
28. Gaydos SP (2003) Passivation of aerospace stainless parts with citric acid solutions. Plat Surf Finish 90:20–25

29. Ferré R, Martín I, Trassl R et al (2011) Simultaneous gettering and emitter formation in multicrystalline-Si wafers by annealing phosphorus doped amorphous silicon compounds. Appl Phys Lett 98:10–13. https://doi.org/10.1063/1.3535616
30. Orpella A, Blanqué S, Roiati V et al (2009) N-type emitters passivation through antireflective phosphorus doped a-SiCxNy:H(n) stacks. Span Conf Electron Devices 2009:357–359
31. Aber;e AG, Glunz S, Warta W (1993) Field effect passivation of high efficiency silicon solar cells. Sol Energy Mater Sol Cells 29:175–182. https://doi.org/10.1016/0927-0248(93)90075-E
32. Aberle AG (2000) Surface passivation of crystalline silicon solar cells: a review. Prog Photovoltaics Res Appl 8:473–487. https://doi.org/10.1002/1099-159X(200009/10)8:5%3c473::AID-PIP337%3e3.0.CO;2-D
33. Richards BS (2004) Comparison of TiO2 and other dielectric coatings for buried-contact solar cells: a review. Prog Photovoltaics Res Appl 12:253–281. https://doi.org/10.1002/pip.529
34. Liao B, Hoex B, Shetty KD et al (2015) Passivation of boron-doped industrial silicon emitters by thermal atomic layer deposited titanium oxide. IEEE J Photovolt 5:1062–1066
35. Liu R, Zheng Z, Spurgeon J, Yang X (2014) Enhanced photoelectrochemical water-splitting performance of semiconductors by surface passivation layers. Energy Environ Sci 7:2504–2517. https://doi.org/10.1039/C4EE00450G
36. Bard AJ (1979) Photoelectrochemistry and heterogeneous photo-catalysis at semiconductors. J Photochem 10:59–75. https://doi.org/10.1016/0047-2670(79)80037-4
37. Gerischer H (1977) On the stability of semiconductor electrodes against photodecomposition. J Electroanal Chem Interfacial Electrochem 82:133–143. https://doi.org/10.1016/S0022-0728(77)80253-2
38. Chen S, Wang LW (2012) Thermodynamic oxidation and reduction potentials of photocatalytic semiconductors in aqueous solution. Chem Mater 24:3659–3666. https://doi.org/10.1021/cm302533s
39. Hasegawa H, Akazawa M, Domanowska A, Adamowicz B (2010) Surface passivation of III–V semiconductors for future CMOS devices—past research, present status and key issues for future. Appl Surf Sci 256:5698–5707. https://doi.org/10.1016/j.apsusc.2010.03.091
40. Zhernokletov DM, Dong H, Brennan B et al (2012) Optimisation of the ammonium sulfide (NH4)2S passivation process on InSb(111)A. J Vac Sci Technol, B: Nanotechnol Microelectron: Mater, Process, Meas, Phenom 30. https://doi.org/10.1116/1.4719961
41. Lebedev MV, Kunitsyna EV, Calvet W et al (2013) Sulfur passivation of GaSb(100) surfaces: comparison of aqueous and alcoholic sulfide solutions using synchrotron radiation photoemission spectroscopy. J Phys Chem C 117:15996–16004. https://doi.org/10.1021/jp401942p
42. Post IRC, Ashburn P, Wolstenholme GR (1992) Polysilicon emitters for bipolar transistors: a review and re-evaluation of theory and experiment. IEEE Trans Electron Devices 39:1717–1731. https://doi.org/10.1109/16.141239
43. Yatabe Z, Asubar JT, Hashizume T (2016) Insulated gate and surface passivation structures for GaN-based power transistors. J Phys D Appl Phys 49:393001. https://doi.org/10.1088/0022-3727/49/39/393001
44. Pariona MM, Teleginski V, dos Santos K et al (2013) Influence of laser surface treated on the characterisation and corrosion behavior of Al–Fe aerospace alloys. Appl Surf Sci 276:76–85. https://doi.org/10.1016/j.apsusc.2013.03.025
45. Jena S, Gupta M, Swain BP (2024) Effect of interlayer thickness on electrochemical properties and corrosion resistance of sputtered TiNxnm/Tixnm multilayer thin films. Mater Sci Semicond Process 182:108729. https://doi.org/10.1016/j.mssp.2024.108729

GPSR Compliance

The European Union's (EU) General Product Safety Regulation (GPSR) is a set of rules that requires consumer products to be safe and our obligations to ensure this.

If you have any concerns about our products, you can contact us on ProductSafety@springernature.com

In case Publisher is established outside the EU, the EU authorized representative is:

Springer Nature Customer Service Center GmbH
Europaplatz 3
69115 Heidelberg, Germany

Batch number: 10704307

Printed by Printforce, the Netherlands